# Biodegradability of Surfactants

# Biodegradability of Surfactants

Edited by

D.R. Karsa
Akcros Chemicals
Eccles
Manchester

and

M.R. Porter
Maurice R. Porter & Associates
Consultants in Speciality Chemicals
Sully
Cardiff

## BLACKIE ACADEMIC & PROFESSIONAL
An Imprint of Chapman & Hall

London · Glasgow · Weinheim · New York · Tokyo · Melbourne · Madras

**Published by**
**Blackie Academic and Professional, an imprint of Chapman & Hall,**
**Wester Cleddens Road, Bishopbriggs, Glasgow G64 2NZ**

Chapman & Hall 2–6 Boundary Row, London SE1 8HN, UK

Blackie Academic & Professional, Wester Cleddens Road, Bishopbriggs, Glasgow G64 2NZ, UK

Chapman & Hall GmbH, Pappelallee 3, 69469 Weinheim, Germany

Chapman & Hall USA, One Penn Plaza, 41st Floor, New York, NY 10119, USA

Chapman & Hall Japan, ITP-Japan, Kyowa Building, 3F, 2-2-1 Hirakawacho, Chiyoda-ku, Tokyo 102, Japan

DA Book (Aust.) Pty Ltd, 648 Whitehorse Road, Mitcham 3132, Victoria, Australia

Chapman & Hall India, R. Seshadri, 32 Second Main Road, CIT East, Madras 600 035, India

First edition 1995

© 1995   Chapman & Hall

Typeset in 10/12pt Times by EXPO Holdings, Malaysia
Printed in Great Britain by TJ Press (Padstow) Ltd., Padstow, Cornwall

ISBN   0 7514 0206 0

A catalogue record for this book is available from the British Library
Library of Congress Catalog Card Number: 94-78346

∞ Printed on acid-free text paper, manufactured in accordance with
ANSI/NISO Z39.48-1992 (Permanence of Paper)

# Preface

The awareness and development of 'biodegradable' surfactants pre-dates current pressures by the environmental movement by nearly three decades, wherein a responsible industry mutually agreed to replace 'hard', non-biodegradable components of household detergents by 'soft', biodegradable alternatives, without course to legislation.

The only requirement at that time was for surfactants used in detergents to exhibit a 'primary biodegradability' in excess of 80%; this referring to the disappearance or removal from solution of the intact surface active material as detected by specified analytical techniques. This proved useful, as observed environmental impacts of surfactants, e.g. visible foam on rivers, are associated with the intact molecule. Test methods for 'primary biodegradability' were eventually enshrined in EU legislation for nonionic surfactants (Directive 82/242/EEC, amended 73/404/EEC) and for anionic surfactants (Directive 82/243/EEC, amended 73/405/EEC). No approved test methods and resultant legislation have been developed for cationic and amphoteric surfactants to date.

The environmental classification of chemical substances, which of course includes surfactants, and associated risk assessment utilises a second criterion – 'ready biodegradability'. This may be assessed by a number of methods which monitor oxygen uptake (BOD), carbon dioxide production or removal of dissolved organic carbon (DOC). Some surfactants which comply with the above Detergents Directive are borderline when it comes to 'ready biodegradability'.

The publication of this book coincides with biodegradability legislation standing at a cross-roads, with some uncertainty as to the way in which it may develop. The European Commission has indicated a desire to develop and introduce ultimate biodegradability legislation or mineralisation into the Detergents Directive, underlining the already existing assessment of ultimate biodegradability by major surfactant end-users regardless of standardised test methods or any legislation being in place. Likewise, attention has also been focused more recently on industrial cleaning applications and some areas of industry where surfactants are used as process aids to assess their biodegradability and potential impact on the environment.

Whereas there is general agreement that surfactants should be subject to some environmental acceptance criteria, there is a growing lobby that suggests that surfactants should no longer require an exceptional role as compared with other chemical compounds released into the environment. Eventually, there may be a valid case to deregulate as far as surfactant biodegradability is concerned and to subject the acceptance of surfactants particularly for use in washing, rinsing and cleaning to an environmental risk assessment which is required as a matter of principle for all new and existing substances (EU directives 93/67/EEC and 793/93/EEC, respectively). Only time will tell as to which approach will be accepted.

Set against such a background, this volume provides a state-of-the-art review of surfactant biodegradability mechanisms, test methods, legislative requirements and individual consideration of the four ionic classifications of surfactant, namely anionics, nonionics, cationics and amphoterics. Each chapter is written by acknowledged experts in their particular field, which should ensure that this book will provide a valuable addition to our knowledge of surfactant biodegradability and become a significant reference work on this subject.

**Acknowledgements**

The editors would like to record their special thanks to the individual authors of each chapter for their time, patience and hard work which has resulted in a volume of substance in which all contributors can take pride. For those working in industry or academia, we would also like to extend our thanks to their individual companies or universities for their support.

Finally, we would thank the publishers for their unlimited patience and understanding, as the gestation period for this particular work proved to be somewhat longer than originally anticipated. We trust the final product has been worth waiting for.

D.R.Karsa
M.R. Porter

# Contributors

**T. Balson**        Dow Europe SA, Bachtobelstr. 3, CH-8810, Horgen,
                     Switzerland

**H. Berger**        Henkel KGaA, TFB/Greb Z33, Henkelstrasse 67,
                     4000 Dusseldorf, Germany

**D. Brown**         Zeneca Brixham Environmental Laboratory, Brixham,
                     Devon TQ5 8BA, UK

**A. Domsch**        REWO Chemische Werke GmbH, Postfach 1160, D-36392
                     Steinam an der Strasse, Germany

**M.S.B. Felix**     Dow Benelux N.V., PO Box 48, 4530 AA Terneuzen, The
                     Netherlands

**H.A. Painter**     FreshField Analysis Ltd., Caxton Villa, Park Lane,
                     Knebworth, Herts SG3 6PF, UK

**N.J. Russell**     Department of Biochemistry, University of Cardiff,
                     PO Box 903, Cardiff CF1 1ST, UK

**J. Steber**        Henkel KGaA, TFB/Greb Z33, Henkelstrasse 67,
                     4000 Dusseldorf, Germany

**C.G. van Ginkel**  Akzo Research Laboratories Arnhem, Corporate Research,
                     Department of Analytical and Environmental Chemistry,
                     Velperweg 76, 6824 BM Arnhem, The Netherlands

**G.F. White**       Department of Biochemistry, University of Cardiff,
                     PO Box 903, Cardiff CF1 1ST, UK

# Contents

**1   Introduction to surfactant biodegradation**                                  **1**

D. BROWN

1.1   Introduction                                                                   1
1.2   Biodegradability and the replacement of soap in detergent
      products                                                                       2
1.3   Biodegradation, biodegradability testing and interpretation of
      results                                                                        5
      1.3.1   Biodegradation                                                         5
      1.3.2   Assessment of biodegradability                                         7
1.4   Sewage treatment processes and their significance for
      surfactants                                                                    8
1.5   Biodegradability test methods for surfactants                                 10
      1.5.1   Biodegradability test methodology                                      11
1.6   The replacement of tetrapropylene benzene sulphonate
      (TPBS)                                                                         15
1.7   Surfactant biodegradability and legislation                                   16
      1.7.1   Anionic surfactants                                                    16
      1.7.2   Non-ionic surfactants                                                  18
      1.7.3   Cationic and ampholytic surfactants                                    18
1.8   Surfactant analysis                                                            18
      1.8.1   Anionic surfactant analysis                                            19
      1.8.2   Non-ionic surfactant analysis                                          20
      1.8.3   Cationic surfactant biodegradability and analysis                      22
      1.8.4   Ampholytic surfactants                                                 23
1.9   Conclusions and thoughts for the future                                       23
References                                                                           25


**2   What is biodegradation?**                                                    **28**

G.F. WHITE and N.J. RUSSELL

2.1   Setting the scene: Microbial nutrition and the carbon cycle                   28
2.2   Surfactants as potential microbial nutrients                                  30
      2.2.1   Accessing the hydrophobic chain                                        31
      2.2.2   Hydrophile degradation in non-ionics                                   37
      2.2.3   Primary versus ultimate biodegradation                                 39
      2.2.4   Anaerobic versus aerobic metabolism                                    40
2.3   Surfactant biodegradation in the environment                                  42
      2.3.1   Pure versus mixed cultures                                             42
      2.3.2   Consortia                                                              44
      2.3.3   Adaptation to surfactants                                              45
      2.3.4   Substrate concentration                                               52
      2.3.5   Mixed substrates                                                       53
      2.3.6   Cometabolism                                                           54
      2.3.7   Retardation by nutrient limitation and predation                       55
      2.3.8   Surfaces and biofilms                                                  56
2.4   Laboratory models of surfactant biodegradation                                58
References                                                                           59

## 3   Biodegradability testing                                      65
H.A. PAINTER

3.1   Introduction                                                    65
      3.1.1   Glossary of terms                                       66
      3.1.2   Early tests                                             68
      3.1.3   Development of the tests                                69
3.2   Nature of biodegradation and influencing factors               72
      3.2.1   Composition of medium                                   72
      3.2.2   Inocula                                                 73
      3.2.3   Physico-chemical factors                                74
      3.2.4   Test substances                                         74
      3.2.5   Control vessels                                         75
      3.2.6   Duration of test                                        76
      3.2.7   Reference compounds                                     77
3.3   OECD and EEC tests for primary biodegradability of
      surfactants                                                     77
      3.3.1   OECD Static test procedure or screening test            77
      3.3.2   OECD Confirmatory test: continuous simulation of
              activated sludge process                                82
3.4   List and synopses of existing methods for ultimate
      biodegradability                                                87
      3.4.1   Ready biodegradability                                  89
      3.4.2   Inherent biodegradability                               94
      3.4.3   Simulation methods                                      97
      3.4.4   Comparison of methods: accuracy and precision           98
3.5   Analytical methods                                             100
      3.5.1   General                                                100
      3.5.2   Anionic surfactants                                    102
      3.5.3   Non-ionic surfactants                                  106
      3.5.4   Cationic surfactants                                   111
References                                                           114


## 4   Testing strategy and legal requirements                      118
H.A. PAINTER

4.1   Selection of tests: strategy of testing                       118
      4.1.1   Primary biodegradability                               118
      4.1.2   Ultimate biodegradability                              119
      4.1.3   Other tests                                            123
4.2   Validation and interpretation of results                      124
      4.2.1   Validity of results                                    124
      4.2.2   Pass levels                                            125
      4.2.3   Interpretation                                         126
4.3   Legal requirements                                            127
      4.3.1   EEC                                                    127
      4.3.2   Other countries                                        129
      4.3.3   USA                                                    129
      4.3.4   EEC Dangerous Chemicals Directives 67/548,
              79/831, 90/C 33/03                                     130
      4.3.5   The future                                             130
References                                                           132

## 5 Biodegradability of anionic surfactants  134
J. STEBER AND H. BERGER

| | | |
|---|---|---:|
| 5.1 | General characteristics of anionic surfactants | 134 |
| 5.2 | Application of anionic surfactants and their environmental relevance | 134 |
| | 5.2.1 Synthetic anionic surfactants and the detergent problem | 135 |
| | 5.2.2 Legal requirements of biodegradability of anionic surfactants | 135 |
| | 5.2.3 Anionic surfactants in the surface waters | 137 |
| 5.3 | Particular structure and application features of anionic surfactants | 138 |
| | 5.3.1 Soaps | 139 |
| | 5.3.2 Alkylbenzene sulfonates | 140 |
| | 5.3.3 Alkane sulfonates | 141 |
| | 5.3.4 $\alpha$-Olefine sulfonates (AOS) | 142 |
| | 5.3.5 $\alpha$-Sulfo fatty acid esters (methyl ester sulfonates, FES) | 143 |
| | 5.3.6 Fatty alcohol sulfates (AS) | 143 |
| | 5.3.7 Alcohol ether sulfates (AES) | 144 |
| | 5.3.8 Sulfosuccinates | 145 |
| | 5.3.9 Alkylphosphates and alkyl ether phosphates | 146 |
| 5.4 | Biodegradation of anionic surfactants | 146 |
| | 5.4.1 Soaps | 146 |
| | 5.4.2 Alkyl benzene sulfonates | 149 |
| | 5.4.3 Secondary alkane sulfonates (SAS) | 159 |
| | 5.4.4 Alpha olefine sulfonates (AOS) | 162 |
| | 5.4.5 $\alpha$-Sulfo fatty acid esters/methyl estersulfonates (FES) | 163 |
| | 5.4.6 Fatty alcohol sulfates (AS) | 167 |
| | 5.4.7 Alcohol ether sulfates (AES) | 171 |
| | 5.4.8 Sulfosuccinates | 175 |
| | 5.4.9 Phosphate ester surfactants | 178 |
| References | | 179 |

## 6 Biodegradability of cationic surfactants  183
C.G. VAN GINKEL

| | | |
|---|---|---:|
| 6.1 | Introduction | 183 |
| 6.2 | Biodegradability of cationic surfactants in OECD Screening tests | 185 |
| | 6.2.1 OECD screening test results | 185 |
| | 6.2.2 Influence of toxicity on the biodegradation | 190 |
| | 6.2.3 Generalizations on biodegradability of quaternary ammonium salts | 191 |
| 6.3 | Behaviour of quaternary ammonium salts in waste water treatment plants | 192 |
| | 6.3.1 Activated sludge plants | 192 |
| | 6.3.2 Absorption of cationic surfactants onto particles | 193 |
| | 6.3.3 Removal of cationic surfactants in activated sludge reactors (bioreactors) | 193 |
| | 6.3.4 Anaerobic biodegradation | 195 |
| | 6.3.5 Influence of cationic surfactants on biological processes | 196 |
| 6.4 | Biodegradation routes of quaternary ammonium salts | 196 |
| | 6.4.1 Tetramethylammonium chloride | 196 |
| | 6.4.2 Ethyltrimethylammonium chloride | 197 |
| | 6.4.3 Alkyltrimethylammonium salts | 198 |
| | 6.4.4 Possible formation of recalcitrant intermediates | 200 |
| References | | 200 |

## 7 Biodegradability of non-ionic surfactants 204

T. BALSON and M.S.B. FELIX

| | | |
|---|---|---|
| 7.1 | Introduction | 204 |
| 7.2 | Structure of polyglycol surfactants | 205 |
| | 7.2.1 Preparation and properties of alkylene oxides and their polymers | 206 |
| | 7.2.2 Initiators used for polyglycol surfactants | 209 |
| | 7.2.3 General properties of polyglycol surfactants | 209 |
| | 7.2.4 Applications of polyglycol surfactants | 210 |
| 7.3 | General biodegradability | 212 |
| | 7.3.1 Primary biodegradation | 213 |
| | 7.3.2 Ultimate biodegradation | 213 |
| | 7.3.3 Inherent biodegradation | 213 |
| | 7.3.4 Ready biodegradation | 213 |
| | 7.3.5 Analytical methods for ready and inherent biodegradation tests | 214 |
| | 7.5.6 The microorganisms | 215 |
| 7.4 | Mechanisms of biodegradation | 215 |
| | 7.4.1 ω-Oxidation | 216 |
| | 7.4.2 β-Oxidation | 216 |
| | 7.4.3 α-Oxidation | 217 |
| | 7.4.4 Alkoxylate chains | 217 |
| 7.5 | Biodegradation of polyglycols | 219 |
| | 7.5.1 Nonyl phenol ethoxylates (NPE) | 219 |
| | 7.5.2 Fatty alcohol ethoxylates (FAE) | 220 |
| | 7.5.3 Fatty alcohol alkoxylates (FAA) | 222 |
| | 7.5.4 Alternative low-foam non-ionics to fatty alcohol EO/PO copolymers | 226 |
| | 7.5.5 Linear block copolymers | 227 |
| | 7.5.6 Fatty amine ethoxylates | 229 |
| | Acknowledgements | 229 |
| | References | 229 |

## 8 Biodegradability of amphoteric surfactants 231

A. DOMSCH

| | | |
|---|---|---|
| 8.1 | Introduction | 231 |
| 8.2 | Structural elements and biodegradation in general | 232 |
| 8.3 | Alkyl betaines | 234 |
| | 8.3.1 Chemical structure | 234 |
| | 8.3.2 Properties, application | 236 |
| | 8.3.3 Primary degradation | 236 |
| | 8.3.4 Ultimate degradation | 237 |
| 8.4 | Alkylamido betaines | 237 |
| | 8.4.1 Chemical structure | 237 |
| | 8.4.2 Properties, application | 238 |
| | 8.4.3 Primary degradation | 238 |
| | 8.4.4 Ultimate degradation | 238 |
| 8.5 | Sulphobetaines and hydroxysulphobetaines | 238 |
| | 8.5.1 Chemical structure | 238 |
| | 8.5.2 Properties, application | 241 |
| | 8.5.3 Primary degradation | 241 |
| | 8.5.4 Ultimate degradation | 242 |
| 8.6 | Alkylamphoacetates | 243 |
| | 8.6.1 Chemical structure | 243 |

|          | 8.6.2 | Properties, application  | 245 |
|          | 8.6.3 | Ultimate degradation     | 245 |
| 8.7      | Polycarboxyglycinates  |              | 245 |
|          | 8.7.1 | Chemical structure       | 245 |
|          | 8.7.2 | Properties, application  | 247 |
|          | 8.7.3 | Primary degradation      | 247 |
|          | 8.7.4 | Ultimate degradation     | 247 |
| 8.8      | Alkylamphopropionates  |              | 247 |
|          | 8.8.1 | Chemical structure       | 247 |
|          | 8.8.2 | Properties, application  | 250 |
|          | 8.8.3 | Ultimate degradation     | 250 |
| 8.9      | Imidazolinium betaines |              | 250 |
|          | 8.9.1 | Chemical structure       | 250 |
|          | 8.9.2 | Degradation              | 251 |
| Acknowledgements |       |                  | 252 |
| References |       |                            | 252 |

**Index**                                                   **255**

# 1   Introduction to surfactant biodegradation

D. BROWN


## 1.1   Introduction

The term surfactant, or surface active agent, is applied to organic molecules whose function is to promote mixing or dispersion between phases of a mixture by lowering the interfacial tension between these phases. For most applications one of these phases is water and the other phase is hydrophobic. Thus, surfactants can be used in a whole range of technical and industrial products where it is required to disperse hydrophobic materials in water or *vice versa*. The major use of surfactants, in terms of the quantities used, is as a component in cleaning preparations or detergents. Such cleaning operations usually result in the discharge of an aqueous effluent, and it is the biodegradability of the surfactants in that effluent, which is the subject of this book.

The earliest known manufactured surfactants are soaps which are the sodium salts of natural, saturated and unsaturated fatty acids formed from the alkaline hydrolysis of animal and plant triglycerides (fats and oils). Recipes for soap manufacture have been found on papyri and clay tablets from ancient civilisations in Egypt and the Tigris/Euphrates (Bock and Stache 1982) and the science of soap-making has progressed through the soap boilers' guilds of the Middle Ages to the sophisticated products of present day commerce.

$$
\begin{array}{l}
CH_2OCOR \\
| \\
CH_2OCOR \quad + \quad 3NaOH \quad \longrightarrow \\
| \\
CH_2OCOR
\end{array}
\qquad
\begin{array}{l}
CH_2OH \\
| \\
CHOH \quad + \quad 3RCOONa \\
| \\
CH_2OH
\end{array}
$$

Triglyceride                                      Glycerol          Soap

Soaps and other synthetic surface active agents used for cleaning purposes are molecules in which there is a hydrophobic group and a hydrophilic group. The nature of the hydrophilic group and the balance between the hydrophilic and hydrophobic parts of the molecule determine which particular surfactants are appropriate.

An often used broad classification of surfactants is based on the charge characteristics of the hydrophilic part of the molecule. Anionic surfactants have a negatively charged hydrophile and include soaps, sulphonates, sulphates; non-ionic surfactants have an uncharged hydrophile often a polyglycol; cationic surfactants are often based on a quaternary ammonium hydrophile while the fourth main class is the amphoterics where the hydrophile contains both positive and negative charges, e.g. an amino carboxylic acid. A more comprehensive description of the different surfactant types is contained in subsequent chapters of this book and has been described by Porter (1991).

As previously mentioned, the major use of surfactants is as a component of cleaning preparations and in this application and several others, essentially all the surfactant is discharged to drain. The major end-uses have been reviewed by Richtler and Knaut (1988) and, as well as home use for personal hygiene, washing and cleaning, these uses include industrial cleaning, textile and leather auxiliaries, emulsifiers, paint additives, oilfield chemicals, etc. These same authors give detailed statistics and trends in consumption and point out the major difference in annual use of cleaning compounds in Western Europe and the USA (10 kg/person per year) compared with the world average (4 kg/person per year).

The estimated 1987 total surfactant consumption (Richtler and Knaut, 1988) for USA (45%), Western Europe (38%) and Japan (17%) is 6.6 M tonne/year with an approximately 1:1 split between household and other uses. Six surfactants together make up approximately 60% of this total consumption, soap (1.5 M tonne/year), linear alkylbenzene sulphonates (1 M tonne/year), alcohol ethoxylates (0.5 M tonne/year), alkyl phenol ethoxylates (0.47 M tonne/year), alcohol ether sulphates (0.35 M tonne/year) and alcohol sulphates (0.25 M tonne/year). Clearly without biodegradation the environmental burden would be enormous.

## 1.2   Biodegradability and the replacement of soap in detergent products

As mentioned in the introduction, soaps have been manufactured and used by mankind for thousands of years and are still very widely employed for personal hygiene and other washing purposes. The statistics above show that soap is the major surfactant in Western Europe, USA and Japan and is even more dominant in the developing countries. Why then did the biodegradability of detergents only become an issue in the middle of the 20th century when the use of soap as the main surfactant in domestic detergents began to be replaced by alternative products?

From the viewpoint of the 1990s, where awareness of environmental issues is sharply focused, two general answers to this question may be made. Firstly, the biodegradability of all substances released to the environment (including soap) is an important factor in defining the levels of a substance in the environment and hence assessing its potential for causing environmental damage (environmental risk assessment). Secondly, based on the so-called precautionary principle, where very major quantities of a substance are released to the environment, regardless of whether the anticipated levels in the environment appear likely to cause harm, the substance should be biodegradable to safeguard against the possibility of future harm due to build-up in the environment.

However, in earlier times and specifically for synthetic surface active agents in the post-World War II period, environmental problems were tackled on an *ad hoc* basis and in general attention was paid only where problems were manifest. The problem which became all too manifest with certain synthetic surface-active agents was foam. Foam at sewage works, foam on rivers and indeed in certain localised areas, below waterfalls, even foam spreading from the rivers to the street (Standing Technical Committee on Synthetic Detergents (STCSD), 1958).

This problem of foam due to the synthetic surface agents being introduced into domestic detergents was certainly not confined to any one country and major investigations into the cause of the problem, delineation of adverse effects other than foam and ways to rectify the situation were made in the USA, Germany, France, the UK and other industrial countries. If fact, this problem was probably the first example of a specific environmental problem attributable to a particular type of material as opposed to general air or water pollution problems caused by the discharge of domestic or industrial wastes direct to the environment.

Two particular attributes of soap, as opposed to the synthetic surface active agents introduced post-World War II, were quickly identified as being the explanation of why foaming problems had not been caused by soap formulations. Soap, as the sodium salt of a fatty acid, reacts with the calcium and magnesium ions in natural waters to form insoluble calcium and magnesium soap (seen as 'scum' in hardwater areas), a major drawback to its use in domestic laundry applications where large amounts of polyphosphate are necessary to prevent deposition of these insoluble soaps in the fabrics being washed. This property of soap, in forming water-insoluble materials with bivalent ions, also means that, once released into water either direct or via a sewage works, it will be immediately and effectively removed as a surface active material. Furthermore, based on our present knowledge of what is, or is not, likely to be easily biodegraded, soap as the sodium salt of fatty acids is expected to be well degraded and indeed a high level of degradation of soap has been demonstrated (Swisher, 1987). Interestingly, however, Swisher makes the comment that the lack of any foaming problems attributable to soap is more due to the insolubility of the calcium and magnesium salts than to biodegradability. To put this Swisher comment into context, it is not being suggested that soap is poorly biodegradable, but rather that even if it were, the chemistry of soap is such that foaming would not be a problem. This, in a more generic sense, is useful in drawing our attention to the fact that biodegradability is not the sole environmental property of consequence in assessing environmental risk.

The sulphation and sulphonation chemistry necessary to replace soap with anionic surfactants which did not suffer the drawback (to the user, if not to the environment) of forming insoluble calcium and magnesium salts began to be developed in the latter part of the 19th century. The sulphation of unsaturated

oils such as castor oil by reaction with concentrated sulphuric acid to give the
so-called Turkey Red oil was developed around 1860 (Fieser, 1950) although
this product was not technically suitable as a domestic detergent ingredient.

$$CH_3(CH_2)_5—CH—CH_2=CH(CH_2)_7COOH$$
$$|$$
$$OSO_3H$$

Turkey Red oil

Further developments in Germany and the USA prior to World War II gave sulphated and sulphonated products primarily based on natural alcohols and oils.
However, as the availability of hydrocarbon feedstocks and the ability of synthetic chemists and chemical engineers increased so did the availability of a particularly cost-effective replacement for soap in the domestic detergent market,
namely the material known as tetrapropylene benzene sulphonate (TPBS) or
sometimes more generally as alkyl benzene sulphonate (ABS). The use of this
product became very widespread in most industrialised countries around 1950
and in the United States, at this time, TPBS was used for most laundry detergents (Swisher, 1987).

   TPBS is made by the alkylation of benzene with propylene tetramer using
Friedel–Crafts type catalysts followed by sulphonation of the alkyl benzenes so
produced. This process gives a whole mixture of isomeric and homologous
materials. A gas chromatogram of the alkylbenzene indicates at least 100 components (Kaelbe, 1963) and the sulphonation process, although predominantly at
the *para* position, will also give other isomers. Faced with this large mixture of
chemical entities, a situation also found with many other synthetic surfactants
including those in use today, scientists wishing to investigate the biodegradability and environmental levels of the product had to adopt approaches rather different from those of classical organic analysis where single compounds are
studied.

   The analytical methods used to follow the biodegradation of surfactants (or
indeed any other organic substance) play a fundamental role in defining what is

$$C_{12}H_{25}$$

$$SO_3Na$$

Tetrapropylene benzene sulphonate (TPBS).

meant by 'biodegradability' in any particular test method or environmental situation. In a historical context, it is important to appreciate that the commonly used MBAS (methylene blue active substance) and BiAS (bismuth active substance) analytical methods for anionic and non-ionic surfactants, respectively, are limited both in the types of anionic and non-ionic surfactant which respond to those methods, and also limited in terms of what is meant by 'biodegradability' when surfactants which do respond are assessed by these methods. These remarks will be elaborated further in following sections of this chapter.

## 1.3   Biodegradation, biodegradability testing and interpretation of results

The three questions, what is biodegradation, how do you measure it and how do you interpret the results, form an exponential cascade of complexity in terms of the difficulty with which answers can be given. This section is aimed at giving the reader at least some insight into the issues for surfactants (and other organic materials) which those questions attempt to address.

### 1.3.1   Biodegradation

This, in its simplest definition, is the breakdown of an organic substance by living organisms but for most purposes, and certainly for the purpose of this book, this definition is narrowed to the breakdown of an organic substance by microorganisms. Even with this rather narrower definition, four other questions immediately open up, namely 'breakdown to what?', 'under what conditions?', 'with what microorganisms?' and 'at what rate?'.

Within the science of biodegradation the 'breakdown to what' question is usually answered in one of two, or possibly three, ways.

'Primary' biodegradation is effectively defined as the breakdown of the substance as measured by a substance-specific analytical method. To take a simple example, an analytical method such as gas chromatography (GC) or high performance liquid chromatography (HPLC) might be set up to monitor the primary biodegradation of a substance such as phenol by the disappearance of the phenol from the chromatograph. With phenol, as with many other single compound substances, it is possible to measure 100% primary biodegradation using compound-specific analytical methods. With complex mixtures, other analytical methods, which will be described later for surfactants, may be used. Alternatively, loss of a key property, such as surface activity/foaming potential, may be used as an indicator of primary biodegradation.

'Ultimate' biodegradation is the second main way in which the 'breakdown to what' question may be answered and 'ultimate' biodegradation may be defined as the complete breakdown of an organic substance to wholly inorganic materials and natural cellular material. That is, to show the complete ultimate degradation of phenol, it would be necessary to demonstrate not only the disappearance of the specific analytical response of phenol, but also that no other organic metabolites resistant to the particular biodegradation conditions are being formed. This analytical task is extremely difficult even for a simple substance such as phenol, and to be carried out with full scientific rigour requires the synthesis of carbon-14 labelled material and rather elaborate experimentation. In practice, ultimate biodegradability is generally assessed in an aerobic system by measuring carbon dioxide produced, by measuring the oxygen consumed during the biodegradation process ('biochemical oxygen demand' or BOD) or by measuring the level of organic carbon remaining in solution during the time course of the biodegradation study. Each such method has its own intrinsic problems leaving aside any experimental problems associated with the measurements themselves.

The measurement of carbon dioxide production or the expression of BOD is a positive indication that some measure of ultimate degradation has occurred. However, microorganisms, like higher organisms, use biodegradable organic materials not only as an energy source but also as a food source to build up their cellular mass. Depending on the nature of the substance in question and the food supply available to the microorganisms (the more food, the more they 'put on weight'), only approximately 60–70% of the theoretical carbon dioxide production or BOD will be found for even a very easily biodegradable substance such as sodium acetate.

The measurement of dissolved organic carbon (DOC) in solution as an indication of ultimate degradation can both in principle and in practice indicate high levels of biodegradation since any of the substrate converted into cellular biomass will be removed by filtration or centrifugation. However, it is not applicable to substances which are sparingly soluble in water or strongly sorptive, and can also be misleading where sparingly soluble or sorptive metabolites are formed. A combination of a high level of ultimate biodegradation as shown by DOC removal (say 95%+) and a high level of ultimate degradation as indicated by carbon dioxide production or BOD (say 60%+) does provide confidence that a high level of ultimate degradation has taken place.

A third way in which the 'breakdown to what' question may be answered is sometimes answered explicitly or implicitly in terms of 'environmentally acceptable' biodegradation. In practice, the methods developed to determine the biodegradation for surfactants implicitly define the biodegradation measured as 'environmentally acceptable'. However, before dealing with the ways in which the biodegradability of surfactants has been and is assessed, some more general elaboration of what is meant by 'environmentally acceptable' will be given.

Many substances of interest and use to mankind are considerably more complex in chemical structure than simple molecules such as phenol, sodium acetate, etc. which can relatively easily be shown to undergo essentially complete and rapid ultimate biodegradation in simple test systems. Furthermore, many of these more complex substances are multi-compound mixtures and, as has already been mentioned, many surfactants of commercial interest fall into this category of multi-compound mixtures of relatively complex molecules. For this type of substance, although it may be relatively easy to demonstrate a high level of primary degradation, a high level of ultimate degradation may only be obtained over a relatively long time-frame and may be difficult, if not impossible, to demonstrate experimentally.

In this situation, biological experiments coupled onto the biodegradation experiment can be used to demonstrate whether the biodegradation process has resulted in metabolites which appear to be more or less toxic than the original material. For surfactants, which in general show a significant toxicity to aquatic organisms such as fish, *Daphnia*, algae, etc. there are a number of reported studies (e.g. Brown, 1976; Reiff, 1976; Kimerle and Swisher, 1977; Neufahrt *et al.*, 1987) which show that the biodegradation of a number of surfactants proceeds with loss or reduction of their toxicity to aquatic organisms.

Having given some indication of the implications of three terms, 'primary', 'ultimate' and 'environmentally acceptable', which may be used to describe biodegradation, the other three questions stemming from the definitions 'under what conditions', 'with what microorganisms' and 'at what rate' will be considered in the context of biodegradability test methods.

### 1.3.2 Assessment of biodegradability

For someone wishing to sell or use a particular surfactant the usual questions asked are either 'is it biodegradable' or 'what is its biodegradability'. If these questions are put in the context of a specified test method which is applicable to that surfactant and for which there is an accepted pass/fail criterion, the answers can be simple and unequivocal; e.g. Yes 94%, or No 18%. In asking these questions and using the answers it is important to understand that the answer is correct only in the context of a test result obtained from a specified test. Different methods could give very different results (e.g. the exemplified answers in the sentence above could be those found for the same surfactant tested by different methods) and, because of the inherent variability of many biodegradability test methods, a test result found in one study could well show a difference when repeated by the same or a different laboratory.

In making the above comments it is not the intention to give the reader the impression that a biodegradability test result has no validity, but to caution that it is not a simple measurement, such as a melting point, for which there is only one correct answer. If then, one is concerned with risk assessment rather than a specific test result, one may wish to understand how well a substance is likely to biodegrade in a particular environmental situation, or to evaluate the relative biodegradability of different substances in a specified test procedure, or to establish whether the substance being investigated has the potential to be broken down by microorganisms. Unfortunately, in the area of biodegradability testing, these distinctions in the objectives of testing are not always well defined and this can lead the unwary to very wrong conclusions.

To take two opposite ends of this spectrum, a substance may be termed 'biodegradable' when, after months of difficult work, a scientist has managed to culture organisms which when fed the substance as the sole carbon source manage to grow and break the substance down. On the other hand a substance may be termed 'not readily biodegradable' because in a rather artificial test for 'ready biodegradation' (OECD, 1993) it does not meet very stringent rate requirements to achieve a particular level of degradation within a 10-day window. The first substance may, in practical terms, be essentially non-degradable while the second may reach very high levels of biodegradation in a real environmental situation.

Having defined three possible objectives of biodegradability assessment, there are also the questions 'under what conditions', 'with what microorganisms' and 'at what rate' posed above. In practice, these questions are effectively linked since the conditions essentially define what microorganisms will be present and viable and the rate is usually defined by the test protocol.

## 1.4  Sewage treatment processes and their significance for surfactants

Before further discussing surfactant biodegradation, an outline of the main sewage treatment processes and their significance for surfactants will be given. Modern society largely takes for granted the amenity and health benefits of a water-borne waste system. The expectation is that virtually all domestic waste can be flushed down the drains and the resulting mixture of human excrement, food waste, soap, detergents and other wastes, both solid and liquid, will somehow be dealt with. In the industrial arena a similar view was long held, although, increasingly, environmental laws and pressures are leading to a better control and understanding of what can and cannot be discharged via the drains as a liquid effluent.

If significant quantities of water-borne sewage are discharged without treatment, gross pollution is the inevitable result. In most developed countries a high proportion of sewage effluent is treated before discharge to fresh water and treatment prior to discharge into estuaries and the sea is beginning to be required by European Community legislation (EEC, 1991).

A sewage treatment works has a primary treatment stage in which solids are removed from the total sewage flow by screens and settling tanks. The settled sewage, which is still highly polluting, may be discharged after primary treatment, but more usually goes forward to a secondary biological treatment process where a high degree of purification takes place.

The basis of the secondary treatment process is the same as the natural treatment processes which would eventually occur if the sewage were to be discharged untreated. That is, aerobic bacteria present in the natural water and the sewage biodegrade the degradable organic components of the sewage. This process requires oxygen and one reason why the discharge of untreated sewage is polluting is that where high levels are discharged, the receiving water becomes anoxic and can no longer support a fish, or aquatic invertebrate, population.

For the secondary sewage treatment process, a high level of bacteria is provided to increase the rate of the purification process and, at the same time, a high level of aeration to maintain oxygen levels. Two main methods of secondary treatment are in common use.

For the 'trickling filter' method, the settled sewage is sprinkled over filter beds containing stones or other types of aggregate. These filter beds are constructed to give a chimney effect so that air is also drawn through the bed. As the sewage percolates through the bed it comes into contact with the bacteria and other microorganisms which grow on the stones.

The effluent leaving the bottom of the bed contains a relatively high level of suspended solids which is predominantly the bacterial biomass which has to slough off the filter beds to prevent the clogging phenomenon known as 'ponding'. The suspended solids are removed in a further settlement tank ('humus' tank) to give a treated effluent usually of a quality suitable for discharge.

The 'activated sludge' treatment process, which is more widely used in large modern treatment plants involves treatment of the settled sewage by mixing it with a high concentration of bacterial biomass (activated sludge) which is held in suspension by the air necessary to keep the system aerobic. This air is supplied either by surface aeration or by submerged diffusers.

The activated sludge concentration in this type of system is usually in the range 2–4 g dry solids per litre and this is removed from the treated sewage in settling tanks before discharge. Most of the settled activated sludge is returned to the activated sludge treatment tanks, but because the degradation process also produces more biomass a proportion is wasted to maintain a balance. The ratio of activated sludge retained to that wasted is essentially the 'sludge age' which is usually expressed in days. The significance of 'sludge age' is picked up later in this chapter.

In the description of primary treatment, both the trickling filter and activated sludge secondary treatment processes, the reader will have noted that sludge solids are being removed and the costs of sludge handling and disposal can be a very significant part of the total costs of sewage treatment.

A widely used sludge treatment process is heated anaerobic digestion in which the sludge is held at approximately 35°C for an average period of say 3 weeks. During this time a large proportion of the organic content is reduced to predominantly methane (used as a fuel for the site) and the resulting sludge is essentially pathogen-free, has a much lower odour and is considerably easier to dewater. Both digested and undigested sludge may be used as a fertiliser on agricultural land and hence provide a route whereby substances contained in the sludge could reach crops and/or cattle. In practice, studies have shown that significant contamination of sludge-treated agricultural land by surfactants does not appear to occur (Holt et al., 1989).

Having described the sewage treatment processes in some detail, it is timely to return to the question of surfactants and the significance of those processes for the treatment of surfactants. As has already been mentioned, domestic sewage is a rather complex mixture. Statistics for water used in the industrialised countries vary, but a water use figure of 200 l/head per day is often quoted. The undissolved solids content of sewage is likely to be around 300–400 mg/l (expressed as dry weight) and the chemical oxygen demand (COD) of settled sewage is 200–500 mg/l (Brown et al., 1986) corresponding to an organic matter content of, say, 100–300 mg/l.

Statistics given in the reports of the UK Standing Technical Committee on Synthetic Detergents (STCSD) indicate that the level of anionic surfactants (excluding soap) in UK sewage is of the order of 10–20 mg/l and data from Germany (Brown et al., 1986) gave anionic surfactant levels in settled sewage in the approximate range 6–17 mg/l and of non-ionic surfactants in the range 3–10 mg/l. From these data, it is apparent that surfactants form a relatively small proportion of the total organic content of sewage but are nevertheless likely to be the major class of man-made organic substances in sewage.

From the point of view of the overall efficiency of the sewage treatment process in removing surfactants from the aqueous phase, sorption and biodegradation mechanisms may both play a role. However, the varying flows, concentrations and nature of the substrate make for great difficulty in defining precisely the relative contributions of these two removal mechanisms. What is clear for anionic and non-ionic surfactants is that, when poor biodegradation is experienced, the levels of surfactant in the final effluent are unacceptable. For cationic surfactants where sorption processes are likely to play a major role in removal from the aqueous phase, it is possible that, as with soap (referred to earlier), high levels of removal from the aqueous phase could be obtained regardless of whether or not the substance is biodegradable (Matthijs et al., 1994).

In the historical context of this chapter, very considerable analytical efforts have been put to monitoring the removal of anionic surfactants during sewage treatment. The analytical method used has almost always been the 'methylene blue active substance' (MBAS) method which, as will be described later, gives an analytical response to alkyl benzene sulphonates and alkyl sulphates but not anionic surfactants such as soap or alkyl phosphates. MBAS is also a convenient non-specific methodology for estimating anionic surfactants in environmental samples although with the advance of analytical science, specific procedures are becoming available for individual anionic surfactants.

## 1.5   Biodegradability test methods for surfactants

In practice, most biodegradation test methods start from the premise that aerobic biodegradability at a sewage treatment works and/or aerobic biodegradability in natural water (usually fresh) is of prime interest and for substances such as surface active agents whose main route to the environment is via a sewage works, this approach is very reasonable.

Other test methods addressing biodegradation in anaerobic systems (sewage sludge, digestion, anoxic sediments) are beginning to be developed and there are established methods for soil scientists to examine the biodegradability of organic substances in soil. However, for surfactants used in detergent formulations, where the prime requirement is for satisfactory sewage works biodegradation, anaerobic and soil test methods are a secondary consideration and will not be further discussed in this chapter.

As already indicated, one important factor in a biodegradability test method is the analytical procedure used to follow the course of the biodegradation; the other is, naturally, the actual test procedure. Both these issues are dealt with in more detail in a later chapter although both are important in an historical context.

### 1.5.1 Biodegradability test methodology

A number of types of test were and are used for assessing the biodegradability of surface active agents. Many organisations contributed to the development of test methodology including in the UK the Standing Technical Committee on Synthetic Detergents (STCSD), in Germany the essentially equivalent group the Hauptauschuss Detergentien (HAD), individual companies, and a number of trade associations such as the US Soap and Detergent Association (SDA).

However, the Organisation for Economic Co-operation and Development (OECD) took a lead through its Environmental Directorate in establishing standard methods which have a high level of international acceptability. This work of the OECD in defining agreed tests for assessing the biodegradability of surfactants was essentially completed in 1976. Since that time, the OECD has continued to be active through its chemicals testing programme in developing a whole range of test guidelines for evaluating physical/chemical, toxicological and environmental properties of chemicals (OECD, 1981) in general. These latter test guidelines continue to be updated (OECD, 1993) and the reader is cautioned not to confuse these general test guidelines with the surfactant specific biodegradability methods referred to below:

*(i) 'Die-away tests'*. The first type, exemplified by the OECD so-called 'static' or 'screening' test (OECD, 1976) is operated with a low concentration of microorganisms on the 'die-away' principle. In essence, the test substance at a specified concentration is placed in a vessel with inorganic nutrients and a bacterial inoculum. The test concentration is measured periodically over the period of the test. For the OECD static test for surfactants, the initial concentration is 5 mg/l and the test period is up to 19 days. The MBAS (methylene blue active substance) and BiAS (bismuth active substance) analytical methods are used for anionic and non-ionic surfactants, respectively. The biodegradability result is

the % biodegradation at 19 days or a plateau level of degradation if this occurs earlier. If this biodegradability result exceeds 80% the surfactant is considered, in the context of this test, to be 'biodegradable'. If the result is less than 80%, the surfactant is further tested in the sewage works simulation test described below.

*(ii)   Sewage works simulation test.*   A second type of test exemplified by the OECD 'simulation' or 'confirmatory' test procedure (OECD, 1976) is considerably more complex in design and operation. In outline, the surfactant at a test concentration of 10 mg/l (anionic surfactants) or 5 mg/l (non-ionic surfactants) is fed in a synthetic sewage to a laboratory simulation of an activated sludge sewage treatment works. The concentration of the surfactant in the effluent is monitored periodically (MBAS or BiAS) during a running-in period until such time as the level of biodegradability appears to have reached a plateau, when an assessment period of 21 days is started. During this assessment period, which must commence within 42 days of the start of the experiment, the average level of biodegradation is calculated from at least 14 separate results. In the context of this test, an average result in excess of 80% is considered to be 'biodegradable' and in practice most anionic and non-ionic surfactants used in domestic detergent formulations commonly give results in this test in excess of 90%.

Two main experimental variants of the above, also known as the continuous activated sludge (CAS) method, are used. The first is based on the Husmann apparatus and is described in the OECD method (OECD, 1976). The second is based on the porous pot method (Water Research Centre (WRC), 1978) in which, by contrast to the Husmann apparatus where gravity separation is used, a porous polyethylene sleeve separates the final effluent from the activated sludge.

Historically, some work was carried out on the biodegradability of anionic surfactants using a laboratory simulation of both trickling filter and activated sludge sewage works using a laboratory-produced detergent-free sewage based on human excreta (Truesdale *et al.*, 1959). Although this sewage was clearly likely to be more representative of real sewage than the synthetic sewage used in the OECD methods, for hygiene and aesthetic reasons tests based on real sewage have not been widely used and have no regulatory status. Similarly, studies using laboratory apparatus designed to simulate the sewage works trickling filter process, although still used in some treatability studies (can a particular waste be treated satisfactorily), are not generally used for the determination of the biodegradability of a specific substance.

*(iii)   Other biodegradability methods and terms.*   The two general methods noted above (the die-away test and the activated sludge sewage works simulation test) are both the major methods of historical importance for investigating the biodegradability of surfactants and also the basis of current European legislation controlling the biodegradability of surfactants to be used in detergents. However, surfactants are but a subset of organic chemicals and the methods and

terms used opposite the biodegradability of organic chemicals in general need some understanding in order to avoid possible confusion.

*Primary biodegradation.*   This has already been defined as breakdown of the substance as measured by a substance specific, or loss of key property, analytical method. A number of biodegradation test methods can be used to measure primary biodegradation and the strictly correct statement of a result is '*X%*' primary biodegradation using biodegradation test method 'A' and analysis by method 'B'.

*Ultimate biodegradation.*   This again has already been defined as complete breakdown to wholly inorganic materials (also called 'mineralisation') and cellular matter. Like 'primary biodegradation', ultimate degradation is defined by the analytical methodology as well as the specific biodegradability test method and the results should be similarly expressed.

*Ready biodegradation.*   The term 'ready' in this context has a very specific meaning as originally defined by the OECD (see OECD, 1981, 1993) and now incorporated into an EEC Directive (EEC, 1992a). Readily biodegradable in this specific context, means that the substance reaches a specified level of ultimate biodegradability (60% as BOD/COD or $CO_2$; 70% on carbon removal), within a specified time-frame and using one of a suite of 'ready biodegradability' test methods. The test methods are all of the 'die-away' type, use a relatively low concentration of bacterial inoculum and the test substance as the sole carbon source. The principle of these tests, which will be described in more detail in a later chapter, is that they are so stringent that a substance judged to be 'readily biodegradable' will biodegrade in all situations where biodegradation is possible. It is important to note that the converse is not true. A substance which does not meet the 'ready' criteria may in practice be very satisfactorily degraded and in the common English usage of the word 'ready' be readily biodegradable.

*Inherently biodegradable.*   The use of this term tends to be both test method and test result related and can be used opposite both primary biodegradation and ultimate biodegradation. As with 'ready' biodegradability, the concept was introduced by the OECD (see OECD, 1981, 1993) and has been taken on in, for example, the EEC Technical Guidance Document on the Risk Assessment of New Substances (EEC, 1993a). As mentioned above, substances which are not 'readily biodegradable' may in practice be biodegradable and OECD took account of this through 'inherent biodegradability'.
The OECD 'pass' level, for a substance to be considered 'inherently biodegradable' on a primary degradation basis, is 20%. This is based on a philosophy that if a substance will break down 20% in a certain test period (and 20% is probably the level which can be determined with sufficient analytical confidence), then there is no reason, in principle, why it should not break down

100% given sufficient time. That is, it is 'inherently' biodegradable. This concept is reasonable for a single compound where primary degradation is being considered, but requires caution for a mixture of compounds (e.g. most surfactants).

Where ultimate degradation is the end-point, the OECD have proposed 70% as evidence for inherent ultimate biodegradation.

As mentioned above, the term 'inherently biodegradable' tends to be both test result and test method related. Thus, although a material giving a primary degradation test result above 20% in a die-away test could be regarded as inherently biodegradable, the term is usually reserved for a result from a so-called 'inherent' test as defined by the OECD.

The inherent test methodology allows for a high level of bacterial inoculum and an extended contact period for acclimatisation of the microorganisms. There are two main types of inherent test, one based on a die-away or 'static' principle in which the test material is the sole carbon source, and the other on a semi-continuous feed principle in which the test substance forms only part of the carbon source.

The most common inherent 'static' test is that based on the Zahn–Wellens test (Zahn and Wellens, 1974; OECD, 1981, 1993). It has found little or no application in the assessment of biodegradability of surfactants, a major problem being that, unless primary degradation is the end-point, a very high test substance concentration is required. This can cause problems due to toxicity of the test substance to the microorganisms and specifically with surfactants, major foaming problems in the early periods of the test.

An alternative static test is the so-called MITI II (OECD 1981, 1993). This test has a very complex sludge preparation procedure and has found little application outside Japan.

Of considerably more interest in the evaluation of surfactants is the SCAS test (OECD, 1981, 1993). This was originally developed by the US Soap and Detergent Association (SDA, 1965) for evaluating the degradability of branched alkyl benzene sulphonate (ABS) and linear alkylbenzene sulphonate (LABS). The SCAS (semi-continuous activated sludge) or 'fill and draw' method for assessing biodegradability involves feeding an aerated activated sludge on a semi-continuous basis (usually every 24 h) with the test substance in either a real or synthetic sewage. At the end of each time period, the aeration is temporarily suspended and a volume of the supernatant, equal to the volume fed, is drawn off for analysis. This analysis may be either for primary biodegradation, or if a $^{14}$C-labelled compound is available, for ultimate biodegradation using the evolution of $^{14}CO_2$ and/or the level of $^{14}C$ in the aqueous sample. By running a parallel control experiment on the sewage/synthetic sewage alone, it is also possible to run SCAS studies on non-labelled substrate to assess ultimate biodegradation on the basis of the dissolved organic carbon (DOC) levels in the effluent.

The technically important difference between the SCAS type of activated sludge test and the continuous activated sludge (CAS) test as exemplified by the sewage works simulation test used for detergents is that of sludge age. The importance of sludge age or sludge retention time (SRT) in biodegradability/treatability studies has been discussed by Birch (1984). In simple terms a long sludge age, corresponding in sewage treatment practice to a low wastage rate of the activated sludge, allows the maximum opportunity for 'competent' organisms (those capable of breaking down the substance) to develop and multiply. A short sludge age, on the other hand, encourages only those organisms which grow rapidly on easily degraded substances.

The SCAS test, in principle, is operated with an infinite sludge age and thus gives the maximum opportunity for biodegradation of more difficult compounds. The CAS test on the other hand should involve controlled wastage of solids on a regular basis. In practice a major drawback of the Husmann apparatus and its operation as originally described is that uncontrolled wastage of solids can occur giving a variable sludge age and for more difficult substrates variable biodegradability results (Painter and Bealing, 1989). This and other factors need to be taken into account in consideration of the OECD Coupled Units Simulation Test 303A (OECD, 1993).

## 1.6 The replacement of tetrapropylene benzene sulphonate (TPBS)

Having so far in this chapter provided the reader with what is hopefully a useful introduction to some of the complexities of biodegradability testing and interpretation, it is timely to return to the historical position which in the early pages of this chapter was essentially left at the point in the early 1950s when TPBS also known as (branched) alkyl benzene sulphonate (ABS) first began to be used in major quantities in domestic detergents. As was noted earlier, this product is superior to soap in technical performance in that it does not form insoluble salts with bivalent metal ions, but it rapidly demonstrated a very major drawback in failing to biodegrade satisfactorily. The effect of this poorly degradable or 'hard' surfactant was all too manifest as unacceptable foam levels at activated sludge sewage treatment works and in receiving waters.

In response to this problem essential parallel initiatives started in a number of countries including the United Kingdom (STCSD) and Germany (HAD). Manufacturers of TPBS and the detergent formulators were, of course, heavily concerned, collectively to understand the problem, and individually as manufacturers to provide an acceptable alternative.

There are many reviews and papers representing the work carried out to replace ABS, the development of test methodology to assess the biodegradability of proposed replacements and the environmental monitoring carried out to check that the replacements were degrading satisfactorily in the field as well as in the laboratory. Malz (1973) has provided one such review from a German perspective, Eden *et al.* (1967) and Price (1980) from a UK perspective and the US SDA (SDA, 1965) presents a US view.

In essence the net result of very intensive work around the period 1955–1965 may be summarised as follows. Poorly biodegradable ('hard') TPBS for use in domestic detergents was largely phased out in industralised countries and replaced with the biodegradable ('soft') alternative linear alkyl benzene sulphonate (LABS). (The acronym LAS is widely used for linear alkyl benzene sulphonate by many authors. LABS will be used for consistency within all chapters of this book.) Laboratory biodegradability test methods were developed to differentiate the biodegradability of TPBS and LABS and field monitoring activity was started to confirm the biodegradability of anionic surfactants in actual sewage works and river situations.

In the area of monitoring, many authors and organisations have made important contributions. In the UK, the Standing Technical Committee on Synthetic Detergents (STCSD 1958–1979) through their reports provided a historical collection of data from a number of UK river systems and a series of papers from Shell Research Ltd. (Mann, 1971a–c) provide case studies on the effects of introducing new surfactants at a small sewage treatment works. In Germany, the HAD provides a similar historical perspective to that from the UK STCSD, whilst the Henkel Company has made a major contribution in monitoring the Rhine and other German rivers over a long period (Fischer, 1976, 1980) and also individual sewage treatment works (Fisher, 1984).

## 1.7  Surfactant biodegradability and legislation

### 1.7.1  Anionic surfactants

As indicated above and in earlier sections of this chapter, the replacement of 'hard' TPBS by 'soft' LABS was facilitated by using laboratory biodegradation tests to assess whether the new surfactants were more degradable than the old. Also noted was the role of the OECD (OECD, 1976) in standardising these test methods. In Germany, the degradability of the anionic surfactants used in detergents was controlled by legislation, whereas in the UK it was found that a voluntary agreement between the government and the detergent industry was effective in achieving the same ends (STCSD, 1977).

In 1973, the EEC issued two directives (73/404/EEC and 73/405/EEC) relating to the degradability of surfactants used in detergents (EEC, 1973a,b).

Directive 73/404/EEC is the so-called 'frame' Directive and states in Article 2 'that member States shall prohibit the placing on the market and the use of detergent where the average level of biodegradability of the surfactants contained therein is less than 90% for each of the following categories: anionic, cationic, non-ionic and ampholytic.' Two particular points are of note from this Article. First, it is not a general directive on the biodegradability of surfactants but only applies to surfactants used in detergents (cleaning preparations). Second, the biodegradability of individual surfactants is not regulated but the collective biodegradability of the classes specified (anionic, non-ionic, etc.). This second point may, on first consideration, appear to be somewhat unusual. However, the legislation is framed in this way for reasons which become more apparent when the second Directive (73/405/EEC) is examined.

The Directive 73/405/EEC specifies the procedures by which the biodegradability of anionic surfactants contained in detergents shall be determined. These procedures are in two parts, the first relating to the extraction of the anionic surfactant from the detergent preparation and the second to the actual biodegradability measurement which is based on the MBAS method of analysis. It is clear, then, that the regulatory control has to be exercised on the detergent product not on the individual ingredients and furthermore the analytical method can only assess total MBAS substance and not discriminate between individual surfactants. It therefore follows why the frame Directive specifies broad classes of surfactant for regulatory control. In practice, however, control is effectively exercised by the detergent manufacturer who tests, or has his surfactant supplier test, the individual surfactants which he is using in his formulations.

The Directive 73/405/EEC has been updated as Directive 82/243/EEC (EEC, 1982a) and the original Directive and its update contain a number of additional points of note which will be commented on in relation to the update Directive.

Article 2 (82/243/EEC) states, *inter alia*, that 'Member States shall prohibit ... a detergent if the biodegradability of the anionic surfactants ... is less than 80% ... .' This 80% pass level, which was also in the earlier Directive, is a contradiction of the frame Directive which as stated above specifies at least 90% degradation. The reason stated in the Directive 82/243/EEC for this contradiction is that 'suitable tolerances for the measurement of biodegradability should be determined (author 'allowed') in order to take due account of the unreliability of test methods ... .'

Article 2 (82/243/EEC) continues by specifying the test methods for determining the biodegradability of the anionic surfactants in a detergent as the OECD method (OECD, 1976) the German method (Bundesgesetzblatt, 1980), the French method (AFNOR, 1981) and the UK method (WRC, 1978).

The OECD method as already described is in two parts, a screening die-away test for which a pass is considered to be a pass for the whole method and a confirmatory sewage works simulation test only carried out in the event of a fail in the screening test. The German method is essentially the same as the OECD confirmatory test, whilst the UK porous pot method uses a different apparatus but is in many essentials the same as the OECD and German methods. The French AFNOR method is an elaboration of the OECD die-away screening test and involves a further dose of the test substance part-way through the test period. It is believed that the latter method is rarely used except, presumably, in France.

With a number of alternative methods available, it is possible that conflicting results may be obtained and Article 3 of Directive 82/243 EEC specifies that the reference method to be used is that given in the Annex to the Directive. This reference method is essentially the same as the OECD confirmatory test, the German method or the alternative UK porous pot method.

### 1.7.2   Non-ionic surfactants

An EEC Directive (EEC, 1982b) on the biodegradability of non-ionic surfactants used in detergent was adopted in 1982 (82/242/EEC). This closely parallels the anionic directive (82/243/EEC) in specifying an 80% pass and the same test methods except in so far as BiAS is the specified analytical method.

The non-ionic Directive does, however, contain an important derogation clause (Article 5) which permitted the temporary use for particular applications of certain non-ionic surfactants not satisfying the biodegradability requirements. This derogation was extended but has now expired. It did, however, represent an important concept, namely that the requirements of public health and industry (the main areas for which the derogation was effective) should be given due weight opposite possible adverse environmental effects and the derogation was not lifted until satisfactory biodegradable alternatives had been developed.

### 1.7.3   Cationic and ampholytic surfactants

Although the frame Directive (73/404/EEC) covers these two classes of surfactant, there are no enabling Directives as yet in place. The reasons for this are almost certainly twofold. The first is that the relatively low tonnage and lack of perceived environmental problems have not prompted biodegradation legislation for these materials, although for cationic surfactants, the increased tonnage via the use of fabric softeners and relatively high aquatic toxicity are certainly live issues. The second relates to the difficulties in analysis of the test substances and particularly with the cationics where their sorptive and biocidal nature presents major difficulties.

## 1.8   Surfactant analysis

As has been indicated earlier in this chapter, the method of analysis used in support of a biodegradability test method has a very important bearing on what is meant by 'biodegradability' and indeed in a legislative sense what is meant by a surfactant.

In the EEC Directives (82/242/EEC, 82/243/EEC) described above, the MBAS method is specified for anionic surfactants and the BiAS method for the non-ionics and it is important to understand the limitations of these methods from both the viewpoint of the biodegradability test methods and the wider perspective of environmental monitoring.

## 1.8.1   Anionic surfactant analysis

The basis of almost all methods used for both the measurement of anionic surfactants in support of laboratory biodegradation experiments and for measurement in environmental samples is the MBAS (methylene blue active substance) method. The method was originally published by Longwell and Maniece (1955) and subsequently modified by Abbott (1962). Other more recent developments for the analysis of environmental samples by the MBAS techniques involve the isolation of the surfactant by the foam sublation technique of Wickbold (1971) and further clean-up by ion exchange (Waters, pers. comm.).

Methylene blue is a cationic dyestuff which forms a chloroform-soluble complex with certain classes of anionic surfactant. By extracting a mixture of anionic surfactant solution and methylene blue with chloroform and measuring the intensity of extracted blue colour, an estimate of the amount of anionic surfactant present may be made. This analytical method depends on two factors. The first is the ability of the anionic hydrophilic group to combine with the methylene blue dyestuff and the second that the overall complex is then sufficiently hydrophobic to be extracted into chloroform from an aqueous medium.

In practice, the first factor appears to limit the type of anionic surfactant response to the MBAS method effectively to sulphates and sulphonates. Soaps, in which the hydrophile is a carboxylic acid, do not respond and in the author's experience neither do anionic surfactants of the phosphate or phosphonate type. Thus, since the EEC Directive 82/243 effectively defines an anionic surfactant as one which responds to the MBAS method, certain important classes of anionic surfactant fall outside its scope. Further, it should be appreciated that although the biodegradability of non-MBAS active anionic surfactants can be measured by alternative methods, these alternative methods do not use an equivalent analytical method to MBAS and hence the results will not be comparable.

The second factor regarding the extractability of the surfactant–methylene blue complex into chloroform has implications in respect of the implicit definition of biodegradability as measured by the method. Aerobic biodegradation of, for example, linear alkylbenzene sulphonates proceeds via oxidative breakdown of the alkyl chain (Swisher, 1963). As the chain becomes smaller and the end-groups more hydrophilic, so the extractability of the methylene blue complex into chloroform will be removed. Insofar as this loss of MBAS appears to be accompanied by loss of foaming properties and also loss of toxicity to aquatic organisms, it appears that biodegradability as measured by MBAS is a satisfactory measure of 'environmentally acceptable' biodegradation. However, it does not necessarily follow that a high level of MBAS degradation is necessarily accompanied by a high level of mineralisation of the molecule or that where mineralisation does take place that it will proceed at the same rate (Wuhrmann and Mechsner, 1974).

### 1.8.2   Non-ionic surfactant analysis

As with the methods for anionic surfactant analysis, the methods used for non-
ionic surfactants have a significant effect on the understanding of biodegradability. The historical methods for the analysis of non-ionics in waters and sewage
have been reviewed by Longman (1975) and by Heinerth (1966) and all involve
essentially two stages. The first stage is the extraction procedure to separate the
non-ionic surfactant into an organic solvent and the second stage is the estima-
tion of the extracted surfactant using chemical methods which, in general,
involve reaction with the hydrophilic (polyglycol) portion of the surfactant.

The extraction procedures used fall into two broad classes, the first standard
solvent extraction procedures and the second, which is now much more widely
used, the solvent sublation procedure of Wickbold which was mentioned a little
earlier in this chapter under anionic surfactants.

A typical solvent extraction procedure is that developed by Patterson *et al.*
(1966) in which the aqueous sample is extracted four times with chloroform
using magnesium sulphate as a salting-out agent. The combined extracts are
washed with acid and then alkali and evaporated to dryness for the subsequent
determination. Such a procedure is very time-consuming, has distinct limitations
in respect of the volume of sample which can be extracted and, with many polluted samples, the separation of the chloroform and the aqueous phase at the
interface can be difficult.

Notwithstanding these criticisms, it should be noted that such solvent extraction procedures may be more effective in extracting the possible degradation
products of non-ionic surfactants than the solvent sublation procedures. This
may be seen both as an advantage if the ultimate (total) degradation of the surfactant is being investigated (Patterson *et al.*, 1968) and also a disadvantage if
primary degradation is the issue of interest. In this context, if comparability with
the MBAS method for anionic surfactant degradation is being sought, then
primary degradation and the solvent sublation extraction procedure are probably
indicated.

The solvent sublation procedure of Wickbold (1971) involves placing the
sample in an aeration column, bubbling ethyl acetate saturated nitrogen up
through the column and collecting the surfactant in a layer of ethyl acetate
floating on the surface of the column. This technique can be used for quite large
volumes of aqueous sample (5 l is probably about the effective limit) and
requires little physical or time input from the operator. More fundamentally, as
indicated above, the sublation principle of this method is reasonably specific for
surface active substances (it has been used for anionic, non-ionic and cationic
surfactants). Thus, the method helps to separate the surface active substances
from the sample and to leave behind some of the other organics which may
interfere with the subsequent analysis.

Having extracted the surfactant sample from the aqueous matrix, a number of
analytical procedures have been used to estimate the non-ionic surfactant

content. As commented above, these all involve reaction with the hydrophilic (polyglycol) part of the molecule and in interpreting the results caution is necessary, particularly with environmental samples, in that the method may not be specific for non-ionic surfactants.

Patterson used a thin layer chromatography (TLC) method with a spray reagent (modified Dragendorff reagent) which reacts with not only the original surfactant but also with any polyglycol degradation products. The TLC conditions were defined to separate the surfactant from degradation products and the surfactant quantified by comparing the intensity of the spot with the intensity of spots from a series of standards.

This Patterson method was widely used in the UK by river authorities for many years to monitor levels in rivers and has found some applicability in monitoring the biodegradation of non-ionic surfactants in sewage treatment works and laboratory experiments. It has the advantage, over virtually all other analytical methods for surfactants of providing information on both parent surfactant and also degradation products. However, the method suffers from two major disadvantages which have severely curtailed its use. Both the extraction procedure and the TLC method are very time-consuming and, more fundamentally, the quantification of the spot intensity is rather subjective and the results can only be described as semi-quantitative.

Hey and Jenkins (1969) describe a variant of the cobaltithiocyanate method originally described by Brown and Hayes (1955). The basis of the original method is the formation of a complex between the non-ionic surfactant and the cobaltithiocyanate ion and the subsequent measurement of the complex by its UV absorbance in an organic solvent at 320 nm. This original method suffers from extraction problems with sewage samples and also interference.

The Hey and Jenkins variation uses solvent sublation for the extraction phase of the analysis, and then destroys the cobaltithiocyanate complex with the surfactant by extraction into acidified water with a subsequent measurement of thiocyanate by absorption at 515 nm. The combination of the solvent sublation step and the change in measurement wavelength was claimed by the authors to increase sensitivity and to reduce interferences. However, the method does not seem to have attracted widespread use, possibly due to the solvent, benzene, suggested for the extraction stage. However, the principles of the methods are in many ways similar to the principles of the Wickbold or BiAS method which will now be described.

The BiAS method as previously mentioned is the analytical method prescribed in the OECD methods for determining the biodegradability of non-ionic surfactants and incorporated into the EEC Directive 82/242/EEC. The method was originally developed by Wickbold and uses foam sublation as described above as the extraction procedure.

The extracted surfactant, after evaporation of the ethyl acetate sublation solvent, is then reacted with a complex bismuth reagent whose chemistry is not dissimilar to the modified Dragendorff spray reagent used by Patterson. This bismuth reagent forms a water insoluble complex with the polyglycol group of a non-ionic surfactant (and also with polyglycol itself if present) which is filtered off. After this precipitate has been washed to remove all traces of soluble bismuth, it is redissolved in water by using ammonium tartrate to form a water soluble bismuth complex. The bismuth content of this aqueous solution is then measured by one of three methods and related to the concentration of either a standard surfactant (nonyl phenol 10 ethoxylate in the EEC Directive) or to the particular surfactant under test in a biodegradation study.

The methods used for measuring the bismuth are potentiometric titration in the OECD/EEC Directive methods, absorbance at 263.5 nm in the presence of EDTA (Longman, 1975) or by direct measurement as elemental bismuth using atomic spectroscopic techniques (the method used in the author's laboratory and elsewhere).

Since the BiAS method has legislative significance, we need to consider its scope and comparability with the MBAS method used for anionics.

In terms of scope the method is stated in the Annex to the EEC Directive 82/242/EEC to be applicable to non-ionic surfactants containing 6–30 alkylene oxide (polyglycol) groups, although in the author's experience this range may be somewhat stretched at both ends. It seems probable that the lower end is governed by the ability of short polyglycol chains to form a complex with the bismuth reagent. At the higher end, since it may be shown that polyethylene glycol itself will react with the bismuth reagent to form an insoluble complex, it seems probable that the specificity of the method is governed by the extraction procedure which is likely to discriminate against the more hydrophilic molecules.

The BiAS method is entirely different in principle to the MBAS method which, as described in the EEC Directive 82/243/EEC, uses a simple chloroform extraction and colorimetric determination of the methylene blue complex. However, as indicated above, the solvent sublation stage of the BiAS method appears to provide discrimination between the surfactant and possible metabolites or other interferences, whereas in the MBAS method, it is the chloroform soluble nature of the surfactant–methylene blue complex which provides the discrimination. Thus, both methods provide for the measurement of the primary biodegradation of the respective surfactant types. For most surfactants, primary degradation is coupled with loss of foaming ability and, as mentioned earlier, loss of toxicity to aquatic organisms.

### 1.8.3    Cationic surfactant biodegradability and analysis

As mentioned previously, although the EEC frame Directive 73/404/EEC prohibits the placing on the market of detergents containing cationic surfactants whose biodegradability is less than 90%, there is no corresponding enabling Directive specifying the test methods to be used. Consideration continues to be given by the European Commission to rectify this omission; however, the definition of suitable test methods for general use does pose considerable technical difficulties.

A number of analytical methods for cationic surfactants are known and have been described by Longman. One, in particular, is effectively the mirror image of the MBAS method for anionic surfactants and involves the formation of a chloroform soluble blue complex of the cationic surfactant with the anionic dyestuff, disulphine blue. This method can be used at high pH to measure the quaternary ammonium cationics and at low pH to measure both quaternary ammonium and tertiary amine cationics which latter are protonated and respond to the analysis at low pH. Thus, there is an analytical method for cationic surfactants equivalent to the MBAS method for anionics and the main problems with defining generally applicable biodegradation test methods for cationics lie elsewhere.

Firstly, cationics have a biocidal nature which tends to preclude the use of a screening type test with a low bacterial inoculum concentration as the test concentration below which the cationics do not inhibit bacterial action tends to be too low for use of a disulphine blue type method. Secondly in a biodegradation study of either a screening test, or, more especially, a confirmatory test nature, absorption is likely to be the prime removal mechanism.

From the viewpoint of measuring environmental levels of cationics both sorption and also the formation of an anionic/cationic complex with anionic surfactants (which are always likely to be present in excess) are factors which need to be taken into account. The paper of Matthijs *et al.* (1994) describes the analytical methodology and monitoring findings at a sewage treatment plant for the cationic fabric softener, ditallow dimethyl ammonium chloride.

### 1.8.4 Ampholytic surfactants

As with cationics, ampholytic surfactants are mentioned in the 73/404/EEC frame Directive as a class of substance whose biodegradability, if present in a detergent, should exceed 90%. There appears to have been little or no work to identify a generally applicable method for the analysis of ampholytics in the context of proposing a biodegradability directive for ampholytics.

## 1.9 Conclusions and thoughts for the future

In this first chapter, giving the historical background to surfactant biodegradation, the aim has been to give the reader an appreciation of what lies behind the term 'biodegradation' and how biodegradation test methods have been used to develop and control the surfactants used in detergents, with particular reference to the EEC Detergent Directives. These measures have, especially in areas where a proper level of sewage treatment is installed, been highly successful in overcoming the evident foam problems due to poorly degraded surfactants.

With this success in mind, the question to be considered is: what of the future for surfactant specific biodegradation legislative test requirements? To address this question some understanding of recent and developing European legislation is required.

The 7th Amendment of the Classification and Packaging of Dangerous Substances Directive (EEC, 1992b) places a heavy emphasis on the need to undertake risk assessment (physical hazard, human health and environmental) for all new substances placed on the European market. The frame methodology for this has been introduced as an EEC Directive (EEC, 1993b) and technical guidance to implement this Directive developed (EEC, 1993c).

A parallel activity for existing substances is currently under way. The Council Regulation on the evaluation and control of risks of existing substances (EEC, 1993d) requires manufacturers and importers of existing substances to report available data on those substances in a standard format and to a time scale defined by tonnage. The high tonnage (>1000 tonnes/annum per manufacturer or importer) substances will be considered first and, based on the data supplied, a priority list of substances for risk assessment drawn up. The frame risk assessment Directive for existing substances and the supporting technical guidance are due to be published during 1994.

In the environmental area, the current main emphasis on risk assessment is in relation to the aquatic environment (clearly a key compartment for detergent surfactant), although the possible effect of substances reaching agricultural land via the use of sewage sludge as a fertiliser is also considered. The basis of the methodology is to compare the measured or predicted environmental concentration (PEC) of the substance under consideration with the predicted no effect concentration (PNEC). PNEC will normally be derived by applying assessment factors to the results of laboratory studies, although, possibly of importance for surfactants, field observations (ecology) can also be used in deriving PNEC.

In the above context biodegradation plays an important, but not a dominant role. In the calculation of PEC the extent of primary degradation is factored into the calculations. Also, considerable attention is paid to whether or not primary degradation is accompanied by a high level of ultimate degradation and within the risk assessment methodology, there is the clear possibility of addressing PEC/PNEC ratios not only for the parent substance, but also any persistent metabolites.

From the above brief description, it is evident that many of the high tonnage surfactants used in detergents are likely to receive a formal environmental risk assessment within the context of the Existing Chemicals Regulation and that this will include consideration of both their primary and ultimate degradability.

There are, however, discussions in progress regarding whether or not the Detergent Directives should be amended to include specified requirements for the ultimate degradability of the surfactant components. If this is done, it raises the question of practical application. The current Detergent Directives are framed on the basis that a control authority can sample detergents at the point of sale or use and, using relatively simple extraction and biodegradation test methods, determine whether or not the surfactants conform to the biodegradability requirements of the Directives. The technical nature of the ultimate biodegradation tests is such that point of sale control would almost certainly not be possible and moreover, the tests themselves are inherently less precise and reproducible than the tests for primary degradation (Brown, 1976).

Whatever the outcome of this discussion on ultimate degradation, there seems to be a clear case for retaining a separate legislation controlling the primary degradation of surfactants used in detergents. In the event of local problems, such as foam, arising from the use of poorly degradable surfactants, environmental analysis would probably indicate the type of surfactant likely to be responsible. Therefore, it would seem desirable to retain the ability for control authorities to sample and test suspect detergent preparations at the point of sale or use.

The overall conclusion reached is that the Directives on the biodegradability of anionic and non-ionic surfactants used in detergents have successfully controlled what is now essentially a historic problem, namely widespread and excessive foam on rivers. They continue to have utility and there seem no obvious grounds for replacement.

The question of whether or not the detergent specific legislation should be extended to cover ultimate degradation remains open. However, any such legislation would almost certainly present considerable experimental difficulties in achieving a reproducible result for a pass/fail boundary unless that boundary were to be set at a very low level, and would almost certainly necessitate departure from the point of sale/use principle which has been the basis of the historic and current achievements.

## Acknowledgment

This chapter was written with the support of ICI Surfactants, PO Box 90, Wilton, Middlesborough, Cleveland TS 90 8JE, UK.

## References

Abbott, D.C. (1962) Colorimetric determination of anionic surfactants in water. *Analyst* **87**, 286.

AFNOR (1981) Association francaise de normalisation (AFNOR). Experimental standard T73-260.

Birch, R.R. (1984) Biodegradation of nonionic surfactants. *J. Am. Oil Chem. Soc.* **61**, 340.

Bock, K.J. and Stache, H. (1982) Surfactants, in *Handbook of Environmental Chemistry*, Vol. 3, part B, ed. O. Hutzinger, Springer-Verlag, Berlin.

Brown, D. (1976) The assessment of biodegradability, in *Proc. VIIth Int. Congress on Surface Active Substances*, Moscow.

Brown D., De Henau, H., Garrigan, J.T., Gerike, P., Holt, M., Keck, E., Kunkel, E., Matthijs, E., Waters, J. and Watkinson, R.J. (1986) Removal of nonionics in a sewage treatment plant. *Tenside Deterg.* **23**, 4.

Brown, E. and Hayes, T. J. (1955) The absorptiometric determination of polyethyleneglycol mono-oleate. *Analyst* **80**, 755.

Bundesgesetzblatt (1980) Part I, p. 706.

Eden, G.E., Truesdale, G.A. and Stennett, G.V. (1967) Pollution by synthetic detergents. Symposium Paper No. 6. Institute of Water Pollution Control, Annual Conference, Torquay.

EEC (1973a) Council Directive of 22nd November 1973 on the approximation of the laws of the Member States relating to detergents. 73/404/EEC. O.J. L347/51.

EEC (1973b) Council Directive of 22nd November 1973 on the approximation of the laws of the Member States relating to methods of testing the biodegradability of anionic surfactants. 73/405/EEC. O.J. L347/53.

EEC (1982a) Council Directive of 31 March 1982 amending Directive 73/405/EEC on the approximation of the laws of the Member States relating to methods of testing the biodegradability of anionic surfactants. 82/243/EEC. O.J. L109/18.

EEC (1982b) Council Directive of 31 March 1982 amending Directive 73/405/EEC on the approximation of the laws of the Member States relating to methods of testing the biodegradability nonionic surfactants. 82/242/EEC. O.J. L109/1.

EEC (1991) Council Directive of 30 April 1991 concerning urban waste water treatment. 91/271/EEC. O.J. L135/40.

EEC (1992a) Annex V Test Methods, Method C4 Determination of Ready Biodegradablity. O.J. L343 A187-225.

EEC (1992b) Council Directive 92/32/EEC of 30 April 1992 amending for the seventh time Directive 67/548/EEC on the approximation of the laws, regulations and administrative provisions relating to the classficiation, packaging and labelling of dangerous substances. O.J. L154/1.

EEC (1993a) Risk Assessment of Notified New Substances. Technical Guidance Document in Support of the Risk Assessment Directive 93/67/EEC, pp. 20–21.

EEC (1993b) Commission Directive 93/67/EEC of 20 July 1993 laying down the principles for assessment of risks to man and the environment of substances notified in accordance with Council Directive 67/548/EEC. O.J. L227/9.

EEC (1993c) Technical Guidance Documents in support of the Risk Assessment Directive (93/67/EEC) for substances notified in accordance with the requirements of Council Directive 67/548/EEC.

EEC (1993d) Council Regulation (EEC) No. 793/93 of 23 March 1993 on the evaluation and control of the risks of existing substances. O.J. L84/1.

Fiesen, L.F. and Fiesen, M. (1950) Organic Chemistry, 2nd Edition. D.C. Heath & Co. Boston, U.S.A.

Fischer, W.K. (1980) Entwicklung der Tensidkonzentrationen in den deutschen Gewässern 1960–1980. Tenside Deterg. 17, 250.

Fischer, W.K. and Gerike, P. (1984) Eine Tensidbilanz im Einzugsgebiet einer kommunalen Kläranlage. Tenside Deterg. 6, 71.

Fischer, W.K. and Winkler, K. (1976) Detergentienuntersuchungen im Stromgebiet den Rheins 1958–1975. Vom Wasser 47, 81.

Heinerth, E. (1966) Zum Problem der Bestimmung geringer Mengen nichtionogener Tenside in Wasser and Abwasser. Tenside Deterg. 3, 109.

Hey, A.E. and Jenkins, S.H. (1969) The determination of nonionic detergents in sewage works samples. Water Res. 3, 887.

Holt, M.S., Matthjis, E. and Waters, J. (1989) The concentrations and fate of linear alkylbenzene sulphonate in sludge amended soils. Water Res. 23, 749.

Kaelbe, E.F. (1963) Detergent component analysis. Soap Chem. Specialties 39(10), 56.

Kimerle, R.A. and Swisher, R.D. (1977) Reduction of aquatic toxicity of LAS by biodegradation. Water Res. 11, 31.

Longman, G.F. (1975) The Analysis of Detergents and Detergent Products, Wiley, New York.

Longwell, J. and Maniece, W.D. (1955) Determination of anionic detergents in sewage, sewage effluents and river waters. Analyst 80, 167.

Malz, F. (1973) Synthetic detergents and water. The solution of the problem in Germany. Prog. Water Technol. 3, 293.

Mann, A. H. and Reid, V.W. (1971a) Biodegradation of synthetic detergents, Evaluation by community trials. 1. Linear alkylbenzene sulphonates. J. Am. Oil Chem. Soc. 48, 588.

Mann, A.H. and Reid V.W. (1971b) Biodegradation of synthetic detergents, Evaluation by community trials. 2. Alcohol and alkylphenol ethoxylates. J. Am. Oil Chem. Soc. 48, 794.

Mann, A.H. and Reid V.W. (1971c) Biodegradation of synthetic detergents, Evaluation by community trials. 3. Primary alcohol sulphates. *J. Am. Oil Chem. Soc.* **48**, 798.

Matthijs, E., Gerike, P., Klotz, H., Kooijman, J.G.A., Korber, H.G. and Waters J. (1994) Removal and mass balance of the cationic fabric softener ditallow dimethyl ammonium chloride in activated sludge sewage treatment plant. *Water Res.*, in press.

Neufahrt, A., Hofmann, K. and Taüber, G. (1987) Biodegradation of nonylphenol ethoxylate and environmental effects of its catabolites, in *Spanish Congress on Detergents and Surfactants*, Barcelona.

OECD (1976) *Proposed Method for the Determination of the Biodegradability of Surfactants used in Synthetic Detergents*, OECD, Paris.

OECD (1981) *OECD Guidelines for Testing of Chemicals*, OECD, Paris.

OECD (1993) *OECD Guidelines for Testing of Chemicals*, Vol. 2, OECD, Paris.

Painter, H.A. and Bealing, D.J. (1989) Experience and data from the OECD activated sludge simulation test. C.E.C. Water Pollution Research Reports No. 18.

Patterson, S.J., Tucker, K.B.E. and Scott, C.C. (1966) Nonionic detergents and related substances in British waters, in *3rd Int. Conf. on Water Pollution Research*, Munich, Section II, Paper 6.

Patterson, S.J., Tucker, K.B.E. and Scott, C.C. (1968) Nonionic detergent degradation III. Initial mechanism of degradation. *J. Am. Oil Chem. Soc.* **47**, 37.

Porter, M.R. (1991) *Handbook of Surfactants*, Blackie, Glasgow.

Price, D.H.A. (1980) The United Kingdom detergent story. *Tenside Deterg.* **17**, 271.

Reiff, B. (1976) The effect of biodegradation of three nonionic surfactants on their toxicity to rainbow trout, in *Proc. VIIth Int. Congress on Surface-Active Substances*, Moscow.

Richtler, H.J. and Knaut, J. (1998) World prospects for surfactants, in *CESIO 2nd World Surfactants Congress*, Vol. I, pp. 3–58.

SDA (1965) Sub-Committee on Biodegradation Test Methods of the Soap and Detergent Association. A procedure and standards for the determination of the biodegradability of alkyl benzene sulfonate and linear alkylate sulfonate. *J. Am. Oil Chem. Soc.* **42**, 986.

STCSD (1958) *Progress Report of the Standing Technical Committee on Synthetic Detergents*, HMSO, London.

STCSD (1958–1979). *Progress Reports 1–19 of the Standing Technical Committee on Synthetic Detergents*, HMSO, London.

STCSD (1977) Standing Technical Committee on Synthetic Detergents. Voluntary Notification Scheme. U.K. Dept. of the Environment.

Swisher, R.D. (1963) The chemistry of surfactant biodegradation. *J. Am. Oil Chem. Soc.* **40**, 648.

Swisher, R.D. (1987) *Surfactant Biodegradation*, 2nd edition, Marcel Dekker, New York.

Truesdale, G.A., Jones, K. and Vandyke, K.G. (1959) Removal of synthetic detergents in sewage treatment process. *Water Waste Treat. J.* November/December.

Wickbold, R. (1971) Enrichment and separation of surfactants from surface waters by transport at the gas/water interface. *Tenside* **8**, 61.

WRC (1978) U.K. Water Research Centre. The 'Porous Pot Test'. Technical Report No. 70.

Wuhrmann, K. and Mechsner,K. (1974) Testing the biodegradability of organic compounds. *EAWAG News*, No. 3.

Zahn, R. and Wellens, H. (1974) Ein einfaches Verfahren zur Prufung der biologischen Abbaubarkeit von Produkten und Abwasserinhaltsstoffen. *Chem. Zeitung* **98**, 228.

# 2 What is biodegradation?

G.F. WHITE and N.J. RUSSELL

## 2.1 Setting the scene: Microbial nutrition and the carbon cycle

In the aerobic atmosphere currently prevailing at the Earth's surface, the most stable form of carbon is its fully oxidised state, $CO_2$. The atmosphere contains $2.6 \times 10^{12}$ tonnes of carbon as $CO_2$, which is in equilibrium with even larger amounts ($1.3 \times 10^{14}$ tonnes) in solution in the rivers, lakes and oceans (Schlegel, 1986). From this pool of oxidised carbon begins a series of exquisitely complex biological processes driven primarily by solar energy, which lead to the reduction of $CO_2$ to C–C, C–H and other bonds in an enormous variety of organic compounds that collectively constitute the biosphere. In effect, photosynthetic organisms, predominantly the green plants and marine algae, are the primary producers which transduce solar energy into chemical energy by converting $CO_2$ into complex assemblies of reduced organic compounds. These materials constitute the base of numerous interrelated food-chains involving herbivores, carnivores and omnivores. During the flux of carbon through these routes, some is oxidised by the consuming organisms to derive energy, while the resulting $CO_2$ is returned to the atmosphere. However, much organic material enters the soils, waters and sediments in the environment either during senescence of plants or the excretions from, or death of, higher organisms. The deposition of organic compounds in the environment provides another pool, this time of reduced carbon compounds which, in the presence of $O_2$, are thermodynamically unstable with respect to $CO_2$ under aerobic conditions.

An integral part of the evolutionary process that has produced this plethora of organic compounds is the evolution of microorganisms that are able to utilise these materials as nutrients. Microorganisms use organic compounds (and other elements such as N and S) in the environment not only as sources of the carbon from which to build cell components, but also for the energy they need to achieve that biosynthesis in order to maintain the low entropy state embodied in the highly organised structures characteristic of biological systems; this energy is also used to accomplish other functions such as motility and bioluminescence. Thus, environmental deposits of organic compounds serve as nutrient and energy sources for microorganisms which ultimately complete the re-oxidation of carbon to $CO_2$ in the atmosphere and surface waters, thereby closing the cycle.

It is easy to see how photosynthesis might be credited with sustaining all life on Earth. However, to maintain a steady state in a cyclic series of events, each contribution to the cycle is equally important. To illustrate the point, one estimate of the rate of $CO_2$ fixation is $1.29 \times 10^{11}$ tonnes per annum (Schlegel, 1986) so that assuming that a linear rate is maintained, the atmospheric $CO_2$ pool, if not replenished, would be exhausted within about 20 years. Although the reservoir of $CO_2$ dissolved in the oceans would buffer the atmospheric depletion, the rate of equilibration between atmospheric and aquatic pools of $CO_2$ is low. However, even assuming rapid and complete equilibration, the global $CO_2$ 'reserve' would be dissipated in about a millenium. Clearly the reoxidation of organic carbon to $CO_2$ is crucial because without it, biogeochemical cycling of carbon would grind to a halt in a few moments of the bio-evolutionary time-scale.

Apart from some relatively minor adjustments for permanent deposition in sediments, the global rate of photosynthetic carbon fixation into organic compounds is balanced by an equally impressive rate of global respiration, of which the major part (about $0.92 \times 10^{11}$ tonnes per annum) is attributable to microorganisms (Figure 2.1). Although there are small amounts of very recalcitrant organic compounds in the humus fractions of soils, the mean residence times for most organics range from a few days in the oceans to a few years in the pedosphere (Hutzinger and Veerkamp, 1981). This rapid turnover means that the concentrations of organic compounds in the environment are generally very low, typically sub-ppm in the oceans and unpolluted freshwaters (Morgan and Dow, 1986). As a result, the majority of microorganisms exist in low-nutrient environments, and in order to survive they must adapt to make best use of the available resources which chemically can be extremely diverse.

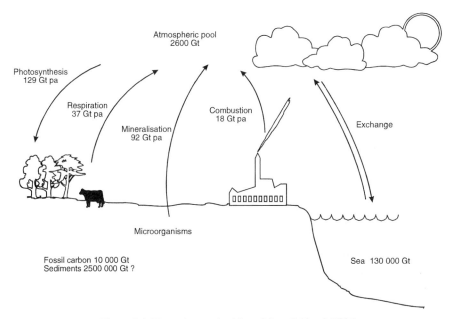

Atmospheric pool
2600 Gt

Photosynthesis
129 Gt pa

Respiration
37 Gt pa

Combustion
18 Gt pa

Mineralisation
92 Gt pa

Exchange

Microorganisms

Fossil carbon 10 000 Gt
Sediments 2500 000 Gt ?

Sea  130 000 Gt

**Figure 2.1** The carbon cycle. Adapted from Schlegel (1986).

Most microorganisms are chemotrophic, i.e. they derive energy from the oxidation of organic compounds. To acquire the necessary metabolic fuels, bacteria have been accredited with a number of adaptations (Morgan and Dow, 1986; Poindexter, 1987), including surface attachment (Dawson *et al.*, 1981; Bachofen, 1986), mobilisation of storage polymer (Dawes, 1976) and chemotaxis (Armitage and Lackie, 1990), together with the development of catabolic diversity to enable them to utilise as wide a range of potential nutrients, including man-made surfactants, as they are likely to encounter (Harder, 1981). During the twentieth century, mankind's mobilisation of the planet's organic chemical resources in plants and mineral deposits through modern agricultural and industrial activities has provided new materials with the potential to sustain microbial growth. Quantitatively the amounts are not very significant, being only *ca.* 0.1% compared with natural carbon compounds (Hutzinger and Veerkamp, 1981), but this contribution is likely to increase as a result of population increases and the spread of industrialisation to developing countries.

The synthetic compounds that will be most readily catabolised will be those that contain chemical structures and motifs in common with those occurring naturally because microbes will already have evolved enzymes to degrade them. Even compounds bearing little structural resemblance to natural products are often biodegradable because analogues such as polycyclic aromatic hydrocarbons formed by diagenesis in sedimentary rocks have also been in contact with microorganisms throughout evolutionary periods of time. Biodegradation is less likely for those compounds containing either individual structures, or combinations of structures, that have never been encountered by living organisms.

## 2.2   Surfactants as potential microbial nutrients

During the relatively short period when synthetic surfactants have been in use, chemists have found that compounds with a wide variety of chemical structures can be produced with the required amphiphilic and detersive properties. Most of this diversity lies with the hydrophilic moieties and their means of linkage to the hydrophobe on which basis surfactants are classified (see Chapter 1). In contrast, the hydrophilic moiety invariably contains medium-to-long alkyl chains, which are potentially excellent substrates for chemotrophic growth because in terms of microbial metabolism the carbon is in a reduced state and so can produce abundant reduced cofactors which in turn are used to yield energy. Therefore, it is sensible to consider biodegradation of surfactants in terms of how bacteria access these sources of reduced carbon.

World production of synthetic (non-soap) surfactants is currently about 7 million tonnes per annum (Hauthal, 1992). Within this total, anionic surfactants predominate with linear alkylbenzene sulphonates (LABS), for example, accounting for 28% of the overall total. Necessarily, therefore, these will provide most of the specific examples used as illustrations in what follows: only general features of microbial biodegradation of surfactants are provided, and the reader is referred to White and Russell (1993) for details of pathways and bio-chemical mechanisms.

### 2.2.1   Accessing the hydrophobic chain

Essentially two mechanisms are used to access the carbon contained in aliphatic hydrocarbon chains: either the hydrophile is first separated from the hydrophobe which is then attacked oxidatively or the alkyl chain is oxidised directly whilst still attached to the hydrophile (Figure 2.2). A surfactant may be attacked by both general mechanisms, either in a single organism or in different biode-graders. Indeed, in a mixture of different organisms in a natural environment it is likely that more than one biodegradative attack is represented. Moreover, for some complex surfactants, consortia of organisms may be necessary for com-plete breakdown of the compound. These points can best be illustrated by describing some examples.

Alkyl sulphates (alcohol sulphates, AS) such as sodium dodecyl (lauryl) sul-phate (SDS or SLS) are readily broken down by bacteria which contain alkylsul-phatases; these release sulphate through cleavage of the C–O or O–S bond of the sulphate ester $CH_3(CH_2)_n$–O–$SO_3$ linkage by means of a hydrolytic reaction mechanism to give a primary alcohol as the product (Figure 2.3). There are also enzymes which can hydrolyse secondary alkyl sulphate esters to give the corre-sponding secondary alcoholic products. The evolution of this biodegradative capacity is probably not recent because analogous sulphate esters are found in biological material, e.g. the sulphated lipids in some freshwater algae (Haines, 1973; Mercer and Davies, 1979). The fact that a variety of sulphate esters occurs naturally in biological material in the environment may explain why such a diversity of alkylsulphatase enzymes has been identified in bacteria, sometimes in the same organism.   Individual bacterial isolates have been found which contain up to seven different alkylsulphatase enzymes!They are not only specific

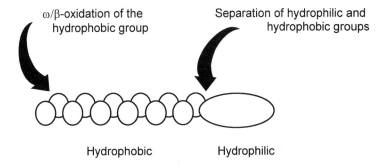

**Figure 2.2** Two strategies for accessing the hydrophobic chain in surfactants.

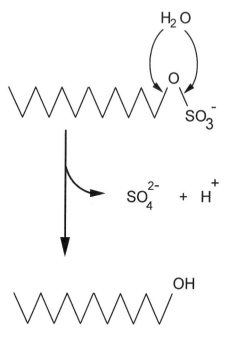

**Figure 2.3** Bacterial primary biodegradation of SDS; an example of hydrophile–hydrophobe separation.

for primary or secondary alkyl sulphates, but also have sharply defined chain length specificities and stereospecificities (Dodgson and White, 1983), and their biosynthesis may be constitutive or inducible or repressible (see Section 2.3.3.3 for an explanation of these terms). It is very easy to isolate alkylsulphatase-containing bacteria from natural environments, even from pristine source waters, whereas in rivers receiving surfactant pollutant inputs the proportion of such bacteria increases with the pollutant loading and may reach 16% of the total population in heavily polluted waters (White *et al.*, 1989).

The biodegradation of the structurally analogous dialkyl sulphosuccinate surfactants, in which the aliphatic alcohols are esterified to succinic acid instead of sulphuric acid as in alkyl sulphates (Figure 2.4), is also initiated by a simple hydrolytic cleavage of the hydrophile from the hydrophobe (Hales, 1993). In both these types of surfactant it is an ester bond which is broken. This is relatively easily achieved because hydrolysis is thermodynamically favourable and no co-factors are required, so the process is energetically inexpensive for the bacteria. Most hydrolase enzymes attack their substrates at the *O*-acid bond of the alcohol–*O*-acid ester linkage. These include most known sulphatases, and all known phosphatases and carboxylesterases of various kinds (Bartholomew *et al.*, 1977). The dialkylsulphosuccinates are therefore most probably hydrolysed at the succinate-carboxyl carbons.

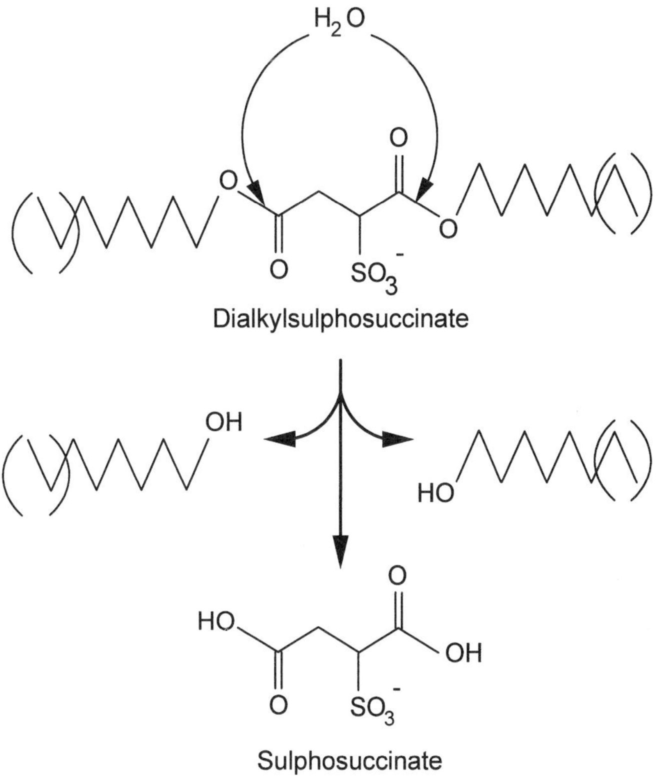

**Figure 2.4** Bacterial primary biodegradation of dialkyl sulphosuccinate; an example of hydrophile–hydrophobe separation.

In comparison, ether bonds are more resistant to chemical attack, so that their presence in surfactants may render the compound less readily biodegradable. For example, the insertion of a polyethylene glycol moiety between the alkyl chain and sulphate in SDS to form dodecyl triethoxy sulphate slows the degradation in pure and mixed cultures. The mechanisms which are utilised must be more elaborate and may require an initial investment of cellular resources, for example in the form of a reduced cofactor. In alkyl ethoxysulphates (alcohol ether sulphate, AES) the hydrophobe is released by etherase enzymes which break the alkyl–glycol (and glycol–glycol) ether bonds to liberate glycols and glycol sulphates (Figure 2.5) and, from the alkyl chain, alkanals and alkanoic acids as well as alcohols. Thus oxidative mechanisms may be needed to break ether bonds, in contrast with the ester bonds where a simple hydrolysis suffices. Ether lipids such as batyl alcohol (3-octadecyloxy-1,2-propandiol) are structurally similar to these surfactants and are possible natural substrates for which the etherase enzymes have evolved. Lipids containing ether bonds are present in

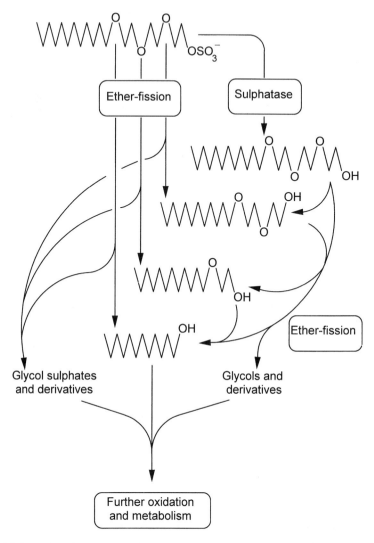

**Figure 2.5** Bacterial primary biodegradation of an alkylethoxy sulphate (alcohol ether sulphate, AES); an example of hydrophile–hydrophobe separation.

several types of bacteria and other microorganisms (Harwood and Russell, 1984), so that bacteria which degrade ether lipids may be expected to be equally common. Whether the ether-lipid degraders will match the ubiquity of the alkyl sulphate degraders remains to be seen.

This trend towards greater stability of the link connecting hydrophobe to hydrophile is seen further when the alkane sulphonate surfactants are considered. These contain a C–S link, which chemically is stable for days in dilute acids, whereas the C–O–S link of most sulphate esters is hydrolysed within minutes. Correspondingly, the simple biological hydrolysis of sulphonates has not been reported and instead bacteria utilise a more complex oxidative mechanism. This pathway is initiated by the input of molecular oxygen together with energy in the form of a reduced cofactor such as NAD(P)H, which are both required by a monooxygenase enzyme; this hydroxylates the sulphonate substrate to give an aldehyde-sulphite adduct as the product which spontaneously eliminates sulphite to leave the aldehyde. This mechanism has been characterised for primary alkane sulphonate surfactants (Thysse and Wanders, 1972, 1974). More recently in our laboratories a similar mechanism has been implicated in the biodegradation of secondary sulphonates (Quick *et al.*, 1994).

In all the biodegradation mechanisms described thus far, separation of the hydrophobe from the hydrophile, whether by hydrolysis or oxidation, yields products which are fatty acids, alcohols or aldehydes (*cf.* Figures 2.3–2.5). These are normal bacterial metabolites and can readily be catabolised via the ubiquitous pathway of fatty acid $\beta$-oxidation which is a cyclical process that releases two-carbon units in the form of the central metabolite acetyl coenzyme A (acetyl-CoA); this compound can in turn be channelled into either energy production or the biosynthesis of cellular components.

The second general strategy of attack on surfactants (Figure 2.2) also employs $\beta$-oxidation, but in this case the surfactant alkyl chain is metabolised with the hydrophilic moiety still attached. However, before $\beta$-oxidation can occur the terminal end (i.e. the $\omega$-position) of the alkyl chain furthest away from the hydrophilic group must first be oxidised by $\omega$-oxidation to give a carboxyl group which can be activated with the coenzyme A that is needed for the $\beta$-oxidation process (Figure 2.6). The $\omega$-oxidation is achieved by a monooxygenase in an energy-dependent reaction that uses NADH and molecular oxygen. This is the same pathway as that used in the microbial breakdown of alkanes and other hydrocarbons which lack a hydrophile (Boulton and Ratledge, 1984). The $\beta$-oxidation of linear alkyl chains may then proceed either to completion or at least until the approaching hydrophilic group is close enough to the site of oxidation to interfere. $\beta$-Oxidation can cope with limited amounts of branching in the alkyl chain, e.g. methyl groups at $\alpha$-positions, but not with geminal dimethyl substituted carbon atoms, i.e. quaternary carbons, in the alkyl chain. This is the structural aspect of tetrapropylene benzene sulphonates (TPS) which makes them very resistant to biodegradation by hydrophobe oxidation: indeed, in the absence of an alternative mechanism of hydrophile cleavage (*vide infra*), these surfactants are very recalcitrant and caused considerable pollution problems when they were first introduced commercially in the 1950s (see Chapter 1).

In principle, the $\omega$-/$\beta$-oxidation pathway could operate for any kind of surfactant with a linear alkyl chain or one containing relatively few branches, provided that the hydrophilic moiety was sufficiently remote from the $\omega$-carbon so as not to interfere with oxidation. Occurrence of this pathway has been most thoroughly studied for the alkylbenzene sulphonates (LABS), for which it is the only

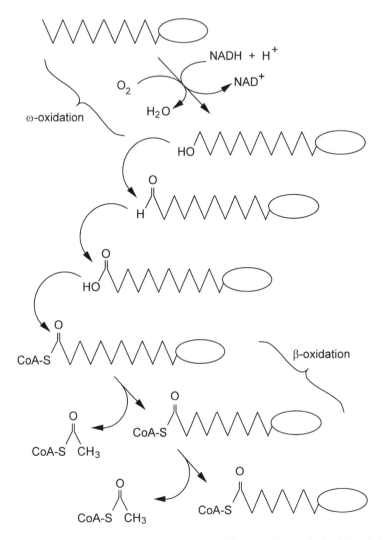

**Figure 2.6** Primary biodegradation by $\omega$-/$\beta$-oxidation; direct attack on the hydrophobic chain.

pathway yet observed (Cain, 1987). Disappearance of the surfactant is accompanied by the formation of sulphophenyl alkanoates with chain-lengths shortened progressively by two-carbon units, and there is no evidence for any hydrophile separation in the early stages of biodegradation. Biodegradation of $\alpha$-sulpho-fatty acid methyl esters is also initiated by $\omega$-/$\beta$-oxidation, with separation of the hydrophilic sulphonate group occurring at a late stage after the hydrophobic group has been extensively degraded (Steber and Wierich, 1989).

Generally speaking, because it requires the investment of energy, the hydrophobe oxidation mode of attack is most often seen for surfactants in which the linkage between the hydrophilic and hydrophobic parts is relatively difficult to break. Thus, for LABS where hydrophile removal would require a difficult disruption of the resonance-stabilised benzene ring, it is the only biodegradative pathway so far observed. For the alkane sulphonates and alkyl ethoxysulphates it occurs as an alternative to hydrophile separation (by oxidative disulphonation and ether cleavage, respectively). At the other end of the 'biodegradability spectrum', alkyl sulphates apparently are not broken down by $\omega$-/$\beta$-oxidation, at least in those competent bacterial isolates which have been studied in laboratories. Energetically it is 'cheaper' to release sulphate by sulphatase action (simple hydrolysis), rather than to expend energy on $\omega$-oxidation, in order to make the alkyl chain available for breakdown by $\beta$-oxidation. Thus, when laboratory isolations of alkylsulphate-degrading bacteria are made from mixed bacterial populations, even if there are representatives capable of using the $\omega$-/$\beta$-oxidation pathway they would be outgrown by those using the sulphatase hydrolytic route.

To summarise, the biodegradation mechanisms for the major anionic surfactants may be correlated, at least qualitatively, with assessments of their chemical stabilities (Table 2.1). In order of increasing chemical stability, the hydrophilic–hydrophobic linkages can be ordered as sulphate/carboxylate esters < alkyl ethers < aliphatic sulphonates and aromatic sulphonates. This order correlates with the shift from simple hydrolytic separation of hydrophile from hydrophobe, through multiple pathways involving oxidation, to exclusive hydrophobe oxidation. There is also a further correlation with the occurrence of natural analogues containing the corresponding hydrophile–hydrophobe link, and, by implication, the enzymes necessary to metabolise them.

### 2.2.2  Hydrophile biodegradation in non-ionics

The highly polar nature of ionic groups such as $-SO_3^-$ in the head-groups of anionic surfactants, gives them a marked propensity for solvation, thus making these groups highly hydrophilic. In contrast, sufficient hydrophilicity in ethoxylate non-ionic surfactants is achieved only by combination of many units (typically 5–25) of uncharged ethylene glycol units. Thus the hydrophilic group in ethoxylated surfactants contains abundant carbon, often more than in the hydrophobic alkyl chain. These moieties are therefore potential sources of bacterial nutrients and this leads to a third possible strategy for biodegradation of these surfactants, namely destruction of the hydrophilic group by successive removal of glycol units from the terminus. Evidence of this contribution to biodegradation is available for a number of non-ionic surfactants, including the linear alcohol ethoxylates (AE), oxo-alcohol ethoxylates (linear and alkyl-branched mixture), and especially the tetrapropylene (highly branched) ethoxylates (Schoberl and Bock, 1980; Schoberl, 1981).

**Table 2.1** Correlation of biodegradability of common anionic surfactants with their chemical stability and the occurrence of natural analogues

| Surfactant Type | Hydrolytic stability | Known natural analogues | Mechanism of biodegradation[a] | | |
| --- | --- | --- | --- | --- | --- |
| | | | Hydrophile/hydrophobe separation | | Direct attack on alkyl chain |
| | | | Hydrolytic | Oxidative | $\omega$-/$\beta$-Oxidation |
| Alkyl sulphate esters | Labile | Many | * | | |
| Alkyl sulpho-succinate (caboxylate) esters | Labile | Many | * | | |
| Alkylethoxy- (ether) sulphates | Labile ester/ stable ether | Some | * | * | * |
| Alkane sulphonates | Stable | Some | | * | * |
| Fatty acid sulphonates | Stable | None | | | * |
| Alkylbenzene sulphonates | Stable | None | | | * |

[a]Asterisks indicate modes of biodegradation which have been observed for each surfactant type.

## 2.2.3  Primary versus ultimate biodegradation

Both hydrophile separation and $\omega$-/$\beta$-oxidation destroy the amphiphilic property of surfactants, a process known as primary biodegradation. This may well leave residues which, from the bacterial perspective, still contain much useful carbon and, from an environmental viewpoint, could be toxic (possibly more so than the parent molecule). The subsequent complete breakdown of residues formed by primary degradation is termed ultimate biodegradation. Sometimes the term mineralisation is used to denote ultimate biodegradation, but this should be reserved for the formation of inorganic residues from the (organic) surfactant—i.e. its metabolism to $CO_2$ and $H_2O$, plus mineral salts such as $SO_3^{2-}$ or $SO_4^{2-}$. Generally there is not complete mineralisation of surfactant, because intermediates of biodegradation are used by the bacteria for the biosynthesis of other cellular components.

Where primary attack is by hydrophile removal, ultimate biodegradation usually follows easily because it occurs via the common oxidative pathways used to break down fatty acids inside bacterial (and other) cells. For example, sulphate removal from primary alkyl sulphates yields an aliphatic alcohol which is readily oxidised by dehydrogenase enzymes found in bacteria to give first the corresponding aldehyde and then the acid which can subsequently be activated and $\beta$-oxidised (*vide supra*). Therefore, in this instance the measurement of primary biodegradation by sulphate release or loss of surfactancy (e.g. with the MBAS test) can be used as an estimate of the complete (ultimate) biodegradability of the compound. However, a separate measurement of ultimate biodegradation must generally be made, for instance by $CO_2$ release.

Complete mineralisation of a surfactant means that any hydrophilic moieties containing carbon must also be metabolised. This process will also mobilise elements such as sulphur for metabolism into other cellular components. For some anionic surfactants primary biodegradation leads directly to full mineralisation of the hydrophile: for example, sulphate and sulphite are released from alkyl sulphate and alkane sulphonate hydrophiles, respectively. With some other surfactants, organic residues may remain. For example, the primary etherase cleavage of the alkylpolyethoxy sulphate surfactant sodium dodecyltriethoxy sulphate (*vide infra*) produces mono-, di- and triethylene glycol monosulphates plus their oxidised derivatives acetic acid 2-sulphate, acetic acid 2-(ethoxysulphate) and acetic acid 2-(diethoxysulphate). Although these intermediates persist in pure cultures (Hales *et al.*, 1986), their accumulation in mixed environmental cultures is only transient (Griffiths *et al.*, 1986), indicating that they do indeed undergo further metabolism.

If the mode of primary attack is by $\omega$-/$\beta$-oxidation, then this will normally proceed unhindered until the alkyl chain becomes too short for further rounds of oxidation. In practice, this means that primary biodegradation produces organic compounds comprised of the hydrophilic group attached to a small residue of the alkyl chain. For example, $\omega$-/$\beta$-oxidation of LABS produces short-chain sulphophenyl alkanoates or sulphobenzoate. Ultimate biodegradation then involves degradation of the aromatic nucleus and separation of the sulphonate group. Biodegradation of benzene sulphonates has been studied extensively, and the current consensus of opinion is that desulphonation immediately precedes and facilitates ring opening via a dioxygenase-catalysed addition of $O_2$ to the benzene nucleus (Cain, 1987; Locher *et al.*, 1989, 1991). With this exception, the ultimate biodegradation of most organic intermediates of primary biodegradation has been poorly studied in comparison with work on the parent surfactants. The reason for this lack of knowledge is that for current legislative purposes it is only necessary to demonstrate an adequate rate of primary biodegradation; ultimate biodegradation and the pathway by which this is achieved need not be elucidated.

### 2.2.4   Anaerobic versus aerobic metabolism

Thus far we have described surfactant biodegradation in terms of the utilisation of the reduced carbon in surfactants by aerobic organisms. The overall process is oxidative and oxidation steps are easily identifiable at some stage in all the pathways, e.g. the $\beta$-oxidation of aliphatic carbon chains. In some of the metabolic reactions there is direct involvement of molecular oxygen from the atmosphere, e.g. in $\omega$-oxidation of alkyl chain. However, there are several situations in which surfactants enter anaerobic environments in which we can anticipate that the mechanisms of surfactant biodegradation may differ from those in aerobic environments.

Surfactants in domestic and industrial waste are most likely to receive some form of biological treatment. Most treatment systems are aerobic, but some are anaerobic (e.g. septic tanks). Even when the main treatment is aerobic, sludges arising from such treatments are often subjected to anaerobic digestion to reduce sludge volume, to eliminate pathogenic organisms and offensive odours, to produce 'bio-gas' (methane) and to convert the solid waste into a form for which disposal is relatively easy, e.g. by land application. Anaerobic digestion is thus an economically important aspect of waste treatment and it is not uncommon for undegraded surfactant to be present in the sewage sludges reaching digesters. Thus the effects of surfactants on the digester, and the biodegradation of surfactants by anaerobic bacteria, are important considerations.

Anaerobic digestion of sludge and other wastes is a two-stage process, involving first the production of short-chain volatile fatty acids (VFA) from the organic waste, and then the production of methane from the VFA and $CO_2$. Bacteria responsible for the second stage are strictly anaerobic and effectively use oxidised carbon ($CO_2$ and VFA) instead of $O_2$ as the terminal electron acceptors. Consequently, the end-product of electron transport is a fully reduced form of carbon, in this instance methane.

Methanogenic bacteria (which produce methane by anaerobic metabolic reactions) are very sensitive to toxic materials (Jackson and Brown, 1970; Surridge *et al.*, 1975) and there is plenty of evidence that surfactants inhibit methane biosynthesis (Osborn, 1969; van der Merwe, 1969; Swanwick and Shurben, 1969). Tetrapropylene benzene sulphonates ('hard' TPS) and the linear alkylbenzene sulphonates ('soft' LABS) are equally inhibitory to digester-gas production and moreover they are not degraded significantly under anaerobic conditions (Bruce *et al.*, 1966; Swanwick *et al.*, 1968; Wood *et al.*, 1970; McEvoy and Giger, 1986). The short-chain sulphophenylalkanoate end-products of $\omega$-/$\beta$-oxidation of LABS are also unlikely to be amenable to anaerobic degradation (Kuhn and Sulfita, 1989). Alkane sulphonates, which undergo aerobic degradation faster than LABS (Hrsak *et al.*, 1981), inhibit methane production and are equally recalcitrant to anaerobic biodegradation (Bruce *et al.*, 1966). Fatty acid ester sulphonates (FES) are also resistant to degradation under anaerobic conditions (Steber and Wierich, 1989). Resistance of these aliphatic and aromatic sulphonated surfactants to anaerobic biodegradation is not at all surprising because these surfactants depend entirely on the involvement of molecular dioxygen ($O_2$) to initiate biodegradation, either by hydrophile removal or for $\omega$-oxidation of the alkyl chain (Section 2.2.1). Without $O_2$, these pathways are impossible.

In contrast, alcohol (alkyl) sulphates (AS) have very little effect on methane production during sludge digestion, and the surfactants are rapidly biodegraded under anaerobic conditions (Maurer *et al.*, 1965; Bruce *et al.*, 1966). This is entirely consistent with the ubiquitous hydrolytic pathway for biodegradation of long-chain alkyl sulphates, which does not require involvement of molecular oxygen (Section 2.2.1).

Microorganisms encounter surfactants in other anaerobic environments besides digesters. For example, river, lake and marine sediments become anaerobic at depths only a few centimetres below the sediment surface, and deeper water in stagnant ponds is also anaerobic. In these circumstances, microbes utilising alternative electron acceptors become important, such as the sulphate-reducing bacteria which reduce $SO_4^{2-}$ to $H_2S$ (Huxtable, 1986), and the denitrifying bacteria which reduce $NO_3^-$ to $N_2$ (Payne, 1981). Several non-fermentative denitrifying bacteria have been isolated which are capable of anaerobic growth on sodium dodecyl sulphate (Dodgson *et al.*, 1984). Cell extracts from aerobic or anaerobic cultures of the most prolific organism contained a single major inducible alkyl sulphatase, strongly implying that the simple hydrolytic mechanism which operates during aerobic biodegradation of alkyl sulphates is also responsible for the biodegradation under anaerobic conditions. Sulphatase-mediated hydrolysis is also the mechanism for anaerobic degradation of sulphate ester surfactants (albeit naturally occurring bile salt sulphate esters) by intestinal bacteria (Huijghebaert *et al.*, 1982).

Thus, from the published literature, it appears that surfactants (e.g. LABS) which are degraded aerobically either solely or mainly by pathways involving $O_2$, will be highly recalcitrant under anaerobic conditions. On the other hand, surfactants which are readily biodegraded via hydrolytic pathways (e.g. sulphate esters) are unlikely to persist under either set of conditions. Whether this generalisation extends to other types of surfactants that undergo hydrolytic primary biodegradation, such as dialkyl sulphosuccinates, remains to be determined.

For ethoxylated surfactants, primary biodegradation by *Pseudomonas* spp. under aerobic conditions is by oxidative separation of the hydrophile (Hales *et al.*, 1982; Griffiths *et al.*, 1986), although it is uncertain whether the process actually requires molecular oxygen (Griffiths, 1985). On the other hand, a denitrifying bacterium (*Alcaligenes faecalis* var. *denitrificans*) is able to grow aerobically and anaerobically on ether-linked surfactants (Grant and Payne, 1983). Surfactants containing primary alcohols attached to 3–7 ethylene oxide units were able to support growth of bacteria respiring nitrate, but surfactants with more ethylene oxide units, and those with secondary alcohols, were unsuccessful. These data indicate that although ether cleavage in this isolate may be oxidative, it probably does not involve $O_2$. Clearly, it would be unwise at present to make generalisations about the anaerobic biodegradability of ethoxylated surfactants from the relatively few available published studies.

## 2.3   Surfactant biodegradation in the environment

Metabolic studies of surfactant biodegradation usually involve bacteria isolated for their capacity to grow rapidly on the surfactant under test (for which they are therefore pre-adapted), acting on the surfactant as a sole source of carbon at concentrations that are often high relative to those normally found in the environment, and in a suspended liquid-phase culture in which nutrients are not limiting. Defining and simplifying the experimental parameters in this way is usually essential in order to achieve the experimental objectives, e.g. establishing metabolic pathways from the sequence of production and disappearance of metabolic intermediates. Almost invariably, the experimental conditions for metabolic and enzymological studies are unrepresentative of the real environments where surfactant biodegradation takes place. Thus, while such studies are invaluable in elucidating the chemical fundamentals of surfactant–bacteria interactions, they must be interpreted cautiously, especially when the results are extrapolated to explain events in the natural environment. It is clear that due account must be taken of a number of other factors when relating metabolic studies of the biodegradation of surfactants to their fate in the real environment.

### 2.3.1   Pure versus mixed cultures

Although the isolates used in laboratory experiments may be derived from such environmentally relevant sources as polluted rivers or activated-sludge plants, they may not individually be truly representative of what happens in the environment where a mixture of many different types of bacteria, and other microorganisms, may all play their part in the ultimate biodegradation of a surfactant. A common experimental solution to this problem is to remove a sample of the whole ecosystem (e.g. riverwater and sediment) to the laboratory and there to establish it as a microcosm. Even then, it is inevitable that there will be qualitative and quantitative changes in the bacterial population. It is well known for example that, when cultures of natural consortia of bacteria are grown continuously in a steady-state chemostat system for any length of time under laboratory conditions in the presence of such compounds as surfactants, a selection process goes on so that after many generations the population can be quite different to that originally taken from the natural ecosystem. Clearly any and all experimental approaches can be criticised on the grounds that they will involve some perturbation of the natural system. In order to construct an accurate perception of what happens in the natural ecosystem, it is therefore important not to rely on any one method but to use a variety of experimental approaches ranging from studies with individual bacterial isolates to consortia, including laboratory modelling, as well as monitoring *in situ* in the environment.

Studies with sodium dodecyltriethoxy sulphate (SDTES) provide a good example of the importance of combining pure and mixed culture studies. Four bacterial isolates were studied for their ability to metabolise SDTES which was broken down by a mixture of sulphate hydrolysis, ether bond cleavage and $\omega$-/$\beta$-oxidation. In *Pseudomonas* spp. DES1 and C12B, which had been isolated originally by enrichment on sulphated surfactants, etherase attack was found to be relatively non-specific in terms of which bond was broken and there was also a significant contribution from sulphatases, whereas in bacterial strains SC25A and TES5, which had been isolated on polyethylene glycols, the etherase was predominant and highly specific for the alkyl ether bond (Table 2.2).

The metabolic pathways of SDTES biodegradation in DES1 are summarised in Figure 2.5 (p. 34), which shows the combination of etherase and sulphatase actions. Up to 20 metabolites were produced because the initial glycol sulphate intermediates could be oxidised to give groups of metabolites but they were not substrates for the etherase; similarly, the sulphatase also acted only on the parent surfactant molecule. Therefore the amounts of each of the three pools of glycol chain-lengths reflected the substrate specificity of the etherase activity (Table 2.2), which seemed to be a single enzyme since growth of DES1 under a wide variety of conditions gave the same result. In both DES1 and C12B there

**Table 2.2** Biodegradation of SDTES in four bacterial isolates

| Bacterial isolate[a] | Percentage metabolism | | | | |
|---|---|---|---|---|---|
| | Ether bond | | | Sulphatase | $\omega$-/$\beta$-Oxidation |
| | 1 | 2 | 3 | | |
| *Pseudomonas* DES1 | 13 | 31 | 22 | 34 | 0 |
| *Pseudomonas* C12B | 16 | 24 | 23 | 38 | 0 |
| Strain SC25A | 86 | 5 | 0 | 9 | 0 |
| Strain TES5 | 82 | 1 | 0 | 6 | 11 |

[a]See Hales *et al.*, 1986

was twice as much etherase activity compared with sulphatase, and the selectivities of their etherases for the three ether links were similar (Table 2.2). In contrast, in SC25A and TES5 there was tenfold more etherase activity compared with sulphatase, and the etherase was highly specific for the alkyl ether bond. In addition in TES5, but not in the other isolates, there was a significant contribution from $\omega$-/$\beta$-oxidation.

These results illustrate that different isolates can degrade a given surfactant by different metabolic pathways and moreover that some isolates use more than one pathway simultaneously.

In tests with natural samples of water or sewage sludge, containing mixtures of bacteria, ether cleavage again predominated over sulphatase attack, and $\omega$-/$\beta$-oxidation was a minor route. However some of the glycol sulphate products of etherase action were difficult to detect and none of the intermediates, which were persistent in the pure culture studies, accumulated in the mixed environmental samples. This was a major difference in the behaviour of the two systems and illustrates the complementarity of pure and mixed culture approaches in metabolic studies of biodegradation.

### 2.3.2   Consortia

Some readily biodegradable surfactants (e.g. alkyl sulphates) are completely biodegraded by single species of bacteria. For more complex surfactants, no one organism may possess a sufficient repertoire of catabolic enzymes to break down all the structural components present. However, in natural environments there will generally be a number of different bacterial species, and other types of microorganisms too, with differing biodegradative capabilities. Mixtures of such organisms collectively may be competent in biodegradation where individual strains fail. Mixed populations of interdependent microbial species which collectively, but not individually, can accomplish the biodegradation of a pollutant, are referred to as consortia. Usually consortia are required for the biodegradation of structurally more complex surfactants. Consortia capable of biodegrading a particular surfactant can be obtained by, for instance, inoculating activated sludge into a chemostat containing growth medium in which the surfactant is the sole carbon source. For example, Jimenez et al. (1991) obtained a four-membered bacterial consortium of three *Pseudomonas* spp. and an *Aeromonas* sp. that was capable of mineralising LABS; pure cultures of a single bacterial species or mixtures of any three did not mineralise LABS although three isolates could carry out primary biodegradation. This study confirmed earlier work which indicated that consortia of bacteria were involved in LABS biodegradation in natural environments (Hrsak et al., 1982; Sigoillot and Nguyen, 1990, 1992). It is unusual to find individual bacterial isolates capable of fully biodegrading 'difficult' surfactants such as LABS, whereas for more readily biodegradable surfactants (e.g. alkyl sulphates) such isolates are commonplace.

The data on LABS illustrate very well how different species cooperate in natural environments to achieve the complete biodegradation of a surfactant. Some members of the consortium will be responsible for primary biodegradation, whilst others utilise the intermediate products to achieve complete biodegradation. This point is illustrated further by the example of SDTES biodegradation discussed above (Section 2.3.1) for which the glycol sulphate intermediates produced by etherase attack persisted in the incubations with pure isolates but not in mixed environmental samples. Similarly dialkylsulphosuccinate undergoes rapid primary biodegradation with extensive utilisation of the alkyl chains (Hales, 1993) but with accumulation of sulphosuccinate. Other species have been isolated which complete the biodegradation of sulphosuccinate (Quick *et al.*, 1994). Another aspect of surfactant biodegradation in the natural environment by consortia, is that commercial surfactant preparations invariably are comprised of mixtures of components varying in, for example, alkyl chain length. For LABS, such chain length isomers are known to biodegrade at different rates (Terzic *et al.*, 1992) and a natural consortium could provide organisms with biodegradative enzymes having a range of chain length specificities.

### 2.3.3    Adaptation to surfactants

*2.3.3.1    What is adaptation?*    The notions of surfactant 'biodegradability' or 'recalcitrance' are based essentially on assessments of rates and extent of biodegradation. There are several experimental designs for measuring these parameters (see Chapter 4) but they invariably involve the introduction of a surfactant into a mixed population of microorganisms, e.g. derived from activated sewage or river water. An important factor in the subsequent speed of events is the previous history of the organisms. It is well established that samples from different soils or sites in a river will give different rates of biodegradation, with sites that have been pre-exposed to pollutant initiating biodegradation faster than those previously unexposed (Wilson *et al.*, 1985). Related phenomena are the apparent 'lag' in the initiation of biodegradation seen in laboratory die-away experiments for many pollutants including surfactants (Griffiths *et al.*, 1986; Anderson *et al.*, 1990; Hales and Ernst, 1991; Russell *et al.*, 1991), and the shortening of the lag following repeated exposure of mixed bacterial populations to test surfactants in the laboratory (Pfaender *et al.*, 1985; Shimp, 1989). These observations are often described as 'adaptation' of the microbial population to the test compound. Adaptation is defined functionally as an increase in the rate of biodegradation with exposure (Aelion *et al.*, 1987; Shimp, 1989) and is observed with samples from diverse environments including soils, river water and estuaries. Its occurrence is demonstrated by (i) comparisons of biodegradation rates achieved by samples pre-exposed and un-exposed to test compounds (Spain and Van Veld, 1983), and (ii) the acceleration achieved by repeated additions of compounds in die-away tests (Larson and Davidson, 1982; Pfaender *et al.*, 1985).

A bacterial population can increase its capacity to biodegrade surfactant in several ways. First, growth of the bacterial population will increase the total cell numbers present and thus potentially the numbers of degraders, and there may be preferential growth of the biodegradation-competent population. A second method is through either of two mechanisms for increasing the amount of enzyme per cell that is biosynthesised in competent bacteria (see Section 2.3.3.3). Third, by a genetic process in which random genetic mutation increases the biodegradation activity or creates a new activity; mutations can occur in either the genes which encode biodegradative enzymes or in the regulatory genes that control the amount of enzyme which is normally produced. Genetic changes can also arise by transfer of the relevant genetic information between members of the microbial community.

*2.3.3.2 Population growth.*   The numbers of competent bacterial cells able to biodegrade a surfactant in a mixed microbial population will usually rise when it is challenged with a surfactant, because those bacterial species which have the capacity to take up and utilise the compound as a carbon/energy source will increase their numbers through growth and cell division. Depending on the presence and nature of other growth substrates in the ecosystem, they may or may not increase as a proportion of the total population. Whether or not they gain such an advantage over other members of the microbial population, the important fact is that the increase in absolute numbers of degraders will raise the overall capacity of the population as a whole for biodegrading the test compound. There are numerous reports of this phenomenon, (e.g. Spain *et al.*, 1984; Russell *et al.*, 1991) where growth of competent members of a mixed population accounts at least in part for the adaptation to a test compound, but it is not necessarily the sole factor.

*2.3.3.3 Increased enzyme biosynthesis.*   The second possible contribution to the adaptation process involves increasing the amount of enzyme per cell by increasing the rate of synthesis of the enzyme(s) responsible for surfactant biodegradation; enzyme level(s) may increase by up to several orders of magnitude from almost negligible background values. This can occur by related but distinct processes, referred to as induction and derepression. These are mechanisms which bring about an increase in the biosynthesis of specific enzymes via controls exerted at the genetic (i.e. DNA) level. Enzymes are proteins and the properties and catalytic activity of an enzyme are determined by the sequence of amino acids which constitute the protein. The amino acid sequence is encoded in the gene for the enzyme which is a section of the DNA in a bacterium. During biosynthesis of the enzyme, the relevant section of DNA is copied (transcribed) into a messenger RNA (mRNA) which serves as a template on which the protein can be assembled (translation) by subcellular units called ribosomes.

The mechanisms of control of gene expression in bacteria have been extensively studied and are now understood in considerable detail. The following paragraphs provide a simplified outline sufficient to show the potential contribution of gene regulation to adaptation in the environment.

In the process of induction, a so-called inducer molecule (usually, but not necessarily, the substrate that is to be biodegraded) interacts with a regulatory protein called the repressor which is normally bound to the DNA of the particular gene encoding the information for the biodegradative enzyme (Figure 2.7, top panel). In the absence of inducer, the binding of repressor to the DNA effectively blocks the gene so that it cannot be copied into mRNA; thus in the absence of inducer, the bacteria do not manufacture the enzyme. Once the inducer has bound to the repressor, the latter's affinity for the gene is reduced and the repressor separates from the DNA (Figure 2.7, centre panel). This activates the gene by allowing it to be copied as mRNA (Figure 2.7, bottom panel) and switches on the sequence of events which culminate in the biosynthesis of many thousands of enzyme molecules because each mRNA can be translated by the ribosome many times into protein.

Once the surfactant (inducer) has been metabolised, the repressor will return to the DNA and switch off the gene. The genes encoding enzymes whose biosynthesis is regulated in this manner are said to be inducible; the fact that they are only made when they are needed helps to conserve cellular resources of energy, carbon and nitrogen.

There is another form of regulation of enzyme biosynthesis, known as derepression, which in a sense is the converse of induction. In this mechanism, a specific gene (or a set of genes) is activated only when the amount of specific regulatory molecule decreases to a certain concentration inside the bacterial cell. The regulatory system also contains a repressor molecule but, in contrast with that found in inducible systems, its affinity for the regulated gene is increased by the presence of the regulatory metabolite. There are essentially two types of metabolic pathways that are regulated by repression: those concerned with biosynthetic pathways and those which produce energy from the breakdown of carbon sources such as glucose and glycerol or, in the present context, surfactant molecules.

For biosynthetic pathways that are regulated by derepression, the regulatory metabolite is usually the final product which can bind specifically to the repressor which in turn recognises the gene encoding the first enzyme in the biosynthetic pathway for the metabolite. The biosynthesis of this enzyme is said to be repressible and in effect the amount of the required product regulates its own biosynthesis via a feedback mechanism. An end-product of biodegradation may also act as a repressor; for example, sulphite may repress the desulphonation of LABS (Cain et al., 1972).

The other system in which derepression operates is when the regulatory metabolite is the starting compound for a catabolic (metabolic breakdown) pathway that produces energy and other useful intermediate metabolites. In this type of control, the availability of readily metabolisable and therefore preferred carbon sources (e.g. pyruvate, succinate) prevents the expression of the genes

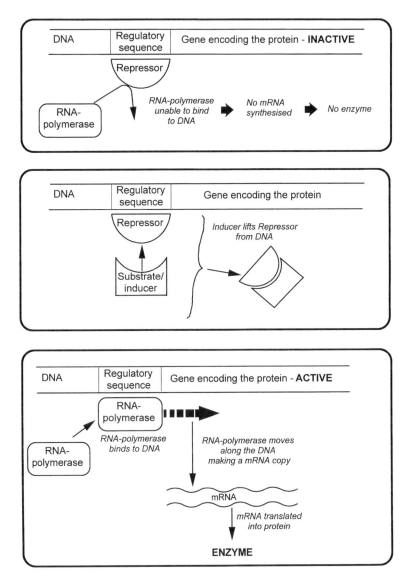

**Figure 2.7** The induction of enzyme synthesis in bacteria; an outline mechanism, showing the major features of the mechanisms involved (details of some other regulatory molecules are omitted for the sake of clarity). Upper panel: repressor protein bound at the regulatory sequence of DNA prevents the binding of RNA polymerase to DNA, thus preventing enzyme synthesis. Centre panel: substrate inducer binds to repressor and the complex detaches from DNA. Lower panel: RNA polymerase now binds to DNA, and makes a copy of mRNA which in turn is translated into enzyme.

for enzymes which would supply carbon from alternative sources (e.g. a surfactant). This system of control of catabolic enzyme synthesis in bacteria is called carbon catabolite repression. It can account for the observation that in some surfactant die-away tests there are very long adaptation or lag periods (when little or no surfactant breakdown occurs), yet once biodegradation eventually begins it proceeds at a rapid rate (Wiggins *et al.*, 1987). The explanation is that an alternative, preferred, carbon source is also present in the die-away flask. This compound (or one of its metabolites) acts as the catabolite repressor for a key gene (or set of genes) specifying surfactant biodegradation enzyme(s) and only when the catabolite repressor concentration has fallen to low enough levels inside the bacteria will the surfactant biodegradative gene(s) be 'switched on' and the requisite enzymes be biosynthesised. Thus, during the lag period the bacteria metabolise the preferred carbon source and bacterial numbers increase; once that compound is exhausted, the enzymes for surfactant breakdown are made and because the bacterial population size has increased there will be a particularly rapid disappearance of the surfactant.

In some organisms, the regulatory system of induction is coupled with carbon catabolite repression so that little or no enzyme is made whilst preferred carbon sources are available, even in the presence of inducer. This prevents wasteful expression of the gene when the enzyme is not needed. Only if the preferred carbon supply is depleted and the inducer is present, is the gene expressed, thus bringing the newly formed enzyme into action. Carbon catabolite repression has been studied in detail only for common carbon sources and there is a paucity of data on the mechanisms by which this system operates in relation to surfactant biodegradation. This information is needed badly because alternative carbon sources are invariably present in natural systems where surfactant biodegradation occurs.

These mechanisms are not used to regulate biosynthesis of all biodegradative enzymes: many are made all the time and such enzymes are said to be produced constitutively. Moreover, a bacterium may simultaneously contain inducible, derepressible and constitutive versions of similar enzymes. For example, *Pseudomonas* strain C12B can synthesise two different primary alkylsulphatases acting on primary alkyl sulphate surfactants such as SDS. One of the enzymes (P2) is inducible by $C_6$–$C_{14}$ alkyl sulphate substrates and analogues such as alkane sulphonates, and has an inducer specificity matching the substrate specificity of the active enzyme. The other primary alkylsulphatase (P1) was originally thought to be constitutive, but there is now considerable evidence that it is subject to repression/derepression control. For example, P1 synthesis is highest when cells are grown on poorly assimilable substrates (e.g. glucose in this particular organism) or when concentrations of readily assimilable substrates (e.g. broth) are kept low (Bateman, 1985). This organism also produces three more enzymes which hydrolyse secondary alkyl sulphates: one of the secondary enzymes is inducible, whereas two are constitutive.

In environmental samples, all three kinds of regulation are to be found. In a survey of a polluted river in South Wales, those bacteria which were able to grow on sodium dodecyl sulphate (SDS) as the sole carbon/energy source and therefore were capable of fully metabolising this surfactant, showed a wide variation in both the number of alkylsulphatase enzymes which they contained and in the pattern of inducibility/repressibility/constitutivity; some contained only one type, whilst others contained a mixture (White *et al.*, 1985, 1989; Anderson *et al.*, 1988). Having enzymes whose production is regulated differently may allow the bacterium to make best use of any alkyl sulphates that it might encounter at concentrations which could be either low (constitutive enzymes important) or high (inducible enzymes important).

Different modes of regulation of surfactant-degrading enzymes may also reflect different physiological roles. The inducible P2 alkylsulphatase mentioned above is thought to offer the cell protection against the harmful effects of SDS on plasma membranes and intracellular structures. Both its induced synthesis in response to the presence of SDS (Cloves *et al.*, 1980) and its periplasmic location in the cell (Fitzgerald and Laslie, 1975; Thomas *et al.*, 1988) are compatible with this idea. On the other hand P1 is repressed by preferred carbon sources and so may serve in a scavenging role for carbon in times of nutrient depletion (Bateman, 1985).

*2.3.3.4 Genetic changes.* The third method of affecting the capacity of a population to biodegrade a surfactant is by a genetic change in one of the genes responsible for the relevant enzyme(s). It is possible that there could be a random mutation in a gene encoding an enzyme which so altered its specificity that it was able to metabolise a new compound (which would probably be related structurally to the original substrate of the enzyme). However, it is more likely that there is a mutation in some regulatory gene which is involved in controlling the amount of the biodegradation enzyme which is produced. The system for gene regulation shown in Figure 2.7 is a simplification in the sense that only the gene which encodes the enzyme protein is shown. In reality, there are usually other regulatory genes involved in a complex multi-component system. Mutations in these regulatory genes may increase or decrease overall rates of biosynthesis of the biodegradative enzyme, and could therefore contribute to the phenomenon of adaptation.

Another way in which genetic change could affect biodegradative enzyme synthesis is through the exchange of genetic information among members of a microbial community. It is a well established fact that the genes encoding the biodegradative enzymes for a number of xenobiotics (man-made compounds) are present not on the main chromosome of bacteria but on a plasmid. Plasmids are pieces of DNA that are separate from and much smaller than the bacterial chromosome; they are replicated independently of the chromosome, can be present in multiple copies, and may be transferred between members of the same or different species within a bacterial population. The potential for genetic transfer via plasmids has been established for the natural environment (Saye *et al.*, 1987) and in such situations as activated sludge (McClure *et al.*, 1990) and laboratory microcosms (Fry and Day, 1989).

In general terms, plasmids encode genes that are responsible for what might be regarded as 'auxiliary' functions, i.e. properties which contribute to the ability of the bacterium to adapt to particular environments rather than the basic growth/cell-division process. Such adaptation includes coping with changes in nutrient availability which can be in the form of a xenobiotic. Among the best understood plasmid systems are the so-called TOL plasmids which encode the genes for the catabolism of toluene and the xylenes (Assinder and Williams, 1990). However, although surfactant biodegradation would appear to be an ideal candidate for a plasmid-encoded set of genes, there is no firm evidence to date that the responsible genes are located on a plasmid rather than the main chromosome.

Cain and co-workers (1987) have investigated the involvement of plasmids in LABS metabolism. The biodegradation of LABS was a likely candidate because it was known already that some of the enzymes needed to open and catabolise the aromatic ring (e.g. those of the so-called *meta*-cleavage and gentisate pathways) were encoded by plasmid genes. These workers found that desulphonating ability of *Pseudomonas testosteroni* PtS-1 was lost under a variety of conditions and at a rate which was compatible with plasmid loss but not chromosomal mutation. Enzyme assays showed that the bacteria lost not only desulphonating ability but also sulphite oxidase and the subsequent enzymes of the *meta*-cleavage pathway. The plasmid DNA was isolated and shown to confer the property of LABS biodegradation when transferred into some other pseudomonads which were normally unable to break down this surfactant. These results not only indicate that the genes are plasmid-encoded but that single species of bacteria may be able to accomplish LABS biodegradation (but *cf.* Section 2.3.2). However, unfortunately no work has been published to confirm these results or characterise the plasmids, and their involvement in LABS biodegradation must be regarded as tentative. In the four-membered consortium of Jimenez *et al.* (1991) which breaks down LABS (see Section 2.3.2), only one of the bacterial species contains a plasmid but whether it was needed for surfactant biodegradation was not determined. Similarly, Breen *et al.* (1992) found that bacterial isolates from wastewaters and ponds were unable to mineralise the LABS ring system, which required a consortium. They also found that, although many of the bacterial isolates contained plasmids, their presence did not correlate with the ability to degrade LABS. Indeed, Smith and Ratledge (1989) obtained evidence that the genes for breakdown of alkylbenzenes in a *Pseudomonas* species were on the bacterial chromosome rather than a plasmid.

Unfortunately, we have to say that very little is known currently about the genetics of surfactant-degrading enzymes. Some progress has been made by Davison and co-workers, who have cloned and sequenced a gene (called *sdsA*) from a *Pseudomonas* species; this gene, which is located on the bacterial chromosome (i.e. not on a plasmid) encodes a primary alkylsulphatase which is responsible for initiating SDS biodegradation in this bacterium (Davison *et al.*, 1990, 1992). A second gene (*sdsB*) was identified close to the *sdsA* gene and its sequence found to be similar to that of a family of bacterial genes (known as *lysR*) which encode a set of regulatory proteins that act by binding to DNA; Davison *et al.* (1992) showed that the *sdsB* gene was involved in the positive regulation of *sdsA*. Currently, these remain the only surfactant-degrading genes to have been isolated.

*2.3.3.5 Overview of adaptation.* Given the possible mechanisms for increasing biodegradative capacity (i.e. growth, physiological regulation and genetic changes) it is relevant to consider their relative contributions to the overall adaptation process. Genetic changes (mutations, plasmid transfer) are usually considered to be too infrequent and random to account for observed adaptations which usually occur in a very reproducible way (Wiggins *et al.*, 1987). Induction of gene expression is far too rapid (minutes to hours) to account for the overall adaptation observed in environmental samples, which in most cases takes days or weeks to become apparent. However, induction of enzymes may contribute in situations where encounters with the xenobiotic are infrequent or previously unknown. At an unpolluted site, constitutive organisms (i.e. those containing constitutively regulated biodegradative enzymes) would be wasting resources by synthesising unnecessary enzymes. They are thus at a selective disadvantage compared with those inducible isolates which only produce the enzyme when required. In contrast, at permanently polluted sites the constitutive organisms are not disadvantaged. Evidence for this comes from a survey of the numbers of alkylsulphatase-containing bacteria in a polluted South Wales river (White *et al.*, 1985, 1989; Anderson *et al.*, 1988). The total number of SDS-degraders was increased at polluted sites along the river compared with the pristine source water; the increase was due largely to a greater number of strains bearing constitutive enzymes, but also the incidence of strains containing multiple alkylsulphatases was much higher at the polluted sites than at the clean source site. Thus, the dominant population at the polluted site is not only constitutive, but also one which contains particularly active and versatile strains.

Generally, most of the adaptation observed during biodegradation of xenobiotics is attributable to growth of the degrading population(s). This is supported by the fact that biodegradation 'die-away' kinetics are often successfully described by mathematical models in which growth is the only factor facilitating an acceleration component in the model (Simkins and Alexander, 1984; Schmidt *et al.*, 1985*a*; Anderson *et al.*, 1990; Hales and Ernst, 1991). In addition, once adaptation has occurred, the capacity to degrade the test compound persists for long periods (Spain and Van Veld, 1983; Shimp, 1989), which is compatible with the formation of a large stable population of degrading organisms.

*2.3.4  Substrate concentration*

Changing the substrate concentration can change the rate of biodegradation in two ways. First, for a given population of cells at fixed cell density, the rate is likely to follow hyperbolic (Michaelis–Menten) kinetics because the rate of the process is usually limited either by the bacterial system responsible for transporting the surfactant into the cell or by one of the enzymic steps in its catabolism; both types of system have a fixed capacity for binding substrate (i.e. they are saturable). At high concentrations of substrate, the rate may be zero-order (i.e. independent of substrate concentration) but at lower concentrations the kinetics approach first-order (i.e. linear dependence of rate on substrate concentration). Second, as biodegradation of the surfactant proceeds, competent bacteria will utilise the available carbon to support cell growth and division (see Section 2.2.1). If the concentration of available substrate is high enough, growth will significantly increase the amount of catalyst, which is effectively equivalent to the number of bacterial cells in the system. In essence the process is autocatalytic and the rate of disappearance of substrate increases as the biodegradation proceeds, leading to the typical sigmoidal curves for disappearance of substrates in laboratory die-away experiments. The kinetics of many of these die-away curves can be described adequately by models that allow the substrate concentration to diminish by first-order or Michaelis–Menten kinetics, under the influence of catalyst (cells) growing according to various types of growth kinetics (Simkins and Alexander, 1984).

If cell density is already high, or the quantity of substrate added is very low, the amount of growth that small quantities of substrate can support may not be significant. In fact, if the substrate concentration is very low, it may even fall below a threshold for growth (Subba-Rao et al., 1982). The existence of such a threshold is evident from both experimental data (van der Kooij et al., 1980) and theoretical considerations (Schmidt et al., 1985b). The priority for bacteria growing under nutrient-depleted (oligotrophic) conditions is the maintenance of existing cells, and population growth must necessarily take second place. A threshold for growth may also amount to a threshold for mineralisation of a pollutant when the competent population is initially so small that it dies before detectable amounts of substrate are metabolised to maintain the cells (Schmidt et al., 1985b). Clearly there is a critical minimum cell density for effective biodegradation to occur (see Section 2.3.7 for further discussion).

Because ionic surfactants are charged molecules, they are associated in solution with counterions. Therefore, depending on the ionic composition of the water, surfactants may exhibit chemical speciation in aquatic systems by forming ion pairs with $Ca^{2+}$, $Mg^{2+}$ or other ions. Although no information is currently available for surfactants, this could affect their biodegradation since chemical speciation of detergent builders such as nitrilotriacetate has been shown to exert a significant effect on their biodegradation rates (Madsen and Alexander, 1985).

### 2.3.5 Mixed substrates

In the natural environment, surfactants would rarely be present in the absence of any other utilisable carbon sources. Mixed microbial populations will develop and grow towards different distributions depending on the nutrient 'mix' available, and this in turn will affect mineralisation of pollutants. The presence of high concentrations of two growth substrates is well known to produce biphasic growth curves of bacteria in batch culture (Harder and Dijkhuizen, 1982), with the substrate that supports the highest growth rate being utilised preferentially, a phenomenon known as diauxie. During the initial growth period, the synthesis of enzymes for the biodegradation of the second substrate may be repressed (Magasanik, 1976), for example, by carbon catabolite repression (see Section 2.3.3.3). At lower, growth-limiting concentrations of both substrates, catabolite repression is relieved and simultaneous utilisation of the various compounds can occur (Schmidt and Alexander, 1985). Thus alternative substrates and uncharacterised dissolved organic carbon in environmental samples may play an important role in controlling the rate and extent of the biodegradation of surfactants.

Although it does not involve surfactants, an example of the effect of a common nutrient on the metabolism of a xenobiotic is demonstrated clearly in the study by LaPat-Polasko et al. (1984) of methylene chloride biodegradation by a Pseudomonas sp. The biodegradation rate of low concentrations (mg/l) of methylene chloride was increased by the presence of higher concentrations (mg/l) of acetate, a readily utilisable substrate; when the substrate concentrations were reversed it was found that acetate utilisation rates were decreased by the addition of methylene chloride. Similar effects have been observed with aromatic pollutants (Schmidt and Alexander, 1985). Therefore, it is clear that when two substrates are metabolised simultaneously their relative concentrations have important effects on the kinetics of their biodegradation. This is an important consideration when analysing biodegradation kinetics for environmental samples from heavily polluted sites containing high organic carbon loads (Russell et al., 1991). It is also particularly relevant to biodegradation in the environment which takes place predominantly in bacterial populations which are attached to surfaces as biofilms (see Section 2.3.8), since LaPat-Polasko et al. (1984) obtained very similar results for both free-living (planktonic) and biofilm bacterial populations.

### 2.3.6   Co-metabolism

The term co-metabolism (sometimes referred to as co-oxidation) describes the situation in which an organism is able to metabolise a compound only in the presence of a second substrate which supports growth (Dalton and Stirling, 1982). This phenomenon has been observed frequently for recalcitrant compounds, e.g. some chlorinated pesticides and aromatic compounds.

On its own the first compound is incapable of supporting growth, but there may be some limited metabolism. Thus, for co-metabolism in the context of surfactants, it is possible that the compound could undergo primary biodegradation in a bacterium, but that mineralisation would occur only in the presence of a growth-supporting substrate.

In situations where there is a mixture of bacteria, one species could provide the growth substrate for another or it could be derived from some natural or industrial discharge. The presence of structurally related molecules may enhance the rates of surfactant breakdown, by inducing enzymes which are capable of metabolising the surfactant in addition; for example, enhancement of polyethyleneglycol biodegradation in sewage sludge has been achieved by exposure to fatty acid diester derivatives (Christopher et al., 1992). However, co-metabolism has not been clearly demonstrated for the biodegradation of a surfactant, although it has been implicated in the breakdown of a branched alkylbenzene sulphonate (Horvath and Koft, 1972). In complex natural consortia of micro-organisms it is likely that such mechanisms would play their part.

### 2.3.7  Retardation by nutrient limitation and predation

It is clear from the foregoing discussion that biodegradation of surfactants is intimately linked to the growth of the bacteria that metabolise them. Growth itself results in utilisation of the carbon in the surfactant and growth also produces more cells with the same biodegradative capabilities, thus accelerating the process. Consequently, factors that limit growth of the competent populations will extend lag periods and/or reduce the rates of biodegradation. Two such factors can be identified.

First, it is important to realise that microbial activities in aqueous systems are often limited not by carbon availability, but by availability of nitrogen or phosphorus. For example, adaptation periods for the biodegradation of p-cresol were longer for samples collected from field sites that were low in inorganic N or P, but were decreased when those samples were amended with N or P (Lewis et al., 1986). The significance of such effects for the biodegradation of surfactants will depend on the particular ecosystem under consideration, because this will determine which elements are limiting for the surfactant-degrading populations present. Moreover, for soil, groundwater and marine systems, long lag periods may not matter as long as the surfactant is eventually degraded, whereas in freshwater rivers and streams adaptation periods of days or even hours could allow surfactants to reach sensitive areas such as freshwater fish (which are particularly susceptible to surfactant toxicity) or municipal water supplies (Lewis et al., 1986).

The second factor limiting growth is predation of bacteria by grazing protozoa (single-celled organisms which include amoebae and Paramecium); these engulf bacteria as a food source and are widely distributed in soils and freshwaters as well as in sewage. Their role in controlling bacterial populations in natural waters has been shown to be significant (Sherr and Sherr, 1987). They are known to delay initiation of biodegradation of synthetic chemicals and to affect the relevant bacterial populations (Wiggins et al., 1987; Zaidi et al., 1989). Their presence and inhibitory effects on pollutant biodegradation by bacteria have recently been demonstrated using specific inhibitors to prevent protozoal (but not bacterial) growth (Ramadan et al., 1990).

The survival of a bacterial population faced with predation by protozoa is likely to be a balance between predation and growth rate of the bacteria; the slower the growth rate of the population, the more likely that protozoa will eliminate that population. In the presence of surfactants, the situation is likely to be further complicated because the presence of surfactants at concentrations above 10 mg/l have been correlated with decreases in protozoal populations in sewage treatment plants (Esteban and Tellez, 1992). Thus biodegradability of a pollutant in the environment is difficult to predict, since it depends on complex interactions of many factors such as the cell density of degraders which in turn depends on previous exposure, and adaptation, to surfactant; on the potential growth rate in presence of surfactant which in turn depends on nutrient status especially with respect to N and P; and on the extent of protozoal grazing which, together with initial cell density and growth rate, will determine whether or not a population survives long enough to complete biodegradation.

### 2.3.8    Surfaces and biofilms

In natural aquatic environments bacterial proliferation is generally limited by the availability of nutrients, and for bacterial populations to survive they must adapt to maximise utilisation of the limiting nutrient source. One strategy for survival, especially relevant to flowing aquatic systems, is the attachment of microorganisms to solid surfaces. For example the attachment of microorganisms to rocks and sediments in rivers offers two advantages. First, because physical and chemical properties at the interface differ from those in the bulk phase (Baier, 1970), the surface tends to accumulate solute molecules. Thus the solid/liquid interface will represent a relatively high-nutrient micro-environment in a low-nutrient macro-environment, and bacteria encountering this surface and staying in close proximity or attaching to it may be at a nutritional advantage in aquatic environments where the flux of carbon can be lower than 1 mg/l per day (Poindexter, 1981). Second, as the nutrients at the surface are utilised they are replenished by adsorption from the flowing water, thus effectively allowing bacteria to access the nutrient in a much larger volume of water than if they remained planktonic and therefore virtually fixed relative to the water column.

The phenomenon of bacterial attachment to solid surfaces in aquatic ecosystems is vitally important because it is well established that by far the greater part of microbial heterotrophic activity in freshwater ecosystems is attributable to the attached (epilithic) bacterial rather than the free-floating (planktonic) populations (Geesey et al., 1978; Costerton et al., 1981; Ladd et al., 1982; Lock et al., 1984; Pignatello et al., 1985). Indeed, Wachtershauser (1988) has speculated that life began at surfaces, the first 'organisms' being surface-adsorbed collections of autocatalytically interacting substances. Perhaps we should not view epilithic bacteria as adapted to growth on surfaces, but rather free-living bacteria as adapted to a planktonic existence!

The formation and properties of biofilms have been the subject of much study, not only because of their environmental relevance (Paerl, 1980; White *et al.*, 1994) but also for their importance in medical (Costerton *et al.*, 1981, 1987; Jacques *et al.*, 1987; Gilbert *et al.*, 1993) and industrial (Characklis and Cooksey, 1983; Bryers, 1993) contexts. Because these aspects have been reviewed extensively elsewhere, only the most salient points are presented here.

Since the earliest observations of bacteria profiting from nutrients adsorbed on solid surfaces (Zobell, 1943), biofilm formation has been considered to occur in two stages. Initial attachment of bacteria is reversible (Marshall *et al.*, 1971) and they can be removed from the surface by gentle washing. This phase precedes irreversible attachment which involves the synthesis of exopolymers and the formation of a hydrated polysaccharide matrix in which the microorganisms are embedded. The long-held view that the biofilm matrix was a homogeneous gel, has recently been radically revised with the advent of non-invasive methods of analysis such as confocal laser microscopy. In the current model (Costerton *et al.*, 1994), biofilm bacteria occur in matrix-enclosed microcolonies interspersed with less dense regions of matrix that include highly permeable water channels.

Although bacteria have been reported to utilise surface-bound molecules (Hermansson and Marshall, 1985), the quantitative significance of adsorbed nutrients in terms of providing carbon/energy for bacterial growth has been questioned (van Loosdrecht *et al.*, 1990). Conversely, the adsorption of organic compounds on a solid surface can slow the rate of biodegradation (Gordon and Millero, 1985). Nevertheless, these various observations point to a close interdependence among the adsorption of organic compounds onto sediments, the attachment of bacteria to sediments and the bacterial biodegradation of adsorbed organics.

Sediment-attachment of organic compounds and of the bacteria that utilise them, is likely to be most significant for compounds that are normally present in water at low concentrations and which adsorb strongly at phase-interfaces. Synthetic surfactants satisfy these criteria (Siracusa and Somasundaran, 1987; House and Farr, 1989; Marchesi *et al.*, 1991), and it is therefore very surprising that the significance of surfactant adsorption at solid surfaces is rarely mentioned in the context of surfactant biodegradation in the environment. Hitherto, three related areas, *viz.* the attachment of bacteria to surfaces, the adsorption of surfactants onto surfaces and the biodegradation of surfactants by bacteria, have been studied independently. However, in the environment all three phenomena will occur simultaneously, and recent investigations (Marchesi *et al.*, 1991; White *et al.*, 1994) have shown that they are in fact closely interrelated. Thus, when river sediment was resuspended in a laboratory microcosm, the attachment of bacteria to sediment particles was stimulated by the addition of those surfactants which are biodegraded by the indigenous population (e.g. SDS). The bacteria remained attached as long as the system contained residual surfactant but when primary biodegradation was complete, the bacteria detached from the surface and resumed their planktonic lifestyle. The effect was not observed if the surfactants used were recalcitrant in the particular system employed (e.g. sulphonated surfactants), or if the compound added was a non-surface-active organic nutrient (pyruvate). The presence of surfactants, and their biodegradability, clearly influence the initial attachment of bacteria to solid surfaces.

Of equal importance is the effect of sediments on the rate of biodegradation of surfactants. When sediment was added to a culture of a known SDS-degrading isolate, the rate of biodegradation was markedly increased compared with the rate in the absence of sediment (White *et al.*, 1994). Again the acceleration was simultaneous with attachment of bacteria to the surface, which was reversed when the surfactant became exhausted. These recent observations have significant implications for the assessment of surfactant biodegradability in environmental situations.

## 2.4   Laboratory models of surfactant biodegradation

A number of different laboratory test methods are used to assess the biodegradability of surfactants and these are discussed in Chapter 4, together with those field tests that are also used. A second type of laboratory test is that used to investigate the biochemistry and chemistry of surfactant biodegradation. Such laboratory-based microcosms are broadly of two types, either closed or open systems. In its simplest form the microcosm is essentially run as a chemostat in which the bacterial suspension or environmental sample is stirred under controlled conditions of nutrient supply, aeration, pH, temperature, etc. Continuous-flow systems to mimic accurately the flow rates in riverine situations are difficult to reproduce in the laboratory because of the large volumes of water required even for reaction vessels with a small capacity, particularly because the water used should ideally be taken from the river being modelled and not recirculated through the reaction vessel. Chemostat systems may suffer from the disadvantage of relatively high washout rates of the biodegradative bacteria, which can be avoided by using membrane diffusion chambers which are permeable to surfactants, gases and low molecular weight nutrients but retain bacteria inside; these can be placed in large-volume tanks containing, for example, lake or river water which can be made into a flow-through system if required to simulate more closely the natural situation (Awong *et al.*, 1990). Other solutions to the problem of bacterial washout in chemostat-based laboratory microcosms involve the use of a recycling loop (e.g. see Murgel *et al.*, 1991).

An important facet of modelling biodegradation in the laboratory is that in the environment the most active bacteria are to be found not free in the water body (planktonic bacteria) but attached to surfaces as a biofilm. It is only comparatively recently that the importance of biofilms in the biodegradation of surfactants and a wide range of man-made pollutants has been appreciated (see Section 2.3.8). The situation is specially complex for compounds such as surfactants because one of their key properties is that they accumulate at surfaces, including those of the bacteria which may biodegrade them as well as mineral particles, rocks or other inert or living surfaces to which the bacteria attach and form a biofilm.

# References

Aelion, C.M., Swindoll, C.M. and Pfaender, F.K. (1987) Adaptation to and biodegradation of xenobiotic compounds by microbial communities from a pristine aquifer. *Appl. Environ. Microbiol.* **53**, 2212–2217.

Anderson, D.J., Day, M.J., Russell, N.J. and White, G.F. (1988) Temporal and geographical distributions of epilithic sodium dodecyl sulfate (SDS)-degrading bacteria in a polluted South Wales river. *Appl. Environ. Microbiol.* **54**, 555–560.

Anderson, D.J., Day, M.J., Russell, N.J. and White, G.F. (1990) Die-away kinetic analysis of the capacity of epilithic and planktonic bacteria from clean and polluted river water to biodegrade sodium dodecyl sulfate. *Appl. Environ. Microbiol.* **56**, 758–763.

Armitage, J.P. and Lackie, J.M. (1990) *Biology of the Chemotactic Response*, Society for General Microbiology Symposium 46, Cambridge University Press, Cambridge.

Assinder, S.J. and Williams, P.A. (1990) The tol plasmids: determinants of the catabolism of toluene and the xylenes. *Adv. Microb. Physiol.* **31**, 1–69.

Awong, J., Britton, G. and Chaudhry, G.R. (1990) Microcosm for assessing survival of genetically engineered microorganisms in aquatic environments. *Appl. Environ. Microbiol.* **56**, 977–983.

Bachofen, R. (1986) Microorganisms in extreme environments. *Experientia* **42**, 1179–1182.

Baier, R.E. (1970) Surface properties influencing bacterial adhesion, in *Adhesion in Biological Systems*, ed. R.S. Manly, Academic Press, London, pp. 15–48.

Bartholomew, B., Dodgson, K.S., Matcham, G.W.J., Shaw, D.J. and White, G.F. (1977) A novel mechanism of enzymic ester hydrolysis. Inversion of configuration and carbon-oxygen bond cleavage by secondary alkylsulphohydrolases from detergent-degrading micro-organisms. *Biochem. J.* **167**, 575–580.

Bateman, T.J. (1985). Primary alkylsulphatase activity in the detergent-degrading bacterium *Pseudomonas* C12B. PhD Thesis. University of Wales.

Boulton, C.A. and Ratledge, C. (1984) The physiology of hydrocarbon-utilizing microorganisms. *Top. Enzyme Ferment. Biotechnol.* **9**, 11–77.

Breen, A., Jimenez, L., Sayler, G.S. and Federle, T.W. (1992) Plasmid incidence and linear alkylbenzene sulfonate biodegradation in wastewater and pristine pond ecosystems. *J. Ind. Microbiol.* **9**, 37–44.

Bruce, A.M., Swanwick, J.D. and Ownsworth, R.A. (1966) Synthetic detergents and sludge digestion: Some recent observations. *J. Proc. Inst. Sewage Purif.* 427–447.

Bryers, J.D. (1993) Bacterial biofilms. *Curr. Opin. Biotechnol.* **4**, 197–204.

Cain, R.B. (1987) Biodegradation of anionic surfactants. *Biochem. Soc. Trans.* **15**, 7S–22S.

Cain, R.B., Willetts, A.J. and Bird, J.A. (1972) Surfactant biodegradation–metabolism and enzymology, in *Biodeterioration of Materials*, eds. H. Walters and E.H. Hueck-Van der Plas, Applied Science Publishers, London, pp. 136–144.

Characklis, W.G. and Cooksey, K.E. (1983) Biofilms and microbial fouling, in *Advances in Applied Microbiology*, ed. A.I. Laskin, Academic Press, London, pp. 93–198.

Christopher, L.J., Holzer, G. and Hubbard, J.S. (1992) Enhancement of polyether degradation in activated sludge following exposure to conditioning agents. *Environ. Technol.* **13**, 521–530.

Cloves, J.M., Dodgson, K.S., White, G.F. and Fitzgerald, J.W. (1980) Specificity of P2 primary alkylsulphohydrolase induction in the detergent-degrading bacterium *Pseudomonas* C12B. *Biochem. J.* **185**, 13–21.

Costerton, J.W., Irvin, R.T. and Cheng, K.-J. (1981) The bacterial glycocalyx in nature and disease. *Annu. Rev. Microbiol.* **35**, 299–324.

Costerton, J.W., Cheng, K.-J. Geesey, G.G., Ladd, T.I., Nickel, J.C., Dasgupta, M. and Marrie, T.J. (1987) Bacterial biofilms in nature and disease. *Annu. Rev. Microbiol.* **41**, 435–464.

Costerton, J.W., Lewandowski, Z., DeBeer, D., Caldwell, D., Korber, D. and James, G. (1994) Biofilms, the customized microniche. *J. Bacteriol.* **176**, 2137–2142.

Dalton, H. and Stirling, D.I. (1982) Co-metabolism. *Philos. Trans. R. Soc. London Ser. B* **297**, 481–496.

Davison, J., Brunel, F. and Phanopoulos, A. (1990) The genetics of vanillate and sodium dodecyl sulphate degradation in *Pseudomonas*, in *Pseudomonas: Biotransformations, Pathogenesis, and Evolving Biotechnology*, eds. S. Silver, A.M. Chakrabarty, B. Iglewski and S. Kaplan, American Society for Microbiology, Washington, DC, pp. 159–164.

Davison, J., Brunel, F., Phanopoulos, A., Prozzi, D. and Terpstra, P. (1992) Cloning and sequencing of *Pseudomonas* genes determining sodium dodecyl sulfate biodegradation. *Gene* **114**, 19–24.

Dawes, E.A. (1976). Endogenous metabolism and the survival of starved prokaryotes, in *The Survival of Vegetative Microbes*, eds. T.R.G. Gray and J.R. Postgate, Society for General Microbiology Symposium 26, Cambridge University Press, Cambridge, pp. 19–53.

Dawson, M.P., Humphrey, B. and Marshall, K.C. (1981) Adhesion: a tactic in the survival strategy of a marine vibrio during starvation. *Curr. Microbiol.* **6**, 195–198.

Dodgson, K.S. and White, G.F. (1983) Some microbial enzymes involved in the biodegradation of sulphated surfactants. *Top. Enzyme Ferment. Biotechnol.* **7**, 90–155.

Dodgson, K.S., White, G.F., Massey, J.A., Shapleigh, J. and Payne, W.J. (1984) Utilization of sodium dodecyl sulphate by denitrifying bacteria under anaerobic conditions. *Fed. Eur. Microbiol. Soc. Lett.* **24**, 53–56.

Esteban, G. and Tellez, C. (1992) The influence of detergents on the development of ciliate communities in activated sludge. *Water Air Soil Pollut.* **61**, 185–190.

Fitzgerald, J.W. and Laslie, W.W. (1975) Loss of primary alkylsulfatase and secondary alkylsulfatase (S-1 and S-2) from *Pseudomonas* C12B; effect of culture conditions, cell-washing procedures, and osmotic shock. *Can. J. Microbiol.* **21**, 59–68.

Fry, J.C. and Day, M.J. (1989) Plasmid transfer in the epilithon, in *Bacterial Genetics in Natural Environments*, eds. J.C. Fry and M.J. Day, Chapman and Hall, London, pp. 55–88.

Geesey, G.G., Mutch, R., Costerton, J.W. and Green, R.B. (1978) Sessile bacteria: an important component of the microbial population in small mountain streams. *Limnol. Oceanogr.* **23**, 1214–1223.

Gilbert, P., Evans, D.J. and Brown, M.R.W. (1993) Formation and dispersal of bacterial biofilms *in vivo* and *in situ*. *J. Appl. Bacteriol.* **74**, 67S–78S.

Gordon, A.S. and Millero, F.J. (1985) Adsorption mediated decrease in the biodegradation rate of organic compounds. *Microb. Ecol.* **11**, 289–298.

Grant, M. and Payne, W.J. (1983) Anaerobic growth of *Alcaligenes faecalis* var. *denitrificans* at the expense of ether glycols and non-ionic detergents. *Biotechnol. Bioeng.* **25**, 627–630.

Griffiths, E.T. (1985). Microbial degradation of surfactants containing ether bonds. PhD Thesis. University of Wales.

Griffiths, E.T., Hales, S.G., Russell, N.J., Watson, G.K. and White, G.F. (1986) Metabolite production during the biodegradation of the surfactant sodium dodecyltriethoxy sulphate under mixed-culture die-away conditions. *J. Gen. Microbiol.* **132**, 963–972.

Haines, T.H. (1973) Halogen- and sulfur-containing lipids of *Ochromonas*. *Annu. Rev. Microbiol.* **27**, 403–411.

Hales, S.G. (1993) Biodegradation of the anionic surfactant dialkyl sulphosuccinate. *Environ. Chem. Toxicol.* **12**, 1821–1828.

Hales, S.G. and Ernst, W. (1991) Biodegradation of nitrilotriacetic acid (NTA) in Weser estuarine water. *Tenside Surfact. Deterg.* **28**, 15–21.

Hales, S.G., Dodgson, K.S., White, G.F., Jones, N. and Watson, G.K. (1982) Initial stages in the biodegradation of the surfactant sodium dodecyltriethoxy sulfate by *Pseudomonas* sp. strain DES1. *Appl. Environ. Microbiol.* **44**, 790–800.

Hales, S.G., White, G.F., Dodgson, K.S. and Watson, G.K. (1986) A comparative study of the biodegradation of the surfactant sodium dodecyltriethoxy sulphate by four detergent-degrading bacteria. *J. Gen. Microbiol.* **132**, 953–961.

Harder, W. (1981). Enrichment and characterisation of degrading organisms, in *Microbial Degradation of Xenobiotics and Recalcitrant Compounds*, eds. T. Leisinger, R. Hutter, A.M. Cook and J. Nuesch, Fed. Eur. Microbiol. Soc. Symposium 12, Academic Press, London, pp. 77–96.

Harder, W. and Dijkhuizen, L. (1982) Strategies of mixed substrate utilization in microorganisms. *Philos. Trans. R. Soc. London Ser. B* **297**, 459–479.

Harwood, J.L. and Russell, N.J. (1984) *Lipids in Plants and Microbes*, George Allen and Unwin, London.

Hauthal, H.G. (1992) Trends in surfactants. *Chim. Oggi* **10**, 9–13.

Hermansson, M. and Marshall, K.C. (1985) Utilisation of surface localised substrate by non-adhesive marine bacteria. *Microb. Ecol.* **11**, 91–105.

Horvath, R.S. and Koft, B.W. (1972) Degradation of alkyl benzene sulfonate by *Pseudomonas* species. *Appl. Microbiol.* **23**, 407–414.

House, W.H. and Farr, I.S. (1989) Adsorption of sulphonates from detergent mixtures on potassium kaolinite. *Colloids Surf.* **40**, 167–180.

Hrsak, D., Bosnjak, M. and Johanides, V. (1981) Kinetics of linear alkylbenzene sulphonate and secondary alkane sulphonate biodegradation. *Tenside Surfact. Deterg.* **18**, 137–140.

Hrsak, D., Bosnjak, M. and Johanides, V. (1982) Enrichment of linear alkylbenzenesulphonate (LAS) degrading bacteria in continuous culture. *J. Appl. Bacteriol.* **53**, 413–422.

Huijghebaert, S.M., Mertens, J.A. and Eyssen, H.J. (1982) Isolation of a bile salt sulfatase-producing *Clostridium* strain from rat intestinal microflora. *Appl. Environ. Microbiol.* **43**, 185–192.

Hutzinger, O. and Veerkamp, W. (1981). Xenobiotic chemicals with pollution potential, in *Microbial Degradation of Xenobiotics and Recalcitrant Compounds*, eds. T. Leisinger, R. Hutter, A.M. Cook and J. Nuesch, Fed. Eur. Microbiol. Soc. Symposium 12, Academic Press, London, pp. 3–45.

Huxtable, R.J. (1986) *Biochemistry of Sulphur*, Plenum Press, New York.

Jackson, S. and Brown, V.M. (1970) Effect of toxic wastes on treatment processes and water courses. *Water Pollut. Control*, **69**, 292–303.

Jacques, M., Marrie, T.J. and Costerton, J.W. (1987) Microbial colonization of prosthetic devices. *Microb. Ecol.* **13**, 173–191.

Jimenez, L., Breen, A., Thomas, N., Federle, T.W. and Sayler, G. (1991) Mineralization of linear alkylbenzene sulfonate by a four-member aerobic bacterial consortium. *Appl. Environ. Microbiol.* **57**, 1566–1569.

Kuhn, E.P. and Sulfita, J.M. (1989) Anaerobic biodegradation of nitrogen-substituted and sulfonated benzene aquifer contaminants. *Hazardous Wastes, Hazardous Mater.* **6**, 121–133.

Ladd, T.I., Ventullo, R.M., Wallis, P.M. and Costerton, J.W. (1982) Heterotrophic activity and biodegradation of labile and refractory compounds by groundwater and stream microbial populations. *Appl. Environ. Microbiol.* **44**, 321–329.

LaPat-Polasko, L.T., McCarth, P.L. and Zehnder, A.J.B. (1984) Secondary substrate utilization of methylene chloride by an isolated strain of *Pseudomonas* sp. *Appl. Environ. Microbiol.* **47**, 825–830.

Larson, R.J. and Davidson, D.H. (1982) Acclimation to and biodegradation of nitrilotriacetate at trace concentrations in natural waters. *Water Res.* **16**, 1597–1604.

Lewis, D., Kollig, H.P. and Hodson, R.E. (1986) Nutrient limitation and adaptation of microbial populations to chemical transformations. *Appl. Environ. Microbiol.* **51**, 598–603.

Locher, H.H., Leisinger, T. and Cook, A.M. (1989) Degradation of *p*-toluenesulphonic acid via side chain oxidation, desulphonation and meta ring cleavage in *Pseudomonas (Comamonas) testosteroni* T-2. *J. Gen. Microbiol.* **135**, 1969–1978.

Locher, H.H., Leisinger, T. and Cook, A.M. (1991) 4-Sulphobenzoate 3,4-dioxygenase. *Biochem. J.* **274**, 833–842.

Lock, M.A., Wallace, R.R., Costerton, J.W., Ventullo, R.M. and Charlton, S.E. (1984) River epilithon: toward a structural-functional model. *Oikos* **42**, 10–22.

Madsen, E.L. and Alexander, M. (1985) Effects of chemical speciation on the mineralization of organic compounds by microorganisms. *Appl. Environ. Microbiol.* **50**, 342–349.

Magasanik, B. (1976) Classical and post-classical modes of regulation of the synthesis of degradative bacterial enzymes. *Progr. Nucleic Acid Res. Mol. Biol.* **17**, 99–115.

Marchesi, J.R., Russell, N.J., White, G.F. and House, W.A. (1991) Effects of surfactant adsorption and biodegradability on the distribution of bacteria between sediments and water in a freshwater microcosm. *Appl. Environ. Microbiol.* **57**, 2507–2513.

Marshall, K.C., Stout, R. and Mitchell, R. (1971) Mechanisms of the initial events in the sorption of marine bacteria to surfaces. *J. Gen. Microbiol.* **68**, 337–348.

Maurer, E.W., Cordon, T.C., Weil, J.K., Nunez-Ponzoa, M.V., Ault W.C. and Stirton, A.J. (1965) The effect of tallow based surfactants on anaerobic digestion. *J. Am. Oil Chem. Soc.* **42**, 189–192.

McClure, N.C., Fry, J.C. and Weightman, A.J. (1990) Gene transfer in activated sludge, in *Bacterial Genetics in Natural Environments*, eds. J.C. Fry and M.J. Day, Chapman and Hall, London, pp. 111–129.

McEvoy, J. and Giger, W. (1986) Determination of linear alkylbenzenesulfonates in sewage-sludge by high-resolution gas-chromatography mass-spectrometry. *Environ. Sci. Technol.* **20**, 376–383.

Mercer, E.J. and Davies, C.L. (1979) Distribution of chlorosulpholipids in algae. *Phytochemistry* **18**, 457–462.

Morgan, P. and Dow, C.S. (1986) Bacterial adaptations for growth in low nutrient environments, in *Microbes in Extreme Environments*, eds. R.A. Herbert and G.A. Codd, Academic Press, London, pp. 187–214.

Murgel, G.A., Lion, L.W., Acheson, C., Shuler, M.L., Emerson, D. and Ghiorse, W.C. (1991) Experimental apparatus for selection of adherent microorganisms under stringent growth conditions. *Appl. Environ. Microbiol.* **57**, 1987–1996.

Osborn, D.W. (1969) Difficulties associated with sludge digestion, with particular reference to synthetic detergents. *Water Pollut. Control,* **68**, 662–663.

Paerl, H.W. (1980) Attachment of microorganisms to living and detrital surfaces in freshwater systems, in *Adsorption of Micro-organisms to Surfaces*, eds. G. Bitton and K.C. Marshall, Wiley, New York, pp. 375–402.

Payne, W.J. (1981) *Denitrification*, Wiley, New York.

Pfaender, F.K., Shimp, R.J. and Larson, R.J. (1985) Adaptation of estuarine ecosystems to the biodegradation of nitrilotriacetic acid: effects of pre-exposure. *Environ. Toxicol. Chem.* **4**, 587–593.

Pignatello, J.J., Johnson, L.K., Martinson, M.M., Carlson, R.E. and Crawford, R.L. (1985) Response of the microflora in outdoor experimental streams to pentachlorophenol; compartmental contributions. *Appl. Environ. Microbiol.* **50**, 127–132.

Poindexter, J.S. (1981) Oligotrophy. Fast and famine existence, in *Advances in Microbial Ecology*, ed. M. Alexander, Plenum Publishing Corporation, New York, pp. 63–89.

Poindexter, J.S. (1987). Bacterial responses to nutrient limitation, in *Ecology of Microbial Communities,* eds. M. Fletcher, T.R.G. Gray and J.G. Jones, Society for General Microbiology Symposium 41, Cambridge University Press, Cambridge, pp. 283–317.

Quick, A., Russell, N.J., Hales, S.G. and White, G.F. (1994) Biodegradation of sulphosuccinate: direct desulphonation of a secondary sulphonate. *Microbiology*, submitted.

Ramadan, M.A., El-Tayeb, O.M. and Alexander, M. (1990) Inoculum size as a factor limiting success of inoculation for biodegradation. *Appl. Environ, Microbiol.* **56**, 1392–1396.

Russell, N.J., Anderson, D.J., Day, M.J. and White, G.F. (1991) Colonisation of biofilms by bacteria capable of biodegrading sodium dodecyl sulphate (SDS) at clean and polluted riverine sites. *Microb. Ecol.* **22**, 85–98.

Saye, D.J., Ogunseitan, O., Sayler, G.S. and Miller, R.V. (1987) Potential for transduction of plasmids in a natural freshwater environment: effect of plasmid donor concentration and a natural microbial community on transduction in *Pseudomonas aeruginosa. Appl. Environ. Microbiol.* **53**, 987–995.

Schlegel, H.G. (1986) *General Microbiology*, Cambridge University Press, Cambridge.

Schmidt, S.K. and Alexander, M. (1985) Effects of dissolved organic carbon and second substrates on the biodegradation of organic compounds at low concentrations. *Appl. Environ. Microbiol.* **49**, 822–827.

Schmidt, S.K., Simkins, S. and Alexander, M. (1985a) Models for the kinetics of biodegradation of organic compounds not supporting growth. *Appl. Environ. Microbiol.* **50**, 323–331.

Schmidt, S.K., Alexander, M. and Shuler, M.L. (1985b) Predicting threshold concentrations of organic substrates for bacterial growth. *J. Theor. Biol.* **114**, 1–8.

Schoberl, P. (1981) Comparative investigations on the microbial metabolism of a nonylphenol and an oxoalcohol ethoxylate. *Tenside Surfact. Deterg.* **18**, 64–72.

Schoberl, P. and Bock, K.J. (1980) Surfactant degradation and its metabolites. *Tenside Surfact. Deterg.* **17**, 262–266.

Sherr, E.B. and Sherr, B.F. (1987) High rates of consumption of bacteria by pelagic ciliates. *Nature (London)* **325**, 710–711.

Shimp, R.J. (1989) Adaptation to a quaternary ammonium surfactant in aquatic sediment microcosms. *Environ. Toxicol. Chem.* **8**, 201–208.

Sigoillot, J.-C. and Nguyen, M.-H. (1990) Isolation and characterisation of surfactant degrading bacteria in a marine environment. *Fed. Eur. Microbiol. Soc. Microbiol. Ecol.* **73**, 59–68.

Sigoillot, J.-C, and Nguyen, M.-H. (1992) Complete oxidation of linear alkylbenzene sulfonate by bacterial communities selected from coastal seawater. *Appl. Environ. Microbiol.* **58**, 1308–1312.

Simkins, S. and Alexander, M. (1984) Models for mineralisation kinetics with the variables of substrate concentration and population density. *Appl. Environ. Microbiol.* **47**, 1299–1306.

Siracusa, P.A. and Somasundaran, P. (1987) Mechanism of hysteresis in sulfonate kaolinite adsorption-desorption systems – chromatographic separation of isomers. *J. Colloid Interface Sci.* **120**, 100–109.

Smith, M.R. and Ratledge, C. (1989) Catabolism of alkylbenzenes by *Pseudomonas* sp. NCIB 10643. *Appl. Microbiol. Biotechnol.* **32**, 68–75.

Spain, J.C. and Van Veld, P.A. (1983) Adaptation of natural microbial communities to degradation of xenobiotic compounds: effects of concentration, exposure time, inoculum, and chemical structure. *Appl. Environ. Microbiol.* **45**, 428–435.

Spain, J.C., Van Veld, P.A., Monti, C.A., Pritchard, P.H. and Cripe, C.R. (1984) Comparison of *p*-nitrophenol biodegradation in field and laboratory test systems. *Appl. Environ. Microbiol.* **48**, 944–950.

Steber, J. and Wierich, P. (1989) The environmental fate of fatty acid *a*-sulfomethyl esters. *Tenside Surfact. Deterg.* **26**, 406–411.

Subba-Rao, R.V., Rubin, H.E. and Alexander, M. (1982) Kinetics and extent of mineralisation of organic chemicals at trace levels in freshwater and sewage. *Appl. Environ. Microbiol.* **43**, 1139–1150.

Surridge, D., Jones, J.C. and Stafford, D.A. (1975) Influence of industrial wastes and retention time on methane production from sewage sludge digesters. *Effluent Water Treat. J.* **15**, 289–291.

Swanwick, J.D. and Shurben, D.G. (1969) Effective chemical treatment for inhibition of anaerobic sewage sludge digestion due to anionic detergents. *Water Pollut. Control*, **68**, 190–201.

Swanwick, J.D., Bruce, A.M. and Vandyke, K.G. (1968) Inhibition of sludge digestion by synthetic detergents. *Water Pollut. Control*, **67**, 91–99.

Terzic, S., Hrsak, D. and Ahel, M. (1992) Primary biodegradation kinetics of linear alkylbenzene sulphonates in estuarine waters. *Water Res.* **26**, 585–591.

Thomas, O.R.T., Matts, P.J. and White, G.F. (1988) Localisation of alkylsulphatases in bacteria by electron microscopy. *J. Gen. Microbiol.* **134**, 1229–1236.

Thysse, G.J.E. and Wanders, T.H. (1972) Degradation of *n*-alkane-1-sulfonates by *Pseudomonas*. *Antonie van Leeuwenhoek* **38**, 53–63.

Thysse, G.J.E. and Wanders, T.H. (1974) Initial steps in the degradation of *n*-alkane-l-sulfonates by *Pseudomonas*. *Antonie van Leeuwenhoek* **40**, 25–37.

van der Kooij, D., Visser, A. and Hihnen, W.A.M. (1980) Growth of *Aeromonas hydrophila* at low concentrations of substrates added to tap water. *Appl. Environ. Microbiol.* **39**, 1198–1204.

van der Merwe, P.H. (1969) The effect of synthetic detergents on sludge digestion at Rondebult sewage treatment works. *Water Pollut. Control* **68**, 669–672.

van Loosdrecht, M.C.M., Lyklema, J., Norde, W. and Zehnder, A.J.B. (1990) Influence of interfaces on microbial activity. *Microbiol. Rev.* **54**, 75–87.

Wachtershauser, G. (1988) Before enzymes and templates; theory of surface metabolism. *Microbiol. Rev.* **52**, 452–484.

White, G.F. and Russell, N.J. (1993) Biodegradation of anionic surfactants and related molecules, in *Biochemistry of Microbial Degradation*, ed. C. Ratledge, Kluwer Academic Publishers, Dordrecht, pp. 143–177.

White, G.F., Russell, N.J. and Day, M.J. (1985) A survey of sodium dodecyl sulphate SDS-resistance and alkylsulphatase production in bacteria from clean and polluted river sites. *Environ. Pollut.* **A37**, 1–11.

White, G.F., Anderson, D.J., Day, M.J. and Russell, N.J. (1989) Distribution of planktonic bacteria capable of degrading sodium dodecyl sulphate (SDS) in a polluted South Wales river. *Environ. Pollut.* **57**, 103–115.

White, G.F., Russell, N.J., Marchesi, J.R. and House, W.A. (1994) Surfactant adsorption, bacterial attachment and biodegradation in river sediment: a three-way interaction, in *Bacterial Biofilms and their Control in Medicine and Industry*, eds. J. Wimpenny, W. Nichols, D. Stickler and H. Lappin-Scott, Bioline, Cardiff, pp. 121–126.

Wiggins, B.A., Jones, S.H. and Alexander, M. (1987) Explanations for the acclimation period preceding the mineralization of organic chemicals in aquatic environments. *Appl. Environ. Microbiol.* **53**, 791–796.

Wilson, J.T., McNabb, J.F., Cochran, J.W., Wang, T.H., Tomson, M.B. and Bedient, P.B. (1985) Influence of microbial adaptation on the fate of organic pollutants in ground water. *Environ. Toxicol. Chem.* **4**, 721–726.

Wood, A.A., Claydon, M.B. and Finch, J. (1970) Synthetic detergents: some problems. *Water Pollut. Control*, **69**, 675–683.

Zaidi, B.R., Murakami, Y. and Alexander, M. (1989) Predation and inhibitors in lake water affect
    the success of inoculation to enhance biodegradation of organic chemicals. *Environ. Sci. Technol.*
    **23**, 859–863.
Zobell, C.E. (1943) The effect of solid surfaces upon bacterial activity. *J. Bacteriol.* **46**, 39–56.

# 3. Biodegradability testing

## H.A. PAINTER

## 3.1 Introduction

The need for tests for biodegradability arose from the large-scale use of synthetic surfactants, or surface-active agents, soon after the Second World War. The most extensively used surfactant, an alkylaryl sulphonate tetrapropylene benzene sulphonate (TPBS) based on tetrapropylene, proved to be only partially (about 50%) removed by sewage treatment and caused widespread foaming on aeration tanks of activated sludge plants and in rivers. Some reports claimed that the efficiency of sewage treatment was impaired, though this was not substantiated, and that the rate of aeration of mixed liquor in the aeration tanks was reduced by the presence of anionic surfactants. Many investigations were set in train by these events and reports, and it was found that branching in the side chain resulted in alkylaryl sulphonates which were less readily attacked by bacteria than were their straight-chain isomers. Eventually the TPBS type was substituted by the straight-chain (linear) varieties (LABS) and the problems associated with TPBS were overcome.

It was almost certainly as a result of the surfactant problem that the term 'biodegradability' and its derivatives came into the language. Hitherto, chemicals were said to be removed, destroyed, attacked, broken down, ruptured, catabolized, metabolized, etc. by microorganisms. However, it is quite certain that the current edifice of testing 'new' and existing chemicals for biodegradability owes much to work done on surfactants.

Scores of tests for the primary biodegradability of surfactants were devised and used in the search for an environmentally acceptable product. These tests were narrowed down by the Environmental Health and Safety division of the organization for Economic Cooperation and Development (OECD, 1971, 1976) to just two — a screening die-away test and a confirmatory test simulating the activated sludge process. These test methods have been embodied in four EC Directives (1973a,b, 1982a,b). Later, the two original tests were adapted to be applied to all organic substances and other tests were introduced, all of which were examined and established by the OECD group (OECD, 1981) and later revised (OECD, 1993a,b). Instead of the disappearance of the parent compound, these later tests allowed the measurement of the extent of mineralization of organic compounds (ultimate biodegradation). The parameters followed are DOC or COD (see Glossary terms, Section 3.1.1) removed, gaseous or dissolved oxygen consumed and carbon dioxide evolved. The test methods have been incorporated into EC legislation, amendments 79/831 and 92/69 to Directive 67/548 (1967, 1979), in the form of three entries in the Official Journal of the European Communities (1984, 1988, 1992). Consequently, all 12 member States will have incorporated the Directives into their national laws, e.g. in the UK as Statutory Instruments (HMSO, 1978, 1984). Also, in some cases test methods have been published in individual countries, for example, in the UK some eight methods have been presented by the Standing Committee of Analysts (SCA) (HMSO, 1983a). More recently, the methods have been extended to a wider audience by the International Organization for Standardization (e.g. ISO, 1985, 1991a–d).

For a clearer understanding a glossary of terms is given, followed by a brief account of the nature of biodegradation. The factors important in testing are then discussed, before descriptions are given of the major test methods for biodegradability and for chemical determination of surfactants. Choice of methods and schemes of testing, as well as interpretation of results and legal requirements are dealt with in Chapter 4.

### 3.1.1 Glossary of terms

It is useful to define some of the more important terms used — see also earlier chapters.

*Biodegradability.* Biodegradability is the capacity of a substance to undergo attack by a biological agent, usually bacteria or fungi.

*Biodegradation.* Biodegradation is the breakdown of a substance by microorganisms and can be (i) primary — a change, or changes, in the chemical structure of a substance resulting in a loss of specific property of that substance, e.g. loss of methylene blue reactivity by anionic surfactants; thus, primary biodegradation is determined by measuring the decrease in the concentration of the specific chemical or group of chemicals; (ii) environmentally acceptable — biodegradation to such an extent as to remove undesirable properties of the substance, e.g. foam caused by surfactants. This will frequently correspond to primary biodegradation (e.g. the loss of methylene blue reactivity with anionic surfactants) but may vary depending on the circumstances under which the products are discharged to the environment and on changes in public opinion as to what is acceptable; (iii) ultimate — mineralization, that is the conversion of an organic substance to fully oxidized, simple molecules, namely $CO_2$, $H_2O$, nitrate, ammonium, etc. and new biomass. In practice, ultimate biodegradation is assessed by measuring the removal of dissolved organic carbon (DOC) or chemical oxygen demand (COD), the production of carbon dioxide ($CO_2$) or the uptake of oxygen or dissolved oxygen.

*Bioelimination.* Bioelimination is the removal of a substance from the liquid phase in the presence of living microorganisms by physico-chemical as well as biological processes.

*BOD.*    Biochemical oxygen demand (BOD) is the amount (mg) of oxygen consumed by microorganisms when metabolizing a test substance, usually under prescribed conditions; expressed as mg oxygen uptake per mg test substance.

*COD.*    Chemical oxygen demand (COD) is the amount (mg) of oxygen consumed during oxidation of a test substance with hot, acidic dichromate; it provides a measure of the amount of oxidizable matter present; expressed as mg oxygen consumed per mg test substance.

*DOC.*    Dissolved organic carbon (DOC) is the organic carbon present in solution, or that which passes through a 0.45 μm filter, or remains in the supernatant after centrifuging at approximately $4000 \times g$ (about 40 000 m s$^{-2}$) for 15 min.

*ThOD.*    Theoretical oxygen demand (ThOD) is the total amount (mg) of oxygen required to oxidize a chemical completely; it is calculated from the molecular formula; also expressed as mg oxygen required per mg test substance.

*ThCO$_2$.*    Theoretical carbon dioxide (ThCO$_2$) is the quantity of carbon dioxide (mg) calculated to be produced from the known or measured carbon content of a chemical when fully mineralized; also expressed as mg carbon dioxide evolved per mg test substance.

*Readily biodegradable.*    Readily biodegradable is an arbitrary classification of chemicals which have passed certain specified screening tests for ultimate biodegradability; the conditions in these tests are so stringent — relatively low density of non-acclimatized bacteria, relatively short duration, absence of other organic compounds — that such chemicals will rapidly and completely biodegrade in aquatic environments under aerobic conditions.

*Inherently biodegradable.*    Inherently biodegradable is a classification of chemicals for which there is unequivocal evidence of biodegradation (primary or ultimate) in any test for biodegradability. No limits are placed on the conditions under which the tests are carried out.

*Treatability.*    Treatability is the amenability of a substance to be removed during biological waste water treatment without adversely affecting the normal operation of the treatment processes. Generally readily degradable substances are treatable but this is not the case for all inherently biodegradable substances. Abiotic processes may also operate.

*Screening tests.*    Screening tests are relatively simple, batch tests which may be used for preliminary assessment of biodegradability; usually correspond to tests for ready biodegradability.

*Simulation tests.*    Simulation tests mimic a given sector of the environment and are designed to predict the rate of biodegradation of a substance under relevant environmental conditions.

*Lag phase.*    Lag phase is the period from inoculation in a screening test until the degraded percentage has increased to about 10%. The lag time is often variable and poorly reproducible.

*Degradation phase.*    Degradation phase is the period from the end of the lag phase to the time when 90% of the maximum level of degradation has been reached.

*10-day window.*    The 10-day window is the period of 10 days immediately following the attainment of 10% biodegradation.

### 3.1.2    Early tests

To assess primary biodegradability of surfactants, the empirical approach was to add a low concentration of the substrate to river water and measure the disappearance of some property of the surfactant. It was thought that if the surfactant degraded in the test then it would also be degraded in the river. Foam height and surface tension were tried as indicators but were unsatisfactory; reactivity of anionic surfactants with the dye, methylene blue, proved more successful. Even so, values for degradation of some chemicals varied with water from other rivers and from the same river on different occasions presumably due to variable mineral and bacterial content. This led to the adoption of BOD 'dilution water' as the medium inoculated with soil extracts, sewage effluents, activated sludge and even dried activated sludge. There were many variants but eventually there was agreement in the form of the OECD Static or Screening Die-Away and Confirmatory tests for anionic and non-ionic surfactants (for details, see Sections 3.3.1 and 3.3.2) which were later incorporated in the EC Detergent Directives for surfactants used in household and industrial cleaning agents.

Meanwhile, attention was being given to modifying the tests so that ultimate biodegradation could be assessed. The reasons for this were twofold: a need to ensure that xenobiotic chemicals were not merely being transformed to other, possibly harmful organic chemicals, thus escaping detection by the specific method, and to avoid having to develop costly analytical methods to determine the test substances. In place of specific analytical test methods, the non-specific or 'summary' parameters DOC and COD were used. Also, indirect methods of assessment of ultimate biodegradability were used by measuring the amounts of carbon dioxide ($CO_2$) evolved, one of the end-products of degradation, or the oxygen consumed in the oxidation. Methods based on measurements of bacterial growth — weight, turbidity, cell density — were largely unsuccessful.

There was a much more systematic approach to the various factors involved in the testing than there had been for primary biodegradation, with the object of arriving at optimal conditions for the various types of tests.

The principle of any test for biodegradation is that the test substance is brought into contact with a mixed population of bacteria in a solution of inorganic nutrients. In tests for primary biodegradability other organic compounds can be present and no blank controls (see Section 3.2.5) are necessary. But for ultimate biodegradability other organic chemicals must largely be absent to make the analyses simpler and to create more stringent conditions in the test. Blank controls are necessary, in which all constituents are present except the test substance, so that the contribution of the inoculum to DOC/COD removal, uptake of oxygen and $CO_2$ production may be measured and allowed for.

### 3.1.3   Development of the tests

As experience with testing accumulated, it was soon found that the size or concentration of the inoculum required in screening tests for complete, or nearly complete, removal of biodegradable anionic surfactants or non-ionic surfactants, that is primary degradation, was extremely small. Normally, as little as 0.5 ml of a paper-filtered sewage effluent from a well-operated waste water treatment plant per litre of medium will result in >90% MBAS or BiAS removal in 7–10 days. (Anionic surfactants are usually expressed as methylene blue active substances (MBAS) and non-ionic surfactants as bismuth active substances (BiAS), see Section 3.5.1.) This inoculum concentration is roughly equivalent to 0.5 to $2.5 \times 10^2$ cells/ml (as estimated on a medium such as casein–peptone–salts). It was found that up to 30 mg dry solids (activated sludge) per litre medium could be used as the inoculum without loss of predictability. That is, with inoculum concentrations of 30 mg solids/l (roughly equivalent to $10^5$ to $2 \times 10^6$ cells/ml) or less, a positive result in the screening test would accurately indicate that the surfactant would degrade in the activated sludge simulation or confirmatory test and hence also in the environment. Above about 100 mg solids/l a small proportion of positive results would be false, while at 1000 mg/l a higher proportion of false positive results would be obtained.

Similarly, experience with tests for ultimate biodegradability (or mineralisation) using DOC removal, $CO_2$ production or oxygen uptake, showed that, in general, the larger the size of the inoculum the shorter was the time required to degrade a biodegradable chemical. Provided 30 mg/l, or less, activated sludge was used as inoculum, positive results from screening tests accurately forecast the behaviour of chemicals in simulation tests and in the environment. Positive results obtained when over about 200 mg solids/l was used could not be relied on to give an accurate prediction; some chemicals passing the test did not degrade in the simulation test. This lack of predictability was found to be due not only to the concentration of microorganisms but also to the duration of the test, the source and treatment of the inoculum, and the presence of other biodegradable organic chemicals.

The duration of the original tests had varied from 14 to 30 days but for uniformity and convenience the duration for all agreed tests was fixed at 28 days.

Since the object of the screening tests is to discover whether a chemical will degrade without difficulty when discharged to the environment, it is important that the microorganisms used as the inoculum should not have been in contact with the test chemical either in the environment or, after collection, in the laboratory. For example, a laboratory percolating filter, receiving almost exclusively TPBS for several weeks, produced effluents which when used as inoculum in batch screening tests satisfactorily degraded the surfactant, whereas normal effluents degraded only low percentages of TPBS.

The last recognized factor inducing an otherwise non-, or poorly, degradable chemical to be degraded is the presence of a degradable chemical, especially if their structures are similar (co-metabolism).

Because the conditions used in this type of test were very stringent

  – only inorganics in the medium,
  – inoculum of cell density not more than about $10^6$/ml,
  – inoculum not pre-exposed to the test chemical,
  – duration of test 28 days,

chemicals which are adequately biodegraded in the test must be easily degradable and have been arbitrarily labelled 'readily biodegradable', and the tests are called 'tests for ready biodegradability'. A further arbitrary condition was applied, namely, that the degradation should occur not only within 28 days, but within 10 days after the initiation of biodegradation, taken to be when 10% degradation was observed. Such chemicals would degrade readily and rapidly when discharged to the aerobic aquatic environment. However, because of the stringent conditions, chemicals which do not degrade adequately within the time limits laid down are not to be taken as non-biodegradable; further testing should be carried out.

It should be pointed out that, for technical and other reasons, the inoculum cell densities in the various tests methods for ready biodegradability vary considerably. This variation has to be accepted because of the arbitrary nature of the classification so that chemicals which pass any of these tests are called readily biodegradable.

If any or all of the conditions of the tests for ready biodegradability are relaxed, the test is designated as one for 'inherent biodegradability'. Some factors which increase the likelihood of degradation are higher cell densities, pre-exposure of the inoculum to the test chemical, longer duration, frequent inoculation. In the two recognized tests for inherent biodegradability (Table 3.1, D–F) the cell densities used are in the region of $10^7$ to $10^8$/ml. Chemicals which degrade by more than 20% in these tests — or in any other lenient test which can be scientifically justified — are classified as being inherently biodegradable. Whereas readily biodegradable chemicals degrade easily and rapidly in the activated sludge simulation test and in the environment, only some inherently

**Table 3.1** List of methods for ultimate biodegradability

---

A *OECD revised methods (1993a) — ready biodegradability*

| | |
|---|---|
| 301A | DOC Die-Away |
| 301B | $CO_2$ Evolution (Modified Sturm Test) |
| 301C | MITI (I) Ministry of International Trade and Industry (Japan) (BOD) |
| 301D | Closed Bottle (BOD) |
| 301E | Modified OECD Screening (DOC) |
| 301F | Manometric Respirometry (BOD) |

B *EEC methods — ready biodegradability*

| | |
|---|---|
| C3 | Modified OECD Screening (DOC) |
| C4 | Modified AFNOR (NF T90/302) (DOC) |
| C5 | Modified Sturm ($CO_2$) |
| C6 | Closed Bottle (BOD) |
| C7 | Modified MITI (I) (BOD) |
| C8 | Biochemical Oxygen Demand (Off. J.E.C. Vol. **251**, 19.9.1984 pp. 160–211) |

C *ISO methods*

(not differentiated between ready and inherent biodegradability)

| | |
|---|---|
| ISO 7827 | Method by analysis of dissolved organic carbon (1985) |
| ISO 9408 | Method by determining the oxygen demand in a closed respirometer (1991a) |
| ISO 9439 | Method by analysis of released $CO_2$ (1991b) |
| ISO/CD 10,707 | Closed Bottle (not yet completed) |
| ISO CD 10,634 | Guidance on evaluating biodegradability of insoluble chemicals (1992). |

D *OECD methods — inherent biodegradability*

| | |
|---|---|
| 302A | Modified semi-continuous activated sludge (SCAS) (DOC) (OECD, 1981) |
| 302B | Modified Zahn–Wellens–EMPA[a] (DOC or COD) (OECD, 1981, 1993b) |

E *EEC methods — inherent biodegradability*

Modified SCAS
Modified Zahn–Wellens
(Off. J.E.C. L133, Vol **31**, 30 May 1988)

F *ISO* (not differentiated)

ISO 9887 SCAS (1991c)
ISO 9888 Zahn–Wellens (1991d)

G *OECD methods*:    *simulation*

| | |
|---|---|
| 303A | Aerobic Sewage Treatment: Coupled Units Test (DOC) (OECD, 1981) |

H *EEC methods*:    *simulation*  (see Ref. in E above)

Activated sludge simulation tests (DOC)

I *ISO*

ISO/TC147/SC5/WG4 N140 (1991) (DOC)
Activated sludge simulation tests

---

[a] EMPA: Swiss Federal Laboratories for Materials Testing and Research.
Sections A–C, D–F and G–I are discussed in Sections 3.4.1, 3.4.2 and 3.4.3, respectively.
Other details of the tests are given in Tables 3.2 and 3.3, and in Table 4.3 of Chapter 4.

biodegradable chemicals will so degrade in the simulation test while at the other extreme some inherently biodegradable chemicals will degrade in the environment only very slowly or not at all, and will not degrade in the simulation test.

Ideally, chemicals would not be classified in this arbitrary manner. Instead, kinetic constants, e.g. specific growth rates, relating to their biodegradation by bacteria, would be measured under conditions prevailing in various parts of the environment — freshwater, estuaries, sea — but this has been found more difficult and costly to achieve than had been earlier thought. The microbial reaction is second-order, so that the concentrations of the competent bacteria as well as those of the chemical undergoing biodegradation are involved.

## 3.2    Nature of biodegradation and influencing factors

Organic substances are removed from the aquatic environment by a variety of mechanisms — autoxidation, adsorption, sedimentation, hydrolysis, photolysis, as well as biological action — but it is accepted that biological processes play the major role. Bacteria are almost certainly the most important of the many types of organisms in metabolizing both natural and synthetic chemicals: bacteria are ubiquitous and their size, high specific rates of growth, metabolic versatility and mode of life make them very suitable for this function. Thus, biodegradation is intimately bound up with bacterial growth, nutrition and metabolism, and factors which affect these bacterial functions will also affect biodegradability assessments.

### 3.2.1    Composition of medium

As indicated earlier, originally uninoculated river water was used as the medium but variations in results with water from different rivers suggested that the composition of the river water, as well as the number and type of bacteria, also varied. For proper growth, microorganisms require a range of elements in addition to carbon: namely, N, P, S, K, Na, Fe, Ca, Mg at 'macro' levels, and probably a number of other elements at 'micro' levels, the so-called trace elements such as B, Co, Cu, Mn, Mo, V, Zn. All synthetic media contain the first series of elements and some, including some in the OECD tests, contain a cocktail of trace elements. The B-group of vitamins, either singly or in various mixtures, are required by some fastidious species and were added, as such or in the form of yeast extract, to some media. However, the recently revised OECD and EC media (OECD, 1993a) do not now contain either the trace elements or vitamin B, except in the Modified OECD Die-Away test (301E; for this and other codes, see Table 3.1). The reason for this is the larger size of inoculum (30 mg/l activated sludge solids), recommended in most tests for ready biodegradability, which will contain sufficient trace elements and B vitamins. In method 301E, however, the inoculum is small for the amount of test substance present and the final medium may be deficient.

The OECD media are buffered to pH 7.0–7.4 with phosphates, although in nature the main buffer is carbonate/bicarbonate. Consequently the concentration of P (116 mg P/l) in most tests is much higher than in sewage (6–25 mg P/l).

All media for screening tests are prepared in distilled or deionized water; for tests simulating sewage treatment tap water is used to prepare the 'synthetic sewage' (see Section 3.3.2.2.1). The water for media must not contain toxic material (e.g. Cu) which would inhibit growth. The buffering capacity of the synthetic sewage depends on the hardness of the tap water used, since the concentration of phosphate added is low. It may be necessary in some soft water areas to add a carbonate buffer to maintain the pH above levels which would inhibit microbial oxidation.

The media are not sterilized and it has been shown that sufficient numbers of bacteria are present in the media to degrade compounds such as acetate, benzoate and aniline within 10 days.

## 3.2.2   Inocula

The nature and quantity of the inoculum play an important role in biodegradability assessments; the inoculum is probably the biggest single factor in the success of the batch test. Pure cultures of extremely versatile single species have been tried but the results were not rewarding. Tests with mixed populations from the environment as inocula gave a larger number of positive results because of their wider range of metabolic activity.

The main sources of inoculum are river water, sewage effluent (unchlorinated), activated sludge (both from a treatment works treating predominantly domestic sewage) and soil, or a mixtures of these. It has been found that sources of inocula other than activated sludge usually yield lower cell densities and give a higher scattering of results.

The OECD Screening test for surfactants (Section 3.3.1), and the Modified OECD DOC Die-Away test (301E) for ultimate biodegradability, use a good quality effluent, filtered through a coarse filter paper, at the rate of 0.5 ml/l medium. The filtration allowed bacteria to pass to the filtrate but removed coarser particles so that adsorption was reduced.

Activated sludge taken from the recirculation line probably needs no treatment; otherwise the sludge is washed with water or medium to remove soluble organic matter and the centrifuged solids are resuspended in the mineral medium. Another alternative is to use the supernatant liquid from centrifuged, homogenized sludge. The Japanese MITI I test uses a special, mixed inoculum prepared by adding the supernatant of a mixture of samples from ten sources — activated sludges, industrial sludge, rivers, lakes, seas — to a glucose/peptone/phosphate medium, and growing activated sludge by the fill-and-draw mode. The sludge is not used as an inoculum until after 1 month's operation, and is discarded after a further 3 months.

In tests for ready biodegradability, the inocula must not have been in contact with the test substance (pre-exposed). Inocula may, however, be pre-conditioned to the experimental conditions by aeration for 5–7 days at the temperature of the test. This sometimes improves the precision of the method by reducing 'blank' values (see Section 3.2.5) of DOC content, $CO_2$ evolution and oxygen uptake.

The bacterial cell density in the medium determines, to a large extent, the length of the lag period and also whether sufficient test substance is degraded within the duration of the test. If the number of cells capable of degrading the test substance is relatively high, the density will soon reach a value which makes a significant reduction in the concentration of the substance. But when the initial cell density is relatively low, the lag period before a significant density is reached may be longer than the 28 days of test. In the surfactant tests, cell numbers were apparently not considered, but in the more general testing rough estimates of cell density were taken into account, using published values for the number of bacteria in sewage effluents, etc. In the Closed Bottle (301D) and Modified OECD Screening (301E) tests the numbers present are in the region of $0.25$ to $2.5 \times 10^2$/ml, derived from the addition of sewage effluent. In the other methods for ready biodegradability the density is $10^4$ to $10^6$/ml, with $10^6$/ml being the maximum. These numbers are reached by the addition of up to 30 mg activated sludge suspended solids per litre or up to 100 ml effluent per litre. It has been shown that when the density is more than about $10^7$/ml, as in the Zahn–Wellens method (302B) for inherent biodegradability, the conditions become so much less stringent that some non-readily biodegradable substances, including surfactants, such as TPBS, are degraded.

### 3.2.3   Physico-chemical factors

The test solutions must be kept aerobic (>2 mg dissolved oxygen per litre) by pre-aeration in the Closed Bottle test (near saturation) (301D), by shaking as in the Modified OECD Screening (301E) and DOC Die-Away (301A) tests, by stirring as in the MITI I (301C) and Manometric Respirometry (301F) tests or by bubble aeration in the Modified Sturm test (301B). The method of aeration also keeps the solids in suspension except in the Closed Bottle method; this is especially important in inherent and simulation tests.

The temperature at which the screening tests are carried out is $22 \pm 2°C$, except in the MITI method which requires $25 \pm 1°C$. It is especially important to keep the temperature within narrow limits in oxygen uptake methods (301C, 301D and 301F).

The Confirmatory test is conducted at a constant room temperature between 18 and 25°C, although for special purposes other temperatures may be chosen.

To avoid algal growth, all the screening tests are conducted in darkened enclosures and to avoid toxic effects the ambient air must be free from solvents and other toxic material.

### 3.2.4   Test substances

Although screening tests for ready biodegradability do not necessarily predict the kinetic rate of biodegradation of a test substance in the environment, it is advisable to use concentrations in the test solutions as near to environmental values as is consistent with being able to detect analytically the changes occurring during the test. Except when $^{14}$C-labelled substances are used, the concentration of test material, to meet this requirement, has to be the equivalent of 10–20 mg carbon/l for DOC Die-Away tests (301A, 301E) and for $CO_2$ production (301B). The Closed Bottle method (301D) can accommodate only 2–5 mg/l of substance, while the respirometric methods (301C, 301F) need much more, 100 mg/l (301C) and 50–100 mg ThOD/l (301F). The concentration should not be so high that microorganisms are inhibited.

Most surfactants are sufficiently soluble in water to allow stock solutions of, say, 1 g/l to be prepared, and aliquots can be used to attain the required concentration. Some cationic surfactants, for example those of the dialkyldimethyl ammonium type, are not very soluble, perhaps below 5 mg/l, so that they cannot be assessed by DOC methods or by the disulphine blue die-away method unless special measures are taken (Sections 3.3.1.6, 3.3.2.7.4). The $CO_2$ evolution and the respirometric methods may be applied and the surfactant may be added either directly as a weighed amount or its dispersion in the test medium may be assisted by sonication, emulsification, dissolution in a volatile solvent followed by evaporation within the test vessel, etc. (ISO, 1993).

Few, if any, surfactants are volatile but such compounds can be tested in respirometers (301C, 301F) modified so as to reduce the head-space volume. They may also be assessed by die-away/closed bottle tests in 'iodine' flasks modified with a side tube fused in the lower side fitted with a mininert valve. The test substance is injected through the valve and samples may be withdrawn via the valve at intervals for specific analysis or dissolved oxygen concentrations may be determined in the flasks by insertion of an oxygen electrode, using a special funnel. The Modified Sturm $CO_2$ method (301B) cannot be used but the, as yet, un-agreed $CO_2$ methods of Struijs and Stoltenkamp (1990) and Birch and Fletcher (1991) can (see Section 4.1.3.2, Ch. 4). These latter two methods can equally well be applied to insoluble and soluble substances.

Substances which strongly adsorb, such as cationic surfactants, may lead to false results so that it is prudent to examine the loss on to suspended solids in the die-away tests (301A, 301E) especially if the inoculum is 30 mg sludge solids per litre. This may be done in a preliminary test or by setting up suitable control vessels (Section 3.2.5) in the biodegradability test.

### 3.2.5 Control vessels

In tests for primary biodegradability there is no need to set up control vessels, unless the possibility of adsorption is being considered. When non-specific or 'summary' parameters (such as DOC or oxygen uptake) are measured to assess ultimate biodegradability, control 'blanks' must be set up since inocula, without test substance, will take up oxygen, evolve $CO_2$ and remove or release DOC. These controls consist of all components of the reaction mixture except the test substance. The blank value thus obtained should be as low as possible so that the final degradation values obtained are of greater precision. These blank values are subtracted from the corresponding values in the presence of the test substance, although it is by no means certain that events (e.g. endogenous metabolism) occurring in the blank vessels due to the inoculum also take place in the presence of the substance, or, if they do, it is not known if they occur to the same extent.

If needed, a vessel may be set up concurrently to determine whether the substance under test inhibits the inoculum at the concentration tested. (Another type of toxicity test may be applied for this purpose independently of a test for biodegradability e.g. OECD (1984), ISO (1986), HMSO (1983b).) The vessel would contain the mineral medium, inoculum, the test substance and a reference substance (see Section 3.2.7), each chemical being added at the same concentrations, respectively, as in the biodegradability test.

Abiotic controls, when required, are set up to check for possible non-biological degradation by sterilizing the reaction mixture containing the test substance but no inoculum. Sterilization can be by filtration through a membrane (0.2–0.45 µm) or by the addition of a suitable toxic substance.

Unless adsorption has been ruled out or is unlikely, DOC die-away tests with activated sludge inocula (301A) should include an abiotic control which is both inoculated and poisoned.

### 3.2.6   Duration of test

The time taken to remove a given concentration of a degradable chemical depends largely on the bacterial cell density. From the academic viewpoint, any attempt to assess biodegradability can be continued for as long as the experimenter wishes, but practically a limit has to be imposed for standardization and legal requirements. The original surfactant die-away tests were designed to remove over 90% in 7–10 days and not longer than 19 days; the volume of the effluent inoculum used was adjusted to get a result in this range.

Before the OECD methods were agreed various incubation times had been used empirically — from 5 days (BOD) to 42 days (Association Francaise de Normalisation, AFNOR 1977). After 'ring-testing' the methods, all tests were standardized at 28 days, but it was also agreed that if degradation had started by the 28th day, it would be sensible to continue for a further short period to ascertain whether a plateau of removal was reached. A chemical requiring this extra time is classified as inherently, not readily, biodegradable.

In addition to the limit of 28 days, there is a second, but controversial limitation: to be classified as readily biodegradable, the removal of the substance must reach the agreed value during a '10-day window' within the 28 days. The 'window' begins when 10% removal is attained and must end before the end of the 28th day. The reason for this limitation is that it was thought it would exclude results from degradation curves which do not reflect normal kinetics. Non-biological removal would, it was agreed, take place at a roughly constant rate throughout the 28 days, so that a slow abiotic removal over 28 days could reach a high level and the test substance would be reported as being readily biodegradable. The 10-day limitation would eliminate such false results. Opponents of this view argue that the shape of the removal curve and the abiotic control values would show up such cases.

### 3.2.7 Reference compounds

In the tests for primary biodegradation of surfactants (OECD, 1976), two standard or reference anionic surfactants are used to check on the activity of the inoculum; one degrades by 90–95%, the other by about 35%. If the readily degradable standard does not attain the expected value, the inoculum activity is judged to be too weak and the test is repeated with an increased volume of inoculum or another source is used. Conversely, if more than 35% of the poorly degradable standard is observed, the test is repeated with a lower volume of inoculum source. Thus, a true check on the inoculum activity is maintained.

In the case of ultimate biodegradability only a readily biodegradable standard is used; sodium acetate, sodium benzoate and aniline have been selected. Initially, the reference substance was thought to test the activity of the inoculum, as with the surfactants, but it was found that these standards, which are more readily degraded than the surfactants, were degraded even when no inoculum was added. It is assumed that the bacteria degrading the standards in this way were present in the water; some samples of deionized water were found to contain in the region of $10^3$ cells/ml. Indeed, the American Society for Testing and Materials (ASTM) specifies that water for use in preparing bacteriological media should contain less than $10^3$ cells/ml, indicating that this is probably a usual value. (These cells, as such, play no part in the subsequent use of the bacteriological media since they are always first sterilized.)

Thus, the use of the standards does not help in deciding whether the bacteria added via the inoculum are sufficiently active; at most their use indicates that the procedures have probably been carried out correctly. If the inoculum is to be tested for activity the medium should be sterilized before inoculation or perhaps less easily degradable standards should be sought.

## 3.3 OECD and EEC tests for primary biodegradability of surfactants

### 3.3.1 OECD Static test procedure or Screening test

This method, applied to both anionic and non-ionic surfactants, is the official method by which surfactants are tested under the EC Directives 73/405 (1973b), 82/242, 82/243 (1982a,b). If surfactants in formulations are to be tested they must first be separated by extraction from constituents which might affect the subsequent degradation assessment, otherwise they may be examined without further treatment. The annexes to the Directives describe the methods of analysis and the method for extracting surfactants from detergent products but not this screening test, which is fully described elsewhere (OECD, 1976). [Note: The method implies that a decrease in concentration of the surfactant is due to biodegradation but abiotic mechanisms could be operative. Separate investigations, e.g. using sterile test media, before the method was agreed, had shown that the mechanism was due to microorganisms.]

*3.3.1.1  Principle.*   A solution of the surfactant at 5 mg active substance per litre in an inorganic medium is inoculated with a relatively small number of aerobic microorganisms from a mixed population. The mixture is aerated at 25 ± 1°C until the concentration of surfactant falls to a constant level, but not for longer than 19 days. Anionic surfactants are determined by the methylene blue method, reported as MBAS, and non-ionic surfactants as bismuth active substances (BiAS). The procedure, including the activity of the inoculum, is checked by means of two standard anionic surfactants.

*3.3.1.2  Preparation.*   The medium based on 'BOD dilution water' is listed in Table 3.2; all constituents are of 'analytical' quality and the water used was distilled or de-ionized and free from toxic substances, in particular copper.
[Note: The buffer is based on phosphate and was originally designed to be at pH 7.2, but, because of a long-undetected error in transposing one text to another, the number of molecules of water of crystallization of the disodium phosphate component was changed from seven to two. This caused the pH of the final medium to be 7.4 instead of 7.2, but it has been decided to retain the dihydrate and leave the pH at 7.4.]
The nutrient salts — magnesium sulphate, calcium chloride and ferric chloride — are made up in three separate solutions and the phosphate buffer plus ammonium chloride in a fourth. The inorganic medium is prepared by adding the stock solutions to water at the rate of 1 ml each per final litre of medium.

*3.3.1.2.1  Standards of biodegradability.*   The method is based on comparing the removal of the test surfactant with that of two 'standard' surfactants — a 'soft' or easily biodegradable surfactant and a 'hard' or relatively non-biodegradable one. The adopted 'soft' material is Marlon A, a commercial linear alkyl benzene sulphonate and the 'hard' standard is the branched tetrapropylene benzene sulphonate type (TPBS). (At the time the method was published, no agreement on standard non-ionics had been reached, so that the same anionic standards are used in both cases.)

*3.3.1.2.2  Stock solutions.*   Stock solutions of the test material and the standards are prepared to contain 1 g active material per litre. Surfactants in formulations are first extracted to separate them from other ingredients which might affect the assessment of biodegradation (OECD, 1976). The stock solutions are

**Table 3.2** Composition of media

| | Normal (g) | MITI (g) |
|---|---|---|
| *Stock solutions* — each made up to 1 l with water[a] | | |
| (a) Potassium dihydrogen orthophosphate | 8.5 | 8.5 |
|     Dipotassium hydrogen orthophosphate | 21.75 | 21.75 |
|     Disodium hydrogen orthophosphate dihydrate | 33.4 | 44.6 |
|     Ammonium chloride | 0.5 (1.7)[b] | 1.7 |
|     pH | 7.4 | 7–7.2 |
| (b) Calcium chloride, anhydrous | 27.5 | 27.5 |
| (c) Magnesium chloride heptahydrate | 22.5 | 22.5 |
| (d) Iron (III) chloride hexahydrate | 0.25 | 0.25 |
|     — add to solution (d) one drop of conc. HCl or 0.4 g | | |
|     disodium ethylenediaminetetra-acetic acid | | Not |
|     (EDTA) to stabilize the solution | | stabilized |

*Normal medium[c]*
Mix 10 ml solution (a) with 800 ml water, then add 1 ml solutions (b), (c), (d) and make up to 1 l

*Closed Bottle and Surfactant Screening tests*
Use 1 ml each of solutions (a), (b), (c) and (d)

*MITI test*
Use 3 ml each of solutions (a), (b), (c) and (d)
Solution (a) contains dodecahydrate of the sodium salt.

[a] Distilled or de-ionized.
[b] 1.7 used in Surfactant Screening test.
[c] In Modifed OECD method (301E) and in other tests where deficiencies may occur, add trace elements and yeast extract at 1 ml/l of each solution:

Trace elements solution mg/l: Manganese sulphate.4$H_2O$ 39.9; Boric acid 57.2; Zinc sulphate.7$H_2O$ 42.8; Ammonium molybate 34.7; Iron chelate ($FeCl_3$-EDTA) 100. Vitamin solution: yeast extract, 15 mg/100 ml; made up freshly.

diluted 1 in 200 with water and the MBAS or BiAS test is applied, as appropriate, to ensure that the final test solutions have concentrations in the range 4.5–5.5 mg/l for the maximum accuracy of the determination.

*3.3.1.2.3 Inoculation.* Although in principle any source of mixed aerobic microorganisms would be appropriate, either secondary effluent or garden soil are the normal sources. Secondary effluent should be collected freshly from a treatment plant dealing with predominantly domestic sewage, avoiding periods when excessive amounts of storm water are present. Between collection and use (on the same day), the effluent is kept aerobic; before use the effluent is filtered through a coarse filter paper and the first 200 ml of filtrate are discarded. Alternatively, 100 g fertile garden soil, free from excessive proportions of clay, sand or organic matter, are made up to 1 l with chlorine-free tap water and settled for 30 min. The supernatant liquid is filtered through course filter paper, discarding the first 200 ml of filtrate. The filtrate, in both cases, is kept aerobic and is used on the same day as prepared. (The use of soil appears in the 1971 edition (OECD, 1971) but not in the 1976 version (OECD, 1976).)

The amount of inoculum required for each inoculation is normally 0.5 ml/l of medium and must be sufficient to remove 90–95% of the 'soft' standard within 14 days, normally 7–10 days. Only about 35% of the 'hard' standard should be removed in the period of the test. The necessary amount is determined in preliminary tests using a range of volumes of the filtrates, and should occasionally be checked, especially if changes in the rate and degree of degradation of the standards are observed.

### 3.3.1.3   Procedure

*3.3.1.3.1   Anionics.*   To 2-l portions of the mineral medium are added 10 ml stock solution of the test material or standards, respectively, together with the appropriate volume of inoculum. These test solutions are allowed to stand so that foam disperses and their MBAS or BiAS content is determined in duplicate; the mean value should lie between 4.5 and 5.5 ± 0.1 mg/l. Portions (900 ml) of each test solution are carefully transferred to two 2-l Erlenmeyer flasks and a plug of loose cotton wool is inserted into the mouth of each vessel. The flasks are placed on a shaker and shaken in the dark at 25 ± 1°C in an enclosure in which the air is free from pollutants and toxic matter, e.g. chlorinated solvents.

After 5 days, and on alternate days from the eighth day, single determinations of MBAS or BiAS in each flask are made until the difference between two values over a period of 4 days within a flask is less than 0.15 mg/l. The beginning of the plateau of the degradation curve (Section 3.3.1.4) is taken to be the first one of these two values. The sampling programme may be varied according to the progress of the test, provided that the point of inflexion is accurately established, but in any case the period of the test should not exceed 19 days. Samples at the beginning can be 10–20 ml, increasing to 100 ml for the last samples. (For treatment of samples, see Section 3.5.1.1.)

*3.3.1.3.2   Non-ionics.*   Because larger samples are required for analysis of non-ionics than for anionics, larger volumes of test solutions have to be prepared. Thus, 5 l (instead of 2 l) of mineral medium plus 25 ml stock solution of the surfactant and appropriate volume of inoculum are used and 1200-ml portions of the test solutions are placed in the 2-l Erlenmeyer flasks. Four Erlenmeyer flasks are used for each non-ionic surfactant. Equal volumes of samples are taken from each of a pair of flasks and combined to give a single sample of sufficient volume for analysis. Samples are taken only on the fifth and 19th days (presumably because of the tedious nature of the analytical method), the respective volumes from each flask being 200 ml and 500 ml giving combined sample volumes for analysis of 400 ml and 1 l. The two anionic surfactant standards are treated in the same way as in the anionic procedure (Section 3.3.1.3.1).

*3.3.1.4  Calculation of results.*    The percentage degradation (*D*) at time *t* is calculated from

$$D_t = (1 - C_t/C_0) \times 100 \text{ per cent,}$$

where $C_0$ is mean initial concentration of surfactant in the test solution (mg/l), and $C_t$ is mean concentration in test solution at time *t* (mg/l).

For anionic surfactants, a graph (Figure 3.1) is drawn of percentage degradation against time and the percentage degradation of the test material is taken to be the mean of the two replicate values at the plateau. With samples for which the degradation curve does not show a plateau, the percentage degradation is taken to be that obtained on the 19th day.

For non-ionic surfactants, the percentage degradation is obtained by taking into the calculation only the mean value of the concentrations on the 19th day, the fifth day value serving to indicate that degradation is progressing at a satisfactory rate. The reason for this is that, in 1976 when the method was published, it appears that fewer biodegradation–time curves for non-ionics had been obtained than for anionics, probably due to the much higher cost and more tedious nature of the bismuth iodide test than for the MBAS test.

*3.3.1.5  Validity.*    The results are valid if the 'soft' standard degrades by 90–95% within 14 days and the 'hard' standard degrades by not more than 35%. Failing this, the whole test series must be repeated making sure that the inoculum is sufficiently active. Samples, which do not reach the required level of

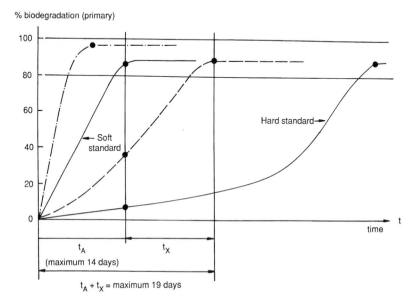

**Figure 3.1** Calculation of biodegradability.

degradation (80%) or for which duplicate tests fail to agree, and one is less than 80%, should be submitted to the confirmatory test.

*3.3.1.6   Note on cationic surfactants.*   The OECD (Cabridenc, 1983) conclud-ed that, although there were suitable methods available for assessing the biodegradability of cationic surfactants based on the use of the disulphine blue method, the adoption of test methods was not urgent because the pollution level resulting from their use was fairly low. This method may be applied to cationic surfactants, but it must be borne in mind that they are adsorbed to a greater extent than the other types of surfactants.

### 3.3.2   OECD Confirmatory test: continuous simulation of activated sludge process

*3.3.2.1   Principle.*   Synthetic sewage containing the required concentration of the surfactant under test is supplied at a constant rate to a vessel in which 3 l of activated sludge is aerated. The mixed liquor passes to an adjoining vessel where it settles and the settled sludge is continuously recycled to the aeration vessel, while the supernatant liquid is collected as effluent. A suitable apparatus is shown in Figure 3.2 and an alternative with no settling or recycling is shown in Figure 3.3. Sewage and effluent samples are analysed for the surfactant over a total period not longer than 9 weeks. The percentage removal of the surfactant is calculated for each sampling time from the concentrations in the sewage and in the correspond-

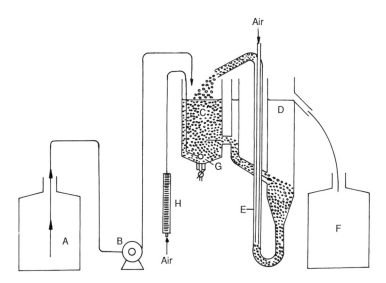

**Figure 3.2** Equipment used for assessment of biodegradability (Husmann unit). A, storage vessel; B, dosing device; C, aeration chamber (3-1 capacity); D, settling vessel; E, air lift pump; F, collector; G, aerator; H, air flow meter.

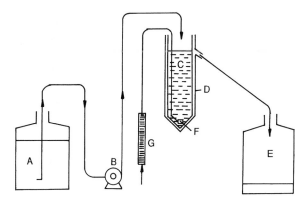

**Figure 3.3** Equipment used for assessment of biodegradability (Porous pot). A, storage vessel; B, dosing pump; C, porous aeration vessel; D, outer impermeable vessel; E, effluent collection vessel; F, diffuser; G, flow meter.

ing effluent, and a graph of percentage removal against time is drawn. The degradability of the test material is taken to be the mean of the values obtained over a 3-week period of steady operation during which removal has been regular. The running-in period before steady operation is attained is limited to 6 weeks.

### 3.3.2.2   Preparation

*3.3.2.2.1   Synthetic sewage.*   The synthetic medium used as 'sewage' consists of the following components in each litre of tap water:
160 mg peptone, 110 mg meat extract, 30 mg urea, 7 mg sodium chloride, 4 mg calcium chloride dihydrate, 2 mg magnesium sulphate heptahydrate, 28 mg dipotassium hydrogen phosphate, plus $20 \pm 2$ mg MBAS or $10 \pm 1$ mg BiAS.

(The original formulation (OECD, 1976) did not contain phosphate as an added salt. It was found (Painter and King, 1978a) that the essential element, phosphorus, was supplied only as an impurity in the peptone. Some peptones contained insufficient phosphate leading to poor settling and not very active sludge and to incomplete degradation; hence the addition of the phosphate salt.)

Uncompounded surfactants are added in the original state but formulated products must first be analysed for surfactants and soap content, and the required surfactant is separated by processes depending on the relative soap content (OECD, 1976).

*3.3.2.2.2   Inoculation.*   The addition of the inoculum should not introduce surfactants to the system. Thus, a freshly collected secondary effluent of good quality (BOD not more than 20 mg/l), kept aerobic until use on the same day as collected, is added at the rate of 1 ml per litre of aeration tank volume, that is 3 ml in all.

[Note: In most locations no inoculation would be necessary for growth on the synthetic sewage, but a few reports surprisingly indicate that in some laboratories inoculation has proved essential.]

*3.3.2.2.3  Apparatus.*   The small activated sludge unit (Husmann apparatus) is shown in Figure 3.2. A storage vessel (A) of capacity at least 24 l contains synthetic sewage and the dosing pump (B) introduces the sewage into the aeration vessel (C), which is about 4 l in volume. The liquor passes into the separator (D) and treated effluent leaves the apparatus to be collected in vessel F. Sludge is returned from the bottom of the separator to the aeration vessel by means of an air-lift pump (E). Aeration of the mixed liquor is effected by the use of a sintered aeration cube(s) and a flow meter (H) indicates the flow rate of the air.

Alternatively, the simpler Water Research Centre (WRC) porous pot apparatus (Painter and King, 1978b) may be used (Figure 3.3). It consists of a porous vessel (C) held inside an impermeable container (D) such that the annular volume is relatively small. A side tube on D is placed so that the porous vessel contains 3 l of activated sludge mixed liquor. Operation is similar to that with the Husmann unit; sewage or synthetic sewage is pumped from the storage vessel (A) into the aeration vessel (C), aerated by means of a diffuser (F). Virtually clear liquid passes through the walls of vessel C and is collected in vessel E. Thus, there is no settlement of sludge or sludge return system, as in the Husmann unit.

*3.3.2.3  Procedure.*   Initially, synthetic sewage is introduced into the system and the height of the separator is fixed so that the aeration vessel contains 3 l. The inoculum is added to synthetic sewage in the aeration vessel and the aerator, air-lift pump and dosing device are then started. The rate of aeration should allow the contents of the aeration vessel to be kept constantly in suspension. The air-lift pump is regulated to cause sludge to be continuously and regularly recycled and to maintain the concentration of dissolved oxygen at above 2 mg/l. The rate of synthetic sewage passing through the aeration vessel should be 1 l/h to give a mean retention time of 3 h. The equipment is kept at room temperature, that is, at a steady value between 18 and 25°C.

Foaming may be prevented by using anti-foaming agents, but these should not contain material which would react positively in the determination of surfactants. Sludge accumulating around the top of the aeration vessel, in the base of the settling vessel or in the circulation circuit should be returned to the circulation at least once per day by brushing, or some other appropriate means. The density of sludge may be increased, if settling is poor, by the addition of 2-ml portions of 5% (w/v) solution of ferric chloride, repeated if necessary.

Effluent from the separator is accumulated in vessel F over a 24-h period, following which a sample is taken after thorough mixing and allowing the foam to disperse.

*3.3.2.4 Analysis.* The appropriate surfactant concentration (in mg/l) of the synthetic sewage is determined immediately before use. The same method is used to determine the concentration of surfactant in the 24-h composite effluents. Samples should be analysed immediately after collection, otherwise they should be preserved, preferably by freezing.

*3.3.2.5 Check on efficiency.* As a check on the performance of the Husmann units, DOC or COD of the filtrate of the effluent is determined at least twice weekly, as well as that in the filtered synthetic sewage in vessel A. The decrease in concentration of DOC or COD in the effluent should level off when a roughly regular daily MBAS or BiAS removal is obtained, at the end of the running-in period (Figure 3.4).

The dry matter content of the activated sludge in the aeration vessel is determined twice a week; if it is greater than 2.5 g/l excess sludge must be discarded.

*3.3.2.6 Calculation.* The percentage removal of MBAS or BiAS is calculated every day from the measured concentration (mg/l) of the surfactant in the synthetic sewage ($C_S$) and in the corresponding effluent ($C_E$) collected in vessel F, that is, removal = $(1 - C_E/C_S) \times 100$ per cent. The removal values are plotted against time (Figure 3.4); there is sometimes a moderately high removal due to adsorption onto the sludge followed by a decrease in removal. If the surfactant is degradable the percentage removal increases, due to growth and/or acclimatization of competent organisms, until a plateau is attained; this acclimatization is called the running-in period. If the surfactant is not very biodegradable, its removal is low and erratic. The running-in period must not be greater than 6 weeks. The plateau period is continued for 21 days and primary degradation is calculated as the mean of the values obtained in this period, provided that

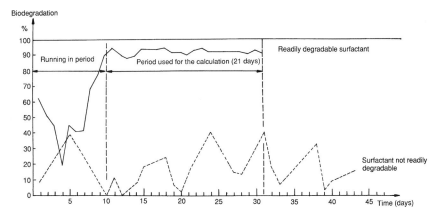

**Figure 3.4** Calculation of biodegradability — dynamic simulation test.

removal has been regular, the operation of the plant has been trouble-free and the removal of DOC or COD reached >80% by 2 weeks from the start of the test.

Individual values are expressed to nearest 0.1% but the final mean value is given to the nearest whole number. In some cases it may be permissible to reduce the frequency of sampling, but at least 14 results collected over the 21 days following the running-in period must be used to calculate the mean.

### 3.3.2.7  Notes on the method

*3.3.2.7.1  Indication for biodegradability.*  Because surfactants adsorb onto sludge, it is not always clear-cut that biodegradation has taken place. The shape of the degradation–time curve can give an indication of what is taking place. A typically shaped biodegradation curve (sigmoidal) has a lag followed by a steady rise in removal over a number of days to a plateau, but if adsorption takes place as well as degradation a positive removal occurs initially followed by a decrease, whereupon the normal biodegradation curve is obtained. If the curve cannot adequately be interpreted, a screening test should be repeated with an exposed inoculum taken from the effluent at the end of the confirmatory test.

*3.3.2.7.2  Synthetic sewage.*  It is accepted that the synthetic sewage has limitations — low carbon/nitrogen, few substrates, low buffering capacity, no continuous inoculation of sewage microorganisms — which tend to give low removals. If nitrification occurs, the pH value of the mixed liquor would fall, possibly interfering with the biodegradation of surfactants. This is more likely to happen with soft tap water and can be corrected by the addition to the synthetic sewage of appropriate amounts of sodium bicarbonate ($NaHCO_3$).

*3.3.2.7.3  Porous pot.*  The UK Porous Pot system (Figure 3.3) (Painter and King, 1978b) may be used instead of the Husmann system, except when two or more laboratories obtain conflicting results in which case the Husmann system has to be used.

*3.3.2.7.4  Cationic surfactants.*  No EC Directives have yet been promulgated for cationics, neither has the OECD produced a method. Some workers (e.g. Gerike *et al.*, 1978; Gerike, 1982) have successfully employed the activated sludge simulation method to follow the primary degradation (disulphine blue) of cationics. But because of the greater adsorption of cationics compared with anionics and non-ionics, it is imperative to determine the content of cationic surfactants on sludge and on the suspended solids leaving the system with the effluents, so that a mass balance may be made.

## 3.4 List and synopses of existing methods for ultimate biodegradability

Some of the present methods have their origins in the OECD tests for surfactants, in which degradation was followed by specific chemical tests, methylene blue reactivity for anionic surfactants and the Dragendorff bismuth reagent for non-ionics. While specific analysis may still be applied in the new methods as an optional addition, the major analytical tools are DOC (or COD), oxygen uptake and $CO_2$ evolution which indicate ultimate biodegradation. When oxygen uptake or $CO_2$ evolution is used, DOC may be applied, for soluble chemicals, at the start and end of incubation as an additional parameter. Some details of the methods are given in Tables 3.1, 3.2 and 3.3 and also in Table 4.3 of Ch. 4.

The media used have been harmonized (Tables 3.2 and 3.3) so that the only differences between the media now are that in the Closed Bottle method the mineral medium is one tenth of the concentration of that used in the other 'ready' tests and the MITI I test employs a slightly different medium at pH 7.0 instead of 7.4. In the Zahn–Wellens inherent test the same stronger medium is used as in most of the tests for ready biodegradability (base-set), while synthetic sewage or domestic sewage is used in the simulation test.

The tests for ready biodegradability are conducted at $22 \pm 2°C$, except for the respirometric method which technically requires a narrower tolerance of $\pm 1°C$ and the MITI test, which is held at $25 \pm 1°C$. The normal duration of the tests is 28 days.

Except for the MITI test, the inocula are drawn from activated sludge, sewage effluent, surface waters or soil extracts, and recommendations are made on the approximate bacterial cell densities to be used (Table 3.3). The inocula may be pre-conditioned to the test conditions by aerating the inocula in the medium in the absence of the test chemical to attempt to reduce the oxygen uptake and $CO_2$ evolution of the 'blank controls'. In the ISO methods only, pre-exposure to the test chemical is also permitted with the object of adapting the population so that it will degrade the chemical within the 28-day test period. The reported results must be accompanied by a description of the details of the pre-exposure and any chemical thus degraded cannot, of course, be described as readily biodegradable. Certain conditions have to be met in these tests to establish their validity. First, a reference chemical (benzoate, aniline, acetate) must have been degraded by 70% DOC (60% ThOD or $ThCO_2$) within specified periods. This condition checks the whole procedure but not the activity of the inoculum since the reference chemicals at present used are so readily attacked that they degrade even in uninoculated media. Next, the control vessels — inoculated, but containing no added organic chemical — should not take up more than 30 mg $O_2$/l or produce more than 40 mg $CO_2$/l in 28 days. Also, in the $CO_2$ evolution test the concentration of inorganic carbon at the start of incubation should not be more than 5% of the total carbon concentration. In the closed bottle test, the blank control oxygen uptake should be less than 1.5 mg/l in 28 days and the concentra-

**Table 3.3** Conditions in the revised OECD tests for ready biodegradability

| Test | DOC Die-Away 301A | $CO_2$ Evolution 301B | Manometric Respirometry 301F | Modified OECD 301E | Closed Bottle 301D | MITI (I) 301C |
|---|---|---|---|---|---|---|
| *Concentrations of test substance* | | | | | | |
| mg/l | | | 100 | | | 100 |
| mg DOC/l | 10–40 | 10–20 | | 10–40 | 2–10 | |
| mg ThOD/l | | | 50–100 | | 5–10 | |
| *Concentration of inoculum* | | | | | | |
| mg/l SS | >30 | >30 | >30 | – | – | 30 |
| ml effluent/l | >100 | >100 | >100 | 0.5 | >5 | |
| Approx. cells/ml | $10^4$–$10^5$>$10^6$ | $10^4$–$10^5$>$10^6$ | $10^4$–$10^5$>$10^6$ | $10^2$–$10^3$ | $10^1$–$10^3$ | $10^4$–$10^5$>$10^6$ |
| *Concentration of elements in mineral medium (in mg/l)* | | | | | | |
| P | 116 | 116 | 116 | 116 | 11.6 | 29 |
| N | 1.3 | 1.3 | 1.3 | 1.3 | 0.13 | 1.3 |
| Na | 86 | 86 | 86 | 86 | 8.6 | 17.2 |
| K | 122 | 122 | 122 | 122 | 12.2 | 36.5 |
| Mg | 2.2 | 2.2 | 2.2 | 2.2 | 2.2 | 6.6 |
| Ca | 9.9 | 9.9 | 9.9 | 9.9 | 9.9 | 29.7 |
| Fe | 0.05–0.1 | 0.05–0.1 | 0.05–0.1 | 0.05–0.1 | 0.05–0.1 | 0.15 |
| pH | 7.4 ± 0.2 | 7.4 ± 0.2 | 7.4 ± 0.2 | 7.4 ± 0.2 | 7.4 ± 0.2 | preferably 7 |
| Temperature (°C) | 22 ± 2 | 22 ± 2 | 22 ± 2 | 22 ± 2 | 22 ± 2 | 25 ± 1 |

DOC, dissolved organic carbon; ThOD, theoretical oxygen demand; SS, suspended solids.

tion of dissolved oxygen in all bottles must not be less than 0.5 mg/l. There are also restrictions on the final pH values of the media.

Finally, the extremes of replicate values of percentage removal should not differ by more than 20%; however, it is found that some chemicals tested with low cell densities often give wider variations than this.

In most of the OECD and EC tests the 'pass' levels have to be reached within 10 days of the removal reaching 10%, called the '10-day window'.

### 3.4.1   Ready biodegradability

The OECD screening test for primary biodegradability of surfactants was developed into a test for ultimate biodegradability of chemicals generally by substituting the determination of DOC for that of MBAS or BiAS; this required the setting up of blank control vessels containing no test chemical (301E; OECD, 1981, 1993a). A development of this was the modified AFNOR method (original 301A; OECD, 1981) in which the removal of DOC was followed, but the inoculum density was much higher, at $5 \pm 3 \times 10^5$/ml and was obtained by re-suspending the suspended solids from membrane-filtered sewage effluent in the test medium. It was the only method in which the number of cells was estimated, by a turbidimetric method; the method has been superseded. Both these methods, 301E and (original) 301A, were adopted first by OECD (1981) and then by EC (Official Journal of the European Communities (Off. J.E.C.), 1984) as C3 and C4. Indeed, there is an agreement between the two organisations that methods developed by them will be adopted by the other in the whole field of biodegradability and ecotoxicology.

Later, these two DOC methods were combined and developed further by ISO (ISO, 1985) by introducing for the first time the optional use of activated sludge inocula (30 mg solids/l, or less). The method has undergone further slight modification in the periodical review by OECD (1993a) and appears as the DOC Die-Away test (new 301A). The Modified OECD Screening (DOC) method has been retained, with the very low cell density (0.5 ml effluent/l) at the request of some member states, as 301E. The EC (EC 1992; Off. J.E.C., 1992) has adopted these two DOC methods along with all the other methods published in the OECD revision (1993a).

Similar changes have occurred with the other two types of tests, $CO_2$ evolution and oxygen uptake. The original test by Sturm (1973) for measuring $CO_2$ as a product of metabolism from surfactants was adopted by the OECD (1981) as 301B and by EC (Off. J.E.C., 1984) as C5. Later, ISO slightly modified the method (ISO, 1991b) by introducing an alternative method for absorbing the $CO_2$ — in NaOH as well as baryta and determination by DIC instead of titration. The 1993 revision by OECD also includes this alternative.

Originally, the MITI method (301C) (enclosed flasks containing air in the head-space) was the only manometric oxygen uptake method in the OECD series (1981) and in the EC group of tests (C7; Off. J.E.C., 1984); it has not been changed in the 1993 OECD revision. Another, very similar method, Manometric Respirometry, 301F, has been introduced by the OECD (1993a) and has as nearly as possible the same conditions as in methods 301A and 301B (1993a). The essential difference between the MITI (301C) and 301F is in the nature of the inocula; the MITI test uses a mixed inoculum which is tedious and time-consuming to produce, while 301F uses activated sludge. ISO has published the Manometric Respirometric method (301F) (ISO, 1991a) and has done so, too (EC 1992: Off. J.E.C., 1992).

Finally, the last agreed test for ready biodegradability is the Closed Bottle method (301D; OECD, 1981, 1993a) based on the 5-day Biochemical Oxygen Demand test (C8; EEC Off. J.E.C., 1984). The oxygen uptake is determined from the measurements of the concentration of dissolved oxygen (in the full bottles) with time. The EC has adopted the method as C6 (Off. J.E.C., 1984) and ISO will soon publish a very similar version (ISO/CD 11 733).

*3.4.1.1   Ready biodegradability: DOC Die-Away test (301A) and Modified OECD (301E).*   The test chemical is added as the sole source of carbon, at 10–40 mg C/l, to a mineral salts medium containing salts of Na, K, Mg, Ca, Fe, $NH_4$, Cl, $SO_4$ and $PO_4$; the medium is buffered by the phosphate salts at pH 7.4. The medium is inoculated to give an increased cell density between about $10^2$–$10^3$/ml (Modified OECD test) and $10^4$–$10^6$/ml (OECD/301A, ISO). At least duplicate vessels are set up containing the chemical plus inoculum and another inoculated pair containing no added test chemical to act as controls. A further inoculated flask is set up containing a reference chemical (aniline, benzoate, acetate) at 20 mg C/l to check the procedure. Abiotic controls are set up, if required, containing a sterilized (e.g. with $HgCl_2$) uninoculated solution of the chemical and a further inoculated vessel containing both the test and reference chemicals is set up if the inhibitory property of the chemical is to be determined.

The flasks are incubated in the dark at $22 \pm 2°C$ (20–25°C, ISO) with shaking. Frequent samples are taken during the 28-day incubation so that an adequate biodegradation curve may be drawn. The samples are either membrane-filtered or centrifuged to remove bacterial cells and DOC is determined in duplicate on the filtrate or supernatant. Filtered samples may be stored at 2–4°C for up to 2 days or below −18°C for a longer period.

The removal of DOC is calculated as:

$$\% \text{ removal} = \frac{\text{DOC at day 0} - \text{DOC at day } t}{\text{DOC at day 0}} \times 100$$

In all cases the DOC values are corrected for those of the blank controls.

A plot is made of the average percentage removal against time (Figure 3.5). The lag time is defined as the time from inoculation until the removal has reached 10% of the starting concentration. This lag time (which is recorded; ISO only) is often highly variable and poorly reproducible. The maximum level of degrada-

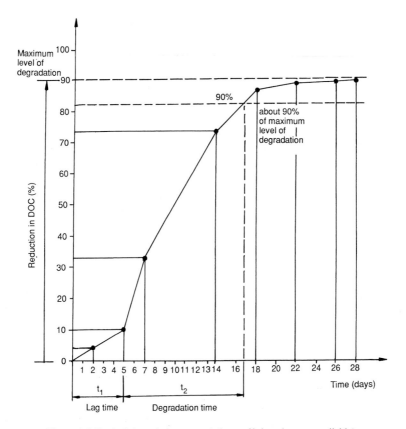

**Figure 3.5** Typical degradation curve (when sufficient data are available).

tion is defined as the approximate level above which no further degradation takes place during the test. Degradation time is defined as the time from the end of the lag period to the time when about 90% of the maximum level of degradation has been reached (ISO only). The 'pass' level (OECD and EEC only) is 70% DOC and must be reached within 10 days of the end of the lag period.

*3.4.1.2   CO₂ evolution (Modified Sturm) test (301B).*   Batches of 3 l of the mineral medium (used in the DOC Die-Away test) contained in enclosed vessels are inoculated normally with 30 mg/l activated sludge, or homogenized sludge at 1% (v/v), giving about $10^5$–$10^6$ cells/ml. The inoculated mixtures are aerated with $CO_2$-free air overnight to purge the system of $CO_2$. The test chemical and reference chemical are added, separately, to give 10–20 mg DOC or TOC/l; insoluble and poorly soluble chemicals may first be treated as described in ISO (1993). Some vessels — control blanks — receive the equivalent amounts of water to equalize the volume in all vessels. As in the DOC die-away test, inhibition and abiotic controls may be set up. Absorber vessels are partially filled with solutions of barium hydroxide or sodium hydroxide.

$CO_2$-free air is bubbled through the liquid in the vessels held at $22 \pm 2°C$ in the dark, at roughly constant rates between 30 and 100 ml/min. The $CO_2$ trapped is determined either by titration of the un-reacted barium hydroxide in the trap nearest to the reaction vessel, or by withdrawing, say, 100 $\mu$l of the sodium hydroxide solution by syringe and injecting into the inorganic part of a carbon analyzer. Determinations of $CO_2$ are made at least at 2- to 3-day intervals in the first part of the incubation and then at least every 5 days. On day 28, 1 ml concentrated HCl is added and the medium is aerated overnight to drive off any residual $CO_2$ into the traps.

Usually 100 ml of 0.125 M barium hydroxide is contained in each trap and the HCl solution is 0.05 M, and hence the mass of $CO_2$ trapped is given by 1.1 $(50 - V)$ mg, when $V$ is the volume of HCl used for the titration. The percentage degradation is given by

$$\frac{\text{mgCO}_2 \text{ in test trap } - \text{ mgCO}_2 \text{ in control trap}}{\text{ThCO}_2} \times 100$$

where

$$\text{ThCO}_2 = 44/12 \times \text{conc. of test chemical C} \times \text{vol. of test solution.}$$

For NaOH in traps,

$$\% \text{ degradation } = \frac{\text{mg IC from test flask } - \text{ mg IC from blank}}{\text{mg TOC added, as test substance}} \times 100.$$

Besides the OECD (1981, 1993a) version, both the EC (Off. J.E.C., 1984, 1992) and ISO (1991b) have published similar texts.

*3.4.1.3   Manometric respirometry (301F).*   The test chemical is added, to give 50–100 mg ThOD/l as the sole source of carbon, to the mineral medium (as used in the DOC Die-Away method). Similarly, a reference chemical is added to another batch of the medium and an equivalent volume of water is added to a third batch to act as a control. If the toxicity of the chemical is to be assessed, a further solution in the medium is prepared containing both the test and reference chemicals at the same concentrations as in the individual solutions. Insoluble and poorly soluble chemicals are added at a later stage. If abiotic degradation is to be investigated, a solution containing the test chemical is sterilized by the addition of a toxic substance (e.g. mercuric chloride).

The specific ThOD (mg $O_2$/mg chemical) is calculated on the basis of the formation of ammonium from N-containing test chemicals, unless nitrification is anticipated when the calculation is based on the formation of nitrate.

Known volumes of each solution are added to the respective respirometer flasks, at least in duplicate, contained in a water bath or incubator at $22 \pm 1°C$, and it is at this stage that insoluble and poorly soluble chemicals are added (see

ISO, 1993). The $CO_2$ absorbent is then added to the absorber compartment and the flasks are inoculated with activated sludge to give a concentration of 30 mg solids/l.

The stirrers are started, the flasks are sealed when they have reached the operating temperature and the measurement of oxygen uptake is begun. With automatic respirometers a continuous record is obtained, so that the lag period and the '10-day window' are easily recognized; with non-automatic respirometers daily readings of the volume/pressure are adequate.

The oxygen uptake is calculated from the readings by the methods given by the manufacturers of the equipment. The specific biochemical oxygen uptake at any time interval is given by specific BOD

$$= \frac{\text{mg } O_2 \text{ uptake by test chemical} - \text{mg } O_2 \text{uptake by blank}}{\text{mg test chemical in respirometer flask}}$$

and the percentage biodegradation is given by

$$\% \text{ThOD} = \frac{\text{specific BOD}}{\text{specific ThOD}} \times 100.$$

COD can be used if ThOD cannot be calculated, but it is a poor substitute since some chemicals do not react fully in the COD test.

If nitrification is expected or thought to have occurred, corrections should be made for the oxygen used in oxidizing ammonium by analysing samples from the respirometer flasks at the beginning and end of incubation for nitrite and nitrate.

Similar texts have been published by ISO (1991a) and by EC (Off. J.E.C., 1992).

*3.4.1.4 MITI test (301C).* The differences between this and the previous method are as follows (see Tables 3.2 and 3.3).

(a) The medium contains less P, Na and K and more Mg, Ca and Fe, and the pH value is 7, not 7.4.
(b) The temperature is $25 \pm 1°C$.
(c) The inoculum is prepared in a relatively complex way. Fresh samples of activated sludges from domestic and industrial waste waters, water and mud from rivers, lakes and seas are collected from ten sites mainly in areas where a variety of chemicals are used and discharged. Equal volumes of these ten samples are thoroughly mixed and the supernatant, after the mixture had been allowed to stand, is adjusted to pH $7 \pm 1$. This supernatant is used to fill a 'fill-and-draw' activated sludge plant, which is then operated on a 24-h cycle (23 h aeration, 1 h settlement) and is fed with 0.1% each of glucose, peptone and phosphate. After operation for 1 month the sludge may be used as a source of inocula for a further 3 months.

(d)  If biodegradation is between 20 and 60% ThOD and the C remaining is not the parent compound, intermediates of the test chemical must be sought.

*3.4.1.5   Closed Bottle test (301D).*   Solutions containing 2–5 mg/l of the test chemical and reference chemical as sole sources of carbon are prepared, separately, in previously aerated mineral medium, in which the concentrations of salts are one-tenth of the concentrations in the medium used in other methods. A blank control solution consists of the aerated mineral medium alone. If toxicity is to be investigated, a solution is prepared containing both the test and reference chemicals. These solutions are then inoculated with secondary effluent or surface water at the rate of 0.05–5 ml/l to give about $10–10^3$ added cells/ml and each well-mixed solution is carefully dispensed into a series of at least ten BOD bottles so that all bottles are completely full. If insoluble and poorly soluble chemicals have been added to the bulk solutions using a method described in ISO/CD 10 634 (1993), ensure that the contents of the containers holding the suspensions are well mixed during the dispensing operation. Otherwise such chemicals may be added directly to the BOD bottles (see ISO/CD 10 634, 1993).

Duplicate bottles of each series are analysed immediately for dissolved oxygen by the modified Winkler or electrode methods and the remaining bottles are carefully stoppered and incubated at $22 \pm 2°C$, or preferably $22 \pm 1°C$, in the dark. Bottles of all series are withdrawn in duplicate for dissolved oxygen analysis at least weekly over the 28-day incubation period. For ensuring identification of the '10-day window', sampling every 3–4 days should be sufficient, but this requires about 20 bottles per series. It is considered that a '14-day window' would suffice for this method.

Corrections for oxygen uptake by nitrification in the case of N-containing chemicals may be made only if the electrode method is used; after analysis for dissolved oxygen, samples are withdrawn from the bottles for analysis of nitrite and nitrate.

The specific BOD and percentage biodegradability are calculated, as indicated in the respirometric method. A similar text has been published by EC (Off. J.E.C., 1984, 1992) and one is to be published by ISO.

### 3.4.2   Inherent biodegradability

Only two methods have been published under this heading; a third method, MITI II, was published but is now not used. Any of the methods in Section 3.4.1 may be used by employing, for example, inocula pre-exposed to the test chemical, longer periods of incubation or, in some cases, higher cell densities.

The test devised by Zahn and Wellens (302B; OECD, 1981) was proposed as a test for ready biodegradability, but was rejected for that purpose because the conditions employed were insufficiently stringent. The inoculum is 200–1000 mg solids/l of activated sludge with 50–400 mg C/l of the test chemical in the mineral medium. Degradation is followed by determining DOC, allowing for DOC in control vessels and for any adsorption of the test chemical on the sludge. It is recommended that sludge taken at the end of a test (pre-exposed) could be used as an inoculum for a second attempt. The EC also published this method (Off. J.E.C., 1988). The OECD has recently (OECD, 1993b) revised the method to bring the mineral medium in line with that used in the revised tests for ready biodegradability; this lowered the concentration of ammonium salts which, in turn, lowered the chance of decreased pH values due to nitrification. Also it was stressed that the ratio of the inoculum to test compound should be kept within certain limits. The ISO version (ISO, 1991d) is very similar to the latest OECD version.

The second agreed test, 302A (OECD, 1981) — the semi-continuous activated sludge method (SCAS) — is based on the USA Soap and Detergent Association (SDA, 1965) 'confirmatory' test for surfactants. The sludge units are operated in the fill-and-draw mode, sewage or synthetic sewage with or without (blank control) the test chemical being added daily. The mean retention time of sewage is 36 h and no sludge is deliberately wasted, thus the conditions are much less stringent than in normal sewage treatment and the test does not truly simulate the activated sludge process. The test is usually operated for 3 months, but it has been found that some chemicals did not degrade until after longer periods, so that it may be advisable to continue the test for up to 6 months. Both the EC (Off. J.E.C., 1988) and ISO (ISO, 1991c) have published methods essentially the same as the OECD version, which has not yet required revision.

*3.4.2.1   Zahn–Wellens — EMPA method (302B).*   It is advisable first to determine the toxicity of the test chemical towards activated sludge so that a non-inhibitory concentration is used in the subsequent test.

A batch of 2 l of the normal mineral medium (Table 3.2) is prepared containing the test chemical at 50–400 mg DOC/l and 200–1000 mg dry solids/l of previously washed activated sludge. The ratio between inoculum and test chemical (as DOC) is kept between 2.5:1 and 4:1. Other batches contain reference chemical and no added chemical, respectively, to act as a control. The suspensions are contained in glass cylinders each equipped with a stirrer and a device for aerating the contents.

The suspensions are aerated with purified, humified air at 20–25°C in the dark or diffuse light for up to 28 days. The mixture is stirred if necessary to keep the sludge in suspension and to maintain the concentration of dissolved oxygen above 1 mg/l. The pH value is frequently monitored and adjusted to pH 6.5–8, if necessary. The first sample is taken at 3 ± 0.5 h after addition of the chemical to estimate any adsorption of the chemical by the sludge. Then, samples are taken on at least four occasions between the first and 27th days, adjusting the frequency to suit the rate of disappearance of the chemical. Finally, samples are taken on the 27th and 28th days, or on the last 2 days of the test run.

The samples are filtered or centrifuged and the concentration of DOC, or COD, is determined on the filtrates or supernatants. The percentage removal is given by

$$\%R_t = \frac{\text{DOC (or COD) at 3 h} - \text{DOC (or COD) at time } t}{\text{DOC (or COD) at 3 h}} \times 100$$

In all cases the values of DOC (or COD) are corrected for the appropriate blank value. A degradation curve is drawn by plotting $\%R$ against time. The shape of the curve, and the difference between the 3-h values and the expected initial value of the DOC (or COD) concentrations, give indications of whether any disappearance is due to biodegradation or physical processes.

*3.4.2.2   Semi-continuous activated sludge (SCAS) test (302A).*   A sufficient number of SCAS units, varying in size between 150 and 1500 ml, are set up so that there is at least one for each test chemical plus a control. The units can be measuring cylinders with a means of aeration or the aeration vessel can be a tube containing a sealed-in air inlet tube and a tap so placed that one-third of the total volume of mixed liquor remains (as settled sludge) in the vessel after draining off the settled supernatant.

The units are filled to the appropriate volume with activated sludge and aeration is begun. After about 23 h aeration is stopped for about an hour to allow the formation of a clear supernatant and a volume equivalent to two-thirds the total volume of mixed liquor is discarded. Sewage or OECD synthetic sewage (Section 3.3.2.2.1) is added to the settled sludge to replace the withdrawn supernatant. The fill-and-draw procedure is repeated daily and the filtered effluents are analysed for DOC (or COD) two or three times per week. When the concentration of DOC attains a constant value, indicating a steady state, the test chemical is added with the sewage to the test unit(s), while only sewage is continued to be added to the control unit. Analysis of the effluents is continued daily, if the DOC value from the test vessels changes significantly, until the difference between the DOC of the control and test effluents remains fairly constant over six consecutive measurements. Otherwise the effluent is analysed two or three times per week; if no removal is observed the analysis is continued for at least 12 weeks but not more than 26 weeks.

The percentage removal is calculated from

$$\%R_t = \frac{\text{Nominal DOC in chemical added} - \text{DOC* at end of aeration}}{\text{Nominal DOC in chemical added}} \times 100$$

where DOC* denotes a value corrected for the DOC in the control effluent, which has been shown to vary from $5.8 \pm 1.9$ mg DOC/l for OECD synthetic sewage and $5.4 \pm 0.8$ mg/l to $13.2 \pm 2.8$ mg DOC/l for domestic sewages.

If the plot of DOC in the test effluent against time has the typical shape of a biodegradability curve with lag and plateau phases, and if the chemical does not adsorb significantly onto sludge, the elimination of the chemical can be reasonably confidently assigned to biodegradation. In case of doubt, a respirometric method using exposed sludge from the SCAS test as the inoculum should differentiate between physical and biological processes.

### 3.4.3 Simulation methods

Simulation tests have been difficult to devise, except for the simulation of the activated sludge process. The only agreed method (OECD, 1981) is based on the OECD Confirmatory test for the primary biodegradation of surfactants (OECD, 1971), which itself was based on the German method (Husmann et al., 1963). To convert the test from one indicating primary to one for ultimate biodegradability, two units rather than one unit have to be used for each chemical; one receives the test chemical, the other receives only sewage (blank control). Also, DOC or COD is determined rather than specific analysis (methylene blue or bismuth iodide). The OECD (1981) version describes only a special mode of operation called 'coupled units' (see Section 3.4.3.1), while the EC version (Off. J.E.C., 1988) describes a number of variations of the method. An ISO version, also describing a number of variations, is under discussion.

#### 3.4.3.1 Activated sludge.
Attempts to simulate the activated sludge process have taken a number of forms and many are probably adequate for the purpose. The OECD (1981) has so far concentrated on one method called the 'coupled-units' test (Fischer et al., 1975), which uses the Husmann apparatus (3-1 aeration vessel) from the OECD Confirmatory test for surfactant biodegradation. The EC (Off. J.E.C., 1988) describes the 'coupled-units' mode of operation and the single or non-coupled mode, as well as the UK Porous Pot system, are also described. The EC method describe a range of a number of operation variables, including type of sewage and inoculum, mean retention time of sewage (3–6 h), mean retention time of sludge (6–10 days), methods of wasting sludge and of adding the test chemical.

Briefly, two units, Husmann (Figure 3.2) or Porous Pots (Figure 3.3), are run in parallel under identical conditions. The test chemical is added to the influent synthetic or domestic sewage to one of the units, while the other receives the sewage alone. In the coupled-units mode sludge is daily interchanged equally between the two units in an attempt to equate the microbial populations in the two sludges.

The concentration of DOC (or COD) in the effluents is determined, but the DOC due to the added chemical in the influent is calculated, not measured. The difference between the mean concentrations of the test and control effluents is assumed to be due to undegraded test chemical. Plots of DOC against time are drawn to show the progression of biodegradation, if any (Figure 3.6). If the plot for the test chemical effluent has a typical sigmoidal shape of a biodegradability curve with lag and plateau phases and if the test chemical does not adsorb

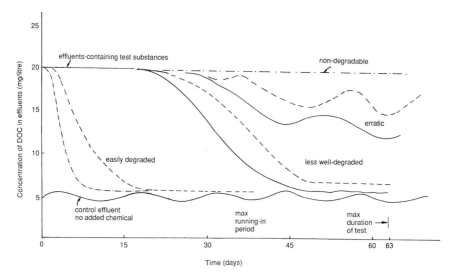

**Figure 3.6** Concentration of DOC in effluents in simulation tests (idealized curves).

significantly on to sludge, the elimination of the chemical can be assumed to be due to biodegradation. This could be confirmed by applying a respirometric method using sludge from the simulation test.

The test normally lasts for no more than 9 weeks; up to 6 weeks are allowed for adaptation of the sludge and 3 weeks for steady operation during which about 14 measurements are made. The mean of these values is used to calculate the percentage elimination from:

$$\%R = \frac{\text{DOC in test effluent} - \text{DOC in control effluent}}{\text{DOC due to test chemical in influent}} \times 100$$

*3.4.3.2  Other simulation methods.*  Methods exist for simulating biological filtration of sewage as well as 'surrogate' methods for simulating rivers, but these have not yet been agreed internationally.

### 3.4.4  Comparison of the methods: accuracy and precision

It will be seen from the previous sections (3.4.1, 3.4.2 and 3.4.3) that the versions of the respective tests published by the three organizations do not significantly differ from one another. Changes made during their evolution have been only minor so that results obtained with any given test will not be influenced by whichever text is followed.

The $CO_2$ evolution method gives the most direct evidence of oxidation of organic carbon during biodegradation; the removal of DOC can be due to processes other than biodegradation and the uptake of oxygen is only an indirect measure for assessing biodegradability. Also, only by using DOC, either in the DOC Die-Away test or as additional determinations in the other two methods, can an indication be obtained of the formation of any recalcitrant intermediate metabolites.

The DOC Die-Away method is obviously limited to soluble chemicals, usually those with a solubility of at least 50 mg/l, and to those which are not significantly adsorbed or are not volatile, although the latter can be assessed by modifying the apparatus. The Respirometric method can accommodate soluble, insoluble and volatile chemicals while the $CO_2$ evolution method can deal with soluble and insoluble but not volatile chemicals. The Closed Bottle method can deal with volatile chemicals if the bottles are modified and although insoluble chemicals can be assessed by this method the values obtained could be falsely low because mixing is extremely deficient. The concentration of test chemicals in this test is limited to 2–5 mg/l because of the low solubility of oxygen in water.

The inherent and simulation tests are obviously suitable for soluble chemicals, and a case can be made for their application to insoluble (but not very volatile) chemicals provided they are suitably dispersed. In inherent tests the resulting exposed sludge could subsequently be used in $CO_2$ evolution or oxygen uptake tests and in the simulation tests the effects of the insoluble chemicals on the performance (BOD removal, nitrification) of the activated sludge units could be ascertained. Volatile chemicals have been examined in simulation tests using enclosed aeration tanks.

The overall accuracy, precision and reproducibility of the various methods are adversely affected by the inconsistency and unpredictability of the inocula which, of necessity, have to be used. For chemicals which are very easily biodegraded, such as the reference chemicals, very high values of %DOC removal approaching the theoretical of 100% are consistently obtained with high precision and reproducibility. However, the %ThOD and %ThCO$_2$ obtained are always lower than %DOC removal (for all chemicals, not just the very easily degradable chemicals) because some of the carbon is converted to biomass. The proportion of the carbon used for cell synthesis varies both between species of bacteria and between chemicals so that the %ThCO$_2$ and %ThOD will vary from test to test and from chemical to chemical, even though simultaneous determinations show that %DOC removal approaches 100%. Hence, there are no 'accurate' values. In these tests the precision with which very easily degradable chemicals are assessed is high but not so reproducible either between tests using inocula from different sources or as for %DOC removals because the division of carbon between cell synthesis and respiration differs between bacterial species. For chemicals which are not so easily degraded and may require longer lag periods, the precision in the various tests is not as high as with chemicals like the reference chemicals, especially when low cell densities are used.

The chemical determinations can be carried out with a greater accuracy and precision than the tests as a whole. The determination of DOC can be made with a precision of $\pm 0.25$ mg/l with a lower limit of about 0.5 mg/l. In the automatic respirometer the 'quantum' of oxygen measured is 0.5–1.0 mg, equivalent to about 1–2 mg DO/l, although this limit could be lowered. In the Closed Bottle test the concentration of dissolved oxygen can be determined to $\pm 0.05$ mg/l. However, measurements of $CO_2$ in the Modified Sturm test are considered to be less precise (than DOC and oxygen in other tests) because the relatively small amount of $CO_2$ evolved in the period between measurements is determined from the difference between two relatively high titration values. For example, a blank titre could be 50 ml and the 'test' titre 45 ml 0.05 M HCl so that the mass of $CO_2$ produced would be

$$1.1 \times (50 - 45) = 5.5 \text{ mg } CO_2.$$

For this reason, Gerike and Fischer (1979, 1981) carried out only one titration for a 28-day test. Others have used sodium hydroxide and determined IC, at frequent intervals, with much greater precision than the determination of $CO_2$ by back titration.

In the SCAS test, the %DOC removal and standard deviation for aniline was $96 \pm 2.6\%$ and for Marlon A $88.8 \pm 9.9\%$ (ISO, 1991c). In the simulation test in the non-coupled mode 1-naphthol was removed by $92 \pm 8.2\%$, pentaerythritol $84 \pm 11\%$ and $18 \pm 19\%$ for sulfanilic acid (Painter and Bealing, 1989). In both tests the standard deviation is often much higher when poorly degradable chemicals are tested.

## 3.5    Analytical methods

### 3.5.1    General

There are many ways by which the progress and extent of primary biodegradation of surfactants may be assessed. In this section methods for determining the surfactants directly, i.e. indicating primary biodegradation, are described. Analytical methods developed for the control of manufacturing processes or the examination of commercial products in bulk quantities are often not applicable to the much lower concentrations (0.1–10 mg/l) of interest in biodegradation and environmental research.

Manual and automatic colorimetric and titrimetric methods which are applicable to a class (anionic, non-ionic or cationic) of surfactants have been used for quantitative determination, since not only are there wide ranges of sub-classes within the three main classes, but the individual commercial products themselves are generally complex isomeric and homologous mixtures. In these cases, estimations on environmental samples are made relative to the responses of relevant arbitrary standard surfactants. For example for anionic surfactants sodium dioctyl sulphosuccinate ('Manoxol OT') is used in the UK, sodium dodecyl sulphate in France, and the methyl ester of dodecyl benzene sulphonic acid is used in Germany. The latter standard is also recommended in the ISO method (ISO, 1984a). For non-ionics the standard chosen is a nonyl phenol ethoxylate containing an average of 8–10 ethoxylate units, such as Synperonic NP8 and Marlophen 810 (ISO, 1984b). These products are chosen since they are available in forms which are of consistent constitution. Normally in biodegradability tests the 'standard' can be the product under test. If a standard is used, concentrations should be reported in terms of that standard. Conversion factors between standards are used, based on molecular weight per active group. For example, the factor for converting Manoxol OT units to lauryl sulphate units is:

$$\frac{\text{molecular weight of sodium lauryl sulphate}}{\text{molecular weight of Manoxol OT}} = 282/444 = 0.635$$

While in biodegradability tests the response to the colorimetric reagent will be almost solely due to the surfactant, in environmental samples other compounds interfere to react with the reagent. Thus, it is realistic in this case to call the material determined in such samples methylene-blue active substances (MBAS) for anionic surfactants, bismuth reagent active substances (BiAS) for non-ionic surfactants.

Details of a number of methods and their application are given by the UK Standing Committee of Analysts (HMSO, 1982) and in wider terms by Swisher (1987).

Chromatographic methods — thin layer, gas, liquid column — have also been developed which can be used for identification both of the groups collectively, separated from other types of surfactants, and for individual isomers and homologues. These methods all require at least a pre-concentration into a solvent by a process such as sublation, adsorption and, usually, clean-up steps to remove interferents and other types of surfactants.

Other methods which can be used for structural and identification purposes, but not primarily for quantification, include infra-red, nuclear magnetic resonance, mass spectrometry and various developments in ionization techniques in GC–MS. Discussion of these and other methods are given by Llenado and Jamieson (1981), Llenado and Neubecker (1983) and Swisher (1987).

*3.5.1.1 Sampling.* Because of the ability of surfactants to adsorb fairly strongly on to surface and suspended solids, extreme care must be taken to obtain representative samples of the material to be tested. For example, the body of liquid from which the sample is to be taken should not contain foam and the sample bottle should be completely filled. Similarly when withdrawing a sub-sample, the contents of the bottle should be carefully stirred with a magnetic stirrer or the bottle should be gently inverted a few times to mix the contents.

In most biodegradability studies, the samples to be tested usually contain little or no suspended solids since the amount of inoculum added is small. If separation is necessary it is better to centrifuge the sample rather than to filter it. If filtration is used it is advisable to use as small a filtration assembly as possible and to reject the first 10–20 ml filtrate before collecting the final volume required for analysis to minimize the effects of adsorption onto the filter material. In activated sludge simulations tests it is usually unnecessary to separate the suspended solids in settled effluent samples. Should material such as activated sludge have to be analysed, the best procedure is to separate the solids by settlement or by centrifugation and to determine the surfactant content in the two phases separately, because the efficiency of extraction and reaction with the reagent (e.g. methylene blue) of surfactants adsorbed onto solids is difficult to determine. Surfactants on solids are first extracted with alkaline methanol (HMSO, 1982) by refluxing, Soxhlet extraction or sonication.

Preservation of samples prior to analysis has been achieved by the addition of mercury (II) chloride or formaldehyde or by freezing. It is advisable to check that the method of preservation does not interfere in the method of analysis subsequently applied. Also, to protect the environment the use of noxious inhibitors should be avoided, if possible.

### 3.5.2   Anionic surfactants

The most common method is the colorimetric method in which the anionic surfactant is reacted with a cationic dye to form a paired ion, or association complex, which unlike the parent dye is soluble in a solvent such as chloroform. The paired ion is extracted under controlled conditions of pH and shaking and the colour is measured spectrophotometrically. Many dyes have been used but methylene blue is by far the most common and various procedures, including automated methods, have been described. Whichever procedure is adopted, it is essential to adhere to a constant procedure throughout a study, since the conditions used — pH, mode and duration of shaking, presence of an inhibitor in the sample — may affect the results.

The specificity of the method is low; all intact sulphonates and sulphates used in commercial preparations react positively. However, as the hydrophobic group is made smaller the ion pair becomes more hydrophilic and less is extracted under the test conditions. For LABS extraction is complete from $C_{16}$ to $C_8$ homologues, while those with fewer carbon atoms in the chain are extracted to a lower extent with little extraction of the $C_4$ homologues. Also, when the hydrophobic group is partially oxidized, e.g. $-CH_2.CH_3$, $\rightarrow -CH_2.COOH$, very much less is extracted than with the parent compound.

A method of differentiating between sulphonates and sulphates is to apply the methylene blue method before and after acid hydrolysis; the difference gives the concentration of alkyl sulphates plus ethoxy sulphates. Sulphonates are stable, whereas the sulphates produce inorganic sulphate plus the corresponding alcohol.

*3.5.2.1  Standard methylene blue method.*  The methylene-blue method for anionic surfactants as adopted by EC (1982b), OECD (1976), ISO (1984a) and UK Standing Committee of Analysts (SCA) (HMSO, 1982) is outlined below. The method is based on improvements of the Abbott (1962) modification of the Longwell and Maniece (1955) method. The range of concentrations tested is 0.1–2.0 mg/l for a 100-ml sample, but can be varied by using different volumes of sample and using 10, 40 or 50 mm optical cells. The limit of detection is 0.02–0.05 mg/l.

*3.5.2.1.1  Principle.*  The surfactant is reacted with methylene blue and the complex is partitioned into chloroform from an alkaline solution of the dye to avoid the negative interference of proteinaceous material present in environmental samples. The chloroform phase is then back-extracted with an acidified methylene-blue solution to remove the interference of those materials, such as inorganic anions, e.g. nitrate, chloride, etc., that form methylene blue complexes of low chloroform extractability. The absorbance of the final chloroform phase is determined at 650 nm.

*3.5.2.1.2  Procedure.*  Briefly, 50 ml water, 10 ml 0.05 M borate or carbonate buffer (pH 10), 5 ml neutral methylene blue solution (250 mg/l) and 10 ml chloroform are shaken for 30 s in a 250 ml separating funnel. The chloroform layer is run off as completely as possible, taking care that none of the aqueous layer is lost. The aqueous layer is rinsed with 2–3 ml chloroform without shaking. The extraction with a further 10 ml chloroform and the rinsing are repeated, and the chloroform extracts are discarded. These steps, which could be carried out in one large separating funnel with sufficient of the dye for all the tests to be made, clean up the methylene blue, which always contains interfering material.

A second separating funnel is used to repeat the alkaline washing of methylene blue, except that 100 ml instead of 50 ml water are added. After discarding the chloroform layers, 3 ml 0.5 M sulphuric acid are added to the funnel and the contents are well mixed.

Up to 100 ml of the sample containing 20–160 $\mu$g of anionic surfactant are added to the first separating funnel followed by 15 ml chloroform. The funnel is shaken evenly and gently twice a second for 1 min, preferably in a horizontal plane. The layers are allowed to separate as completely as possible and the funnel is swirled to dislodge droplets from the sides of the funnel. After 2 min as much as possible of the chloroform layer is run into the second funnel containing the acidified dye solution. The second funnel is shaken and the contents are allowed to separate for 2 min, as before. The chloroform solution is run through a filter funnel containing a freshly chloroform-washed cotton wool plug into a 50 ml volumetric flask. The extraction of the alkaline methylene blue solution in the first funnel is repeated twice with 2 × 15 ml chloroform, the chloroform layer is then transferred to the second funnel and extracted under acid conditions. The extracts are filtered into the volumetric flask and chloroform is added to the 50 ml mark.

For each batch of samples the procedure is applied to a blank and one concentration of standard (e.g. Manoxol OT). Occasionally eight or nine concentrations of standard are used to construct a calibration graph.

The absorbance of blanks, standard(s) and samples is measured in 10 mm cells against a chloroform reference at 650 nm; it is important that the cells are rinsed out three times with the chloroform solutions before taking a reading.

If the absorbance is too low, the 40 mm cell should be used. The number of micrograms in the sample is read from the calibration graph corresponding to the size of cell used, making due allowance for the absorbance of the blank, and the concentration is calculated from this weight and the volume of sample used. Results are expressed in terms of the standard used (Manoxol OT, lauryl sulphate, dodecyl benzene sulphonate) and reported as methylene blue reacting substance (MBAS) unless the identity of the anionic surfactant present in the sample is known.

*3.5.2.2   Comments on the method.*   It is worth mentioning a few points about this method, which is in widespread use.

*3.5.2.2.1   ISO version.*   The ISO version recommends, but does not insist upon, the application of centrifugation followed by Wickbold sublation (see Section 3.5.3) to samples before applying the colorimetric procedure. There is some controversy on this point. The clean-up gives greater specificity but whereas good recovery of added LAS is obtained with clean river waters after the standard two sublations, recovery from sewage samples was poor. Because of this, it is argued that it is unnecessary to modify and complicate a technique which has been universally accepted in order to gain an apparent increase in its specificity (by sublating) for synthetic anionic surfactants at the expense of possibly seriously underestimating the total MBAS in the sample.

*3.5.2.2.2   Methylene blue.*   The purest form of the dye should be used, but even then it appears that other, interfering dyes, are present which have to be removed by alkaline washing just before use. This can be done for each test sample or in bulk for all samples tested on any one occasion.

*3.5.2.2.3   Cotton-wool filter.*   Instead of using pre-washed cotton wool for filtering the final extract, pre-washed Whatman silicone-treated phase-separation paper (1PS) has been found more efficient and convenient (R. Brett, pers. comm.).

*3.5.2.2.4   Chloroform.*   This solvent is a suspected carcinogen; any substitute would have to be tested for its ability to extract the complexes of the various surfactants, including the lower chain length homologues of LABS. At the present time there seems little incentive to establish another solvent, partly because of the introduction of HPLC methods.

*3.5.2.2.5  Automated methods.*   Automated versions of the methylene blue method have been described (e.g. HMSO, 1982). Whereas for homogeneous aqueous samples there is a greater than 90% probability that automated and manual procedures would produce the same value for a given sample, those containing suspended solids give significantly lower MBAS values with the automated procedure. Solids settle out in the sampling cups and the mixing in the coils is not as vigorous as in the manual procedure.

### 3.5.2.3  Chromatographic methods

*3.5.2.3.1  Thin layer chromatography.*   Paper chromatography and TLC are suitable for separating the various types of anionic surfactants, after a preconcentration step, and also for resolution of homologous hydrophobes, but not the various hydrophilic classes. Individual chain-length homologues in alkyl sulphates, alkylaryl sulphonates, $\alpha$-olefine sulphonates and alkane sulphonates were separated on TLC either directly or in some cases after conversion to the sulphonate methyl esters. But these methods can be considered only semiquantitative at best; both GC and HPLC offer much better means of quantification.

*3.5.2.3.2  Gas chromatography.*   GC cannot be applied directly to sulphonates because of their low volatility, so they first have to be converted to volatile derivatives. Originally LABS and other aryl sulphonates were desulphonated by heating with phosphoric acid under standard conditions and the resulting hydrocarbons (alkyl benzenes) were subjected to chromatography, usually with capillary columns. Other derivatives used have been sulphonyl chlorides, methyl esters, fluoride derivatives and thiols. The conditions of the GC separation can be modified, e.g. temperature programming, to achieve various degrees of separation, but complete resolution of all isomers does not appear to have been achieved even with the high efficiency capillary columns. The use of a mass spectrometer as detector (GC–MS) provides a further powerful means of identification.

Secondary alkane sulphonates were separated by GC after conversion to corresponding olefines by alkaline fusion; the peaks were made more distinct by first converting the olefines to alkanes by hydrogenation.

*3.5.2.3.3  High performance liquid chromatography.*   HPLC techniques have been developed such that they are superseding GC methods and are so versatile that the possibilities range from obtaining a single peak representing all the LABS through to single peaks representing nearly all homologues and isomers present. No derivatives have to be formed, but pre-concentration steps are necessary.

A typical HPLC procedure for LABS described by the Standing Committee of Analysts (HMSO, 1993a) is summarized here; it is based essentially on the methods of Matthijs and de Henau (1987) and Holt et al. (1989). A known volume of the unfiltered liquid sample, containing 1–100 $\mu$g of LABS, is evaporated to dryness and the dried residue is carefully extracted three times with methanol. The concentrated extract is anion-exchanged to separate anionics from other surface-active material. LABS is eluted from the resin with methanolic HCl and then concentrated onto a C18 reversed-phase column from a neutralized aqueous solution. Any inorganic salts and highly polar organics are removed before LABS is selectively eluted with methanol. Solid samples, e.g. activated sludge, are first extracted with methanol — reflux or Soxhlet — before clean-up.

The LABS residues are then analysed on a reversed-phase HPLC system, e.g. $\mu$Bondapak C18, with a mobile phase of sodium perchlorate in aqueous methanol and fluorescence detector at 232–290 nm. If a fluorescence system is not available, a linear gradient elution programme may be used with UV detection at 230 nm; the mobile phase comprises varying proportions of aqueous sodium perchlorate and aqueous acetonitrile solutions of sodium perchlorate.

Separations of not only LABS, but also alkyl sulphates, alkyl ethoxysulphates and secondary alkane sulphonates, have been achieved, for example, with a styrene–divinyl benzene copolymer column and a mobile phase of methanol and aqueous ammonium acetate (Irgolic and Hobill, 1987). The detector was a sulphur-specific inductively coupled argon-plasma emission spectrometer. The detection limit was equivalent to about 0.3 ng S/l.

Alkyl sulphates and alkyl ethoxy sulphates were separated by reversed-phase HPLC on columns of Hypersil SAS or ODS with water–acetone mixtures as the mobile phases. Detection was by ion-pairing with acridinium chloride and monitoring the complex (in chloroform) fluorimetrically; the limit of detection was about 1 $\mu$g/l in the original aqueous sample under the conditions of the method.

### 3.5.3   Non-ionic surfactants

As with anionic surfactants, commercial non-ionics of the alkoxylate group (the main type) are not single molecular entities, but are made up of a large number of homologues and ethoxamers. As a result the analytical determination of non-ionics at low concentrations is more complex than that for anionic and cationic surfactants on three counts:

– the polyethoxylates include a wider variety of chemical substances,
– interference by material in environmental samples is more troublesome,
– biodegradation intermediates are not as sharply distinguished from the intact surfactants in their analytical response.

Because of this it is necessary to apply preliminary purification steps certainly for environmental samples, and it is also advisable for biodegradation test samples, in order to concentrate non-ionics in the sample.

There are four methods which are commonly applied (HMSO, 1982):

(a) reaction with the Dragendorff bismuth reagent, e.g. Wickbold (1972), Waters and Longman (1977);

(b) reaction with cobaltothiocyanate, e.g. Crabb and Persinger (1968), Boyer *et al.* (1977);

(c) thin layer chromatography, e.g. Patterson *et al.* (1966);

(d) chemical fission followed by gas chromatography (GC), e.g. Kaduji and Stead (1976), Tobin *et al.* (1976).

The first two methods are colorimetric or titrimetric and depend on the formation of the complexes between the polyethoxylate and the metal Bi or Co; in other similar methods complexes with phosphotungstate or phosphomolydate are formed. These methods determine compounds containing 6–30 ethoxy or propoxy groups.

Since it is recognized that materials (synthetic and natural) other than non-ionic surfactants can positively respond in methods (a) and (b), the entities detected are commonly referred to as bismuth (or Dragendorff) active substances (BiAS) or as cobaltothiocyanate active substances (CTAS).

*3.5.3.1 Wickbold method.* In the Wickbold method a simple solvent (ethyl acetate) sublation step is used to concentrate selectively intact non-ionic surfactants. These concentrates will be free from non-surface active materials such as polyoxyethylene glycols (PEG) and non-ionic surfactant degradation intermediates. For samples from simulation biodegradation tests and for environmental samples (e.g. in community field tests) the next stage is to apply cation-exchange steps to remove other interfering surface-active materials (anionics and cationics).

Briefly, the sample, containing 0.25–1 mg of non-ionic surfactant, is diluted to a standard volume with the addition of sodium bicarbonate and sodium chloride and placed in a Wickbold gas-stripping sublation tube (Figure 3.7). Ethyl acetate is carefully added to the top of the aqueous sample and air or nitrogen gas, saturated with the ester, is passed through the sublation tube as fine bubbles at about 0.6 ml/min for 5 min. The organic layer is discharged via the tap provided into a separating funnel and any water collected is returned to the Wickbold tube. The sublation is repeated and the two 100 ml aliquots are combined in a beaker and evaporated on a steam bath. The two sublation steps extract over 90% of the non-ionic surfactants. The residue is dissolved in 5 ml methanol and, if necessary, cationics are removed at this stage by passage through a microporous cation-exchange resin. Water (40 ml) is added to the methanolic solution together with bromocresol purple, as indicator. The precipitating agent (a freshly prepared mixture of a solution of basic bismuth nitrate plus potassium iodide in acetic acid with a solution of barium chloride) is added at this stage. After stirring for 10 min and standing for at least 5 min the precipitate is filtered through a sintered glass, or Gooch, crucible.

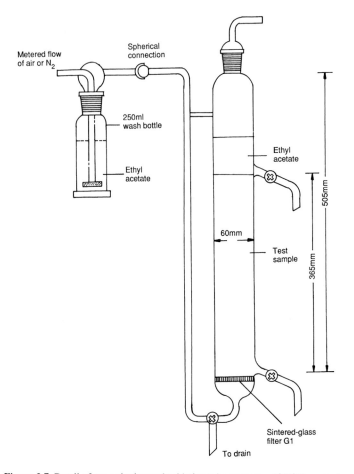

**Figure 3.7** Detail of gas stripping and sublation tube apparatus (OECD version).

The precipitate is thoroughly washed with aliquots of glacial acetic acid and the bismuth content is determined by dissolving in hot ammonium tartrate solution and titrating against carbate solution (pyrrolidine dithiocarbamate). The 'carbate' solution is standardized using copper (II) sulphate solution.

This method, which has found widespread use, has been adopted by France, Germany and the UK and also by the OECD (OECD, 1976), EC (EC, 1982a) and ISO (ISO, 1984b).

Alternatively, the bismuth may be determined in the ammonium tartrate solution by adding an EDTA solution and measuring the absorbance in the UV region (263.5 nm). A further method is to determine bismuth in a nitric acid solution of the precipitate by atomic absorption spectrometry.

Although the Wickbold and the following method are more susceptible than the other two methods to positive interference, this is unlikely to be important in biodegradability assessments since the appropriate control samples are available to correct for any interfering material present.

*3.5.3.2   Cobaltothiocyanate method.*   This method (HMSO, 1982) is favoured by the American Soap and Detergent Association (Boyer *et al.*, 1977) since it is thought to be simpler and faster to operate than the Wickbold titrimetric procedure in both biodegradation and environmental studies. Two to four sublation steps are applied to concentrate the sample (containing 0.2–2 mg) and the ethyl acetate is evaporated. (If ionic surfactants are to be removed, the residue is taken up in methanol and poured down a cationic–anionic resin column, and eluted with methanol in the normal way. The methanol eluate is evaporated to dryness and taken up in dichloromethane.) The residue is taken up in a known volume of dichloromethane and the extracted non-ionic surfactant is reacted with the cobaltothiocyanate reagent (30 g cobalt (II) nitrate.$6H_2O$ plus 200 g ammonium thiocyanate/l by shaking in a separating funnel for a constant time (1 min). After initial separation, the organic layer is further separated by centrifuging. The dichloromethane layer is then transferred by Pasteur pipette to a photometric cell for measurement of absorbance at 620 nm in a spectrophotometer. Normally a calibration graph is prepared using four or five solutions of varying concentrations of a standard non-ionic surfactant (nonyl phenol ethoxylate, Synperonic NP8 or Marlophen 810) and the graph is used to calculate the concentrations of non-ionic surfactant in the tested samples.

In other versions of the method, the surfactant was extracted into ether, the complex was formed in water and then extracted into chloroform (Crabb and Persinger, 1964, 1968). With the latter solvent the response continued to increase at least up to $RE_{19}$ and PEGs were extracted, while with benzene the response diminished above $RE_9$. Petts and Sliney (1981) reported a promising automated version of the method but further investigation is necessary.

*3.5.3.3   Chromatographic methods.*   The thin layer chromatographic (TLC) method (Patterson *et al.*, 1966) and the ether fission–GC method (3.5.3.4) are very much less convenient for biodegradation tests but they offer distinct advantages over the simpler methods. The chromatographic methods allow the determination of much lower concentrations of non-ionic surfactants (down to about 5 $\mu g/l$) and the TLC method gives additional information about the nature of the surfactant and its degradation intermediates. The fission–GC procedures can estimate a wider range of materials, including the short (EO or PO) chain non-ionic surfactants.

The TLC method detects APE and alcohol ethoxylates with chain lengths of at least four ethylene oxide groups. APEs with more than about 20 EO groups form a streak rather than a spot on the TLC plate. In some cases they can be conveniently estimated against standards of a suitable molecular weight polyethylene glycol.

Briefly, a solution of $MgSO_4$ is added to a 250 ml sample of concentration 0.05–2.5 mg non-ionic surfactant/l, to produce a 30% (w/v) solution of the salt. The mixture is extracted with chloroform taking care not to form stable emulsions. Some of the co-extracted impurities are removed by washing the chloroform extract with saline acid and alkaline solutions and the cleaned-up extract is evaporated to dryness. The residue is taken up in a small volume of chloroform and is spotted (2–20 $\mu$l) alongside standards (1–5 ng) on TLC plates coated with silica gel. The plates are run in suitable water–acetic acid–ethyl acetate solvent systems. After the plates have been dried and heated, non-ionic surfactants are finally characterized and visually quantified on the plates by formation of a coloured derivative with a modified Dragendorff spray reagent by matching sample and standard spots.

*3.5.3.4  Chemical fission and GC.*   Samples containing the equivalent of 50–500 ng of nonylphenol-8-ethoxylate are extracted into ethyl acetate by sublation, as for the Wickbold test. To each ethyl acetate extract a known weight of polytetrahydrofuran is added as an internal fissionable standard; it generates 1,4-dibromobutane. After evaporation of the ethyl acetate, the concentrates are transferred to thick-walled fission tubes using dichloromethane to effect the transfer. This solvent in turn is evaporated by heating and applying gentle currents of air. To the dried residue 45% (w/v) hydrogen bromide in glacial acetic acid reagent (0.4 ml) is added and the tube is sealed in a hot flame. The sealed tubes are placed in a safety metal container and heated in an oven at 150°C for 3 h, removed and cooled. The tubes are cut open and the contents are transferred to screw-capped glass bottles using water (4 ml) to complete the transfer. Carbon disulfide (0.5 ml) is added, the bottles are shaken vigorously for 1 min and the layers allowed to separate. Aliquots of the lower, organic layers are subjected to GC.

The ratios of the peak areas of 1,2-dibromoethane (DBE), from the EO groups, to 1,4-dibromobutane (DBB) for each individual chromatogram are calculated. A calibration graph is obtained by plotting DBE/DBB peak area ratios against weight of standard non-ionic surfactant. The data for this graph are obtained by applying the procedure to four amounts (100, 200, 300 and 400 ng) of the surfactant. The weights of surfactant in the tested samples are read off the calibration graph.

*3.5.3.5  High performance liquid chromatography.*   Rapid developments have been seen in recent years for detecting alkyl phenol ethoxylates (APE) in low concentration using GC–MS, fast atomic bombardment and other ionization techniques. However, the simplest and most suitable technique for the quantitative analysis of APEs is high performance liquid chromatography (HPLC), the sensitivity of the procedure being dependent on the method of detection. Preliminary concentration and separation by sublation, ion-exchange and, perhaps, alumina chromatography have first to be carried out.

Normal phase chromatography on amino silica columns with gradient elution coupled with UV or UV fluorescence detection has separated APEs with a detection limit of 1 $\mu$g/l (Ahel and Giger, 1985). The ethoxymers were separated but the hydrophobes were not resolved; the latter was done by using reversed phase chromatography on octyl silica, in which case the ethoxymers were eluted as a single peak. The most sensitive method to date is that of Holt *et al.* (1986) and has been published by the UK Standing Committee of Analysts (HMSO, 1993b). Gradient elution with a solvent system containing methyl *t*-butyl ether, acetonitrile and methanol by normal phase HPLC was used with a UV-fluorescence detector giving a minimum level of detection of 0.2 ng.

There is a difficulty in the direct detection of alcohol ethoxylates (AE) so that after concentrating and clean-up, derivatives have to be prepared and it is usual to use an aromatic reagent. By reacting with phenyl isocyanate (Allen and Linder, 1981), AEs were separated according to their hydrophobic chain length on a C18 reverse phase column, using a water–methanol gradient system (Schmitt *et al.*, 1990). Separation of the ethoxymers was achieved by normal phase HPLC with a hexane/dichloroethane to acetonitrile/isopropanol gradient and a detection limit of 0.1 ppm. Alternative derivatives are 3,5-dinitro benzoate (Nozawa and Ohnuma, 1980) and those produced by 1-anthroyl nitrite (Kudoh *et al.*, 1984); both these methods have been applied to environmental samples.

### 3.5.4 Cationic surfactants

The most common methods for determining cationic surfactants are colorimetric in which the surfactant is reacted with an anionic dye. The ion-association complex so formed is extracted into an organic solvent and the absorbance of the solution is measured spectrophotometrically. If the sample to be analysed contains no other type of surfactant, for example biodegradability test liquors, the test can be made directly on the sample. However, for sewage and other environmental samples direct application is not possible since the cationic surfactant will react preferentially with the anionics present rather than with the dye. In such cases, the residues from pre-concentrated samples are extracted with methanol and the extract is treated on an ion-exchange resin to remove interfering anionic components. The isolated cationic is then reacted with the dye, as before.

It has been established that only long alkyl chain quaternary ammonium compounds react to form stable chloroform-extractable complexes with the intensely coloured disulphine blue (DSB). Alkyl trimethyl ammonium and dialkyl dimethyl ammonium compounds with more than ten C and more than eight C, respectively, in the alkyl groups have been found to form completely extractable complexes. On this basis the commercially important cationic surfactants should be responsive in the DSB reaction. As with anionics and non-ionics, DSB reacts with compounds other than surfactants, e.g. the OECD synthetic sewage used in simulation biodegradability tests gives a value of 0.1–0.3 mg/l, so that concentrations are reported in terms of disulphine blue active substances (DSBAS). But in die-away tests for biodegradability the values measured should give a good indication of the cationic content.

*3.5.4.1　Outline of SCA method (HMSO, 1982).*　Briefly, if the aqueous sample contains no anionic surfactants or other anionic interferants, up to 15 ml sample containing up to 100 $\mu$g cationic surfactant are placed in a centrifuge tube and the volume is made up with water to 15 ml, if necessary. Acetate buffer (2.5 ml of 2 M) plus 1 ml DSB reagent (1.3 mM in 10% ethanol in water) are added to the tube, followed by 10 ml chloroform. The contents of the tube are stirred vigorously with an electric stirrer to mix the two phases completely. The mixture is centrifuged for about 30 s to separate the layers. A rubber teat is fixed to the short arm of an extraction device (Figure 3.8) (Taylor and Fryer, 1969) and the longer arm is introduced through the aqueous layer into the $CHCl_3$ layer. A slight positive pressure is maintained on the teat to produce a slow stream of air bubbles. The bung is screwed into the mouth of the centrifuge tube and the teat is then removed. The $CHCl_3$ extract is transferred directly to a glass 10 mm optical cell by applying pressure with the syringe via the needle. (The cell should have been previously conditioned with a DSB-cationic chloroform stan-

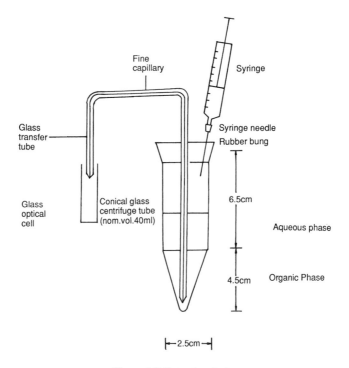

**Figure 3.8** Extraction device.

dard solution.) The cell is washed once with the extract and the absorbance of the extract is then immediately measured at 628 nm against a chloroform reference.

A calibration graph is prepared by applying the procedure to five or six solutions having different concentrations between 0 and 10 mg/l of the cationic surfactant under test, or a 'standard' such as cetyltrimethylammonium chloride.

*3.5.4.1.1 Samples containing anionic surfactants.* Samples containing anionic surfactants, such as effluents from simulated activated sludge systems using domestic sewage, are first evaporated to dryness; up to 200 ml of the unfiltered sample containing 25–1000 $\mu$g of cationic surfactant are used. Similarly, a volume of a standard cationic surfactant solution equivalent to that taken for the test determination is also evaporated to dryness. Anionic surfactant (e.g. Marlon 350; about 1 g/l solution) is added to samples low in anionics to bring the concentration to about 10 mg/l before evaporation. This aids the recovery of the cationic as its neutral association complex from the dry residue.

The sample residues are broken up and extracted three times with 20 ml boiling methanol each time. Each hot extract is filtered into a beaker through a small cotton-wool pad plugged into the neck of a funnel. The pad is washed each time with 10 ml methanol. The combined extracts are evaporated to dryness and then dissolved in 10 ml methanol. The whole extract is transferred to the top of a prepared anionic-exchange resin column and the extract is allowed to pass through the column slowly, <1 ml/min, collecting the eluate. A further two portions of 10 ml methanol are used to ensure quantitative transfer of the sample to the column. The sample is washed through the column with a further 100 ml methanol at 2–3 ml/min and the whole column eluate is evaporated to dryness.

At this stage the residue is taken up in hot water-saturated $CHCl_3$ and the above procedure for forming the coloured complex with DSB and extraction is applied.

Alternatively a separating funnel technique may be used. The final residue is dissolved in 20 ml warm water-saturated $CHCl_3$ and is added to a separating funnel containing 2.5 ml acetate buffer, 15 ml water and 1 ml of the DSB reagent. After shaking on a flask shaker at a known, fixed rate for 5 min, the phases are allowed to separate and the $CHCl_3$ phase is run into a 50 ml graduated flask. The transfer and extraction steps are repeated with a further 20 ml portion of warm, water-saturated $CHCl_3$ and the combined extracts are made up to 50 ml with $CHCl_3$. The absorbance of the solution is measured in a 10 mm cell at 628 nm; further dilution with $CHCl_3$ may be required to bring them into the range of a 0–250 $\mu$g calibration curve.

A drawback of the method is that putative intermediate metabolites of QACs — long chain primary, secondary or tertiary alkylamines — which may be present also react with DSB. This interference may be avoided by removing

QACs from the pre-formed QAC–DSB complexes by cation-exchange chromatography and determining the QACs by TLC. The combined DSB–TLC procedure permits the determination of both DBS-active substances and specific QACs.

*3.5.4.2  Chromatographic methods.*   TLC has also been used directly, using samples previously concentrated by sublation followed by separation from anionic surfactants by ion-exchange, using the Dragendorff reagent for detection (Michelsen, 1978).

Of the HPLC methods, which are less time-consuming and tedious, that of Wee and Kennedy (1982) seems superior in sensitivity, specificity and ease of performance. Clean-up was by extraction with dichloromethane after adding excess LAS to the acidified sample, and a polar column (with cyano and amino groups) was used for separation with a mobile phase of dichloromethane–methanol and a conductivity detector. The detector limit was about 2 $\mu$g/l. In a comparison of methods, the DSBAS procedure was less specific and sensitive than either DSBAS–TLC or HPLC procedures, while HPLC was more sensitive and convenient than the other two methods.

# References

Abbott, D.C. (1962) Colorimetric determination of anionic surfactants in water. *Analyst* 7, 286–293.

Ahel, M. and Giger, W. (1985) Determination of surfactants of the APE type by HPLC. *Anal. Chem.* 57, 2584–2590.

Allen, M.C. and Linder, D.E. (1981) Ethylene oxide oligomer distribution in nonionic surfactants via HPLC. *J. Am. Oil Chem. Soc.* 58, 950–957.

Association Francaise de Normalisation (1977) Method for the evaluation in aqueous medium of the biodegradability of total organic products, T90/302.

Birch, R.R. and Fletcher, R.J. (1991) The application of dissolved inorganic carbon measurements to the study of aerobic biodegradability. *Chemosphere* 23, 507–524.

Boyer, S.L., Guin, K.F., Kelley, R.M., Mausner, L.M., Robinson, H.F., Schmitt, T.M., Stahi, C.R. and Setzkorn, E.A. (1977) Analytical method for nonionic surfactants in laboratory biodegradation and environmental studies. *Environ. Sci. Technol.* 11, 1167–1171.

Cabridenc, R. (1983) Testing of the biodegradability of cationic surfactants. Report III/125/83-EN for the European Commission.

Crabb, N.T. and Persinger, H.E. (1964) The determination of poly EO nonionic surfactants in water at the parts per million level. *J. Am. Oil Chem. Soc.* 41, 752–755.

Crabb, N.T. and Persinger, H.E. (1968) A determination of the apparent molar adsorption coefficients of the cobalt thiocyanate complexes of nonylphenyl EO adducts. *J. Am. Oil Chem. Soc.* 45, 611–615.

EC Directive 67/548 (1967) Dangerous substances — classification, packaging and labelling. *Off. J. E.C.* 196, 1–67.

EC Directive 73/404 (1973a). On the approximation of the laws of the Member States relating to detergents. *Off. J. E.C.* L347, 51–52.

EC Directive 73/405 (1973b) Relating to methods of testing the biodegradability of anionic surfactants. *Off. J. E.C.*, L347, 53–54.

EC Directive 79/831 (1979) Amending 67/548 EC for the sixth time — Dangerous substances — classification, packaging and labelling. *Off. J. E.C.* L259, 10–28.

EC Directive 82/242 (1982a). Relating to methods of testing the biodegradability of non-ionic surfactants. 31 March 1982.

EC Directive 82/243 (1982b). Relating to methods of testing the biodegradability of anionic surfactants. 31 March 1982.

Fischer, W.K., Gerike, P. and Holtmann, W. (1975) Biodegradability determinations via unspecific analyses (COD, DOC) in coupled units of the OECD Confirmatory test — I The test. *Water Res.* **9**, 1131–1135.

Gerike, P. (1982) Biodegradation and bioelimination of cationic surfactants. *Tenside* **19**, 162–164.

Gerike, P. and Fischer, W.K. (1979) A correlation study of biodegradability determinations with various chemicals in various tests. *Ecotoxicol. Environ. Safety* **3**, 157–173.

Gerike, P. and Fischer, W.K. (1981) A correlation study of biodegradability determinations with various chemicals in various tests. II Additional results and conclusions. *Ecotoxicol. Environ. Safety* **5**, 45–55.

Gerike, P., Fischer, W.K. and Jasiak, W. (1978) Surfactant quaternary ammonium salts in aerobic sewage digestion. *Water Res.* **12**, 1117–1122.

HMSO (Her Majesty's Stationery Office) (1978) Statutory Instruments, No. 564. The Detergents (Composition) Regulations 1978. London.

HMSO (1982). Analysis of surfactants in waters, wastewaters and sludges 1980, Methods for the examination of waters and associated materials, London. ISBN 0 11 751605 8. 68 pp.

HMSO (1983a). Assessment of biodegradability 1981, Methods for the examination of waters and associated materials, London. ISBN 0 11 751661 9. 104 pp.

HMSO (1983b). Methods for assessing the treatability of chemicals and industrial waste waters and their toxicity to sewage treatment processes 1982, Methods for the examination of waters and associated materials, London. ISBN 0 11 751959 6. 89 pp.

HMSO (1984). Statutory Instruments No. 1369. The Detergents (Composition) (Amendment) Regulations 1984. London.

HMSO (1993a). Determination of linear alkyl sulphonates (LAS) in sewage, sewage effluent, river water, sewage sludge, river sediment and sludge amended soil samples by High Performance Liquid Chromatography (HPLC) with fluorescence detection, 1993. Methods for the examination of waters and associated materials, London.

HMSO (1993b). Determination of alkylphenol ethoxylates in sewage, sewage effluent and river water by High Performance Liquid Chromatography (HPLC), 1993, Methods for the examination of waters and associated materials, London.

Holt, M.S., McKennel, E.H., Perry, J. and Watkinson, R.J. (1986). Determination of alkyphenol ethoxylates in environmental samples by HPLC coupled to fluorescence detection. *J. Chromatogr.* **362**, 419–424.

Holt, M.S., Matthijs, E. and Waters, J. (1989) The concentration and fate of linear alkylbenzene in sludge-amended soils. *Water Res.* **23 (6)**, 749–759.

Husmann, W., Malz, F. and Jendreyko H. (1963) Removal of detergents from wastewaters and streams. *Forschungsber. Landes Nordrhein-Westfalen* 95–105, **1153**.

Irgolic, K.J. and Hobill, J.E. (1987) Simultaneously inductively coupled argon-plasma atomic emission spectrometer as a sulphur-specific detector for the HPLC analysis of sulphur-containing surfactants. *Spectrochim. Acta* **42B**, 269–273.

ISO 7875/1 (1984a) (International Organisation for Standardization) Determination of surfactants — Part 1 — determination of anionic surfactants by the methylene blue spectrometric method.

ISO 7875/2 (1984b) Determination of surfactants — Part 2 — determination of nonionic surfactants using the Dragendorff reagent.

ISO 7827 (1985) Water quality — evaluation in an aqueous medium of the 'ultimate' aerobic biodegradability of organic compounds — method by analysis of dissolved organic carbon.

ISO 8192 (1986) Water quality — test for inhibition of oxygen consumption by activated sludge.

ISO 9408 (1991a) Water quality — evaluation in an aqueous medium of the 'ultimate' aerobic biodegradability of organic compounds — Method by determining the oxygen demand in a closed respirometer.

ISO 9439 (1991b) Water quality — evaluation in an aqueous medium of the 'ultimate' aerobic biodegradability of organic compounds — method by analysis of carbon dioxide.

ISO 9887 (1991c) Water quality — evaluation of the aerobic biodegradability of organic compounds — semi-continuous activated sludge method (SCAS).

ISO 9888 (1991d) Water quality — evaluation of the aerobic biodegradability of organic compounds — static test (Zahn–Wellens).

ISO 10634 (1993) Water quality — guidance on the preparation of poorly soluble organic compounds before the evaluation in an aqueous medium of their 'ultimate' biodegradability.

Kaduji, I.I. and Stead, J.B. (1976) Determination of polyoxyethylene in small amounts of nonionic detergents by hydrogen bromide fission followed by GC. *Analyst* **101**, 728–731.

Kudoh, M., Ozawa, H., Fudamo, S. and Tsuji, K. (1984) Determination of trace amounts of alcohol and alkyphenol ethoxylates by high-performance liquid chromatography with fluorimetric determination. *J. Chromatogr.* **287**, 337–344.

Llenado, R.A. and Jamieson, R.A. (1981) Surfactants. *Anal. Chem.* **53 (5)**, 174R–182R.

Llenado, R.A. and Neubecker, T.A. (1983) Surfactants. *Anal. Chem.* **55 (5)**, 93R–102R.

Longwell, J. and Maniece, W.D. (1955) Determination of anionic detergents in sewage, sewage effluents and river waters. *Analyst* **80**, 167–171.

Matthijs, E. and de Henau, H. (1987) Determination of LAS. *Tenside Deterg.* **24 (4)**, 193–199.

Michelsen, E.R. (1978) Quantitative determination of quaternary ammonium bases in water and waste water by TLC. *Tenside* **15**, 169–175.

Nozawa, A. and Ohnuma, T. (1980) Improved high-performance liquid chromatographic analysis of ethylene oxide condensates by their esterification with 3,5-dinitrobenzoyl chloride. *J. Chromatogr.* **187**, 261–263.

OECD (1971) (Organization for Economic Cooperation and Development). Pollution by detergents. Determination of the biodegradability of anionic synthetic surface active agents. Paris.

OECD (1976) Environment Directorate. Proposed method for the determination of the biodegradability of surfactants used in synthetic detergents. Paris.

OECD (1981) Chemicals Group. Guidelines for testing chemicals, Paris.

OECD (1984) Section 2. Effects on biotic systems. 209. Activated sludge respiration inhibition test. Paris.

OECD (1993a) Chemicals Group. Revised guidelines for tests for ready biodegradability. Paris.

OECD (1993b) Chemicals Group. Revised Zahn–Wellens/EMPA test for Inherent biodegradability. Paris.

Official Journal of the European Communities (1984) Tests for Degradation. L251, pp. 160–211. Office for Official Publications. L2985, Luxembourg.

Official Journal of the European Communities (1988) Methods for the determination of ecotoxicity. L133, 31 pp. 88–127. Office for Official Publications, L-2985, Luxembourg.

Official Journal of the European Communities (1992) Determination of ready biodegradation. L383A, pp. 187–225.

Painter, H.A. and Bealing, D.J. (1989). Experience and data from the OECD activated sludge simulation test. in Laboratory tests for simulation of water treatment processes. Report No. 18. Commission of the European Communities. pp. 113–138.

Painter, H.A. and King, E.F. (1978a). The effect of phosphate and temperature on growth of activated sludge and on biodegradation of surfactants. *Water Res.* **12**, 909–915.

Painter, H.A. and King, E.F. (1978b). WRc porous pot method for assessing biodegradability. Technical Report TR70 : Water Research Centre, Stevenage, UK.

Patterson, S.J., Hunt, E.C. and Tucker, K.B.E. (1966) The determination of commonly used nonionic detergents in sewage effluents by a TLC method. *J. Proc. Inst. Sewage Purif.* **65 (2)**, 3–11.

Petts, K.W. and Sliney, I. (1981) An automated method for the determination of nonionic surfactants in water. *Water Res.* **15**, 129–132.

Schmitt, T.M., Allen, M.C., Brain, D.K., Guin, K.F., Llenado, R.A. and Osborn, Q.W. (1990). HPLC determination of ethoxylated alcohol surfactants in waste water. *J. Am. Oil Chem. Soc.* **67**, 103–109.

S.D.A. (1965) (Soap and Detergent Association). A procedure and standard for the determination of the biodegradability of alkyl benzene sulfonate and linear alkylate sulfonate. *J. Am. Oil Chem. Soc.* **42**, 986–993.

Struijs, J. and Stoltenkamp, J. (1990) Headspace determination of evolved carbon dioxide in a biodegradability screening test. *Ecotoxicol. Environ. Safety*, **19**, 204–211.

Sturm, R.N. (1973) Biodegradation of non-ionic surfactants: screening test for predicting rate and ultimate biodegradation. *J. Am. Oil Chem. Soc.* **50**, 159–167.

Swisher, R.D. (1987) *Surfactant Biodegradation*. Marcel Dekker, Inc., New York and Basel.

Taylor, C.G. and Fryer, B. (1969) The determination of anionic detergents with iron (II) chelates; application to sewage and sewage effluents. *Analyst* **94**, 1106–1116.

Tobin, R.S., Onuska, F.I., Brownlee, B.G., Anthony, D.H.J. and Comba, M.E. (1976) The application of an ether cleavage technique to a study of the biodegradation of a linear alcohol ethoxylate nonionic surfactant. *Water Res.* **10**, 529–535.

Waters, J. and Longman, G.F. (1977) A colorimetric modification of the Wickbold procedure for the determination of nonionic surfactants in biodegradation test liquors. *Anal. Chim. Acta* **93**, 341–344.

Wee, V.T. and Kennedy, J.M. (1982) Determination of trace levels of quaternary ammonium compounds in river water by HPLC with conductometric detection. *Anal. Chem.* **54**, 1631–1633.

Wickbold, R. (1972) On the determination of nonionic surfactants in river and wastewaters. *Tenside* **9**, 173–177.

# 4  Testing strategy and legal requirements

## H.A. PAINTER

### 4.1  Selection of tests: strategy of testing

#### 4.1.1  Primary biodegradability

The selection of a suitable test for assessing primary biodegradability is straight-forward if the surfactant is sufficiently soluble. The OECD Screening test (Ch. 3: 3.3.1) is first applied and if the percentage removal is more than 80% no

**Table 4.1** Tests for primary biodegradability — obligatory for all surfactants in washing, rinsing and cleaning products. (Test philosophy and practical details are given in EC Directives 73/404, 73/405 (1973b), 82/242, 82/243 (1982a,b), and in OECD (1976))

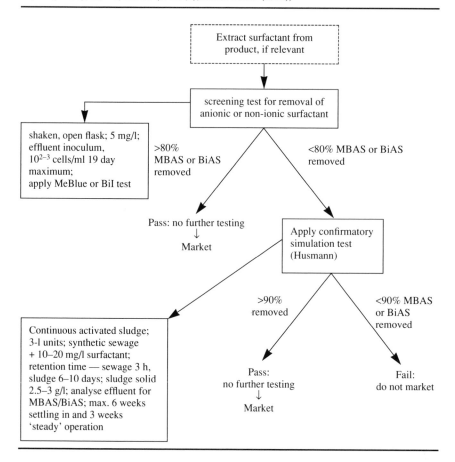

further testing is required (Table 4.1). Surfactants which are degraded by less than 80% are then subjected to the Confirmatory test (activated sludge) procedure (Ch. 3: 3.3.2) and the value of percentage removal obtained is taken as that which would occur in the environment. (USA practice is different, see Section 4.3.3) In the EC a surfactant not reaching >90% in the Confirmatory test may not be marketed (Table 4.1).

When the replacement of the poorly degradable anionic surfactants (TPBS) was under consideration, a few likely candidates were tested even further in field and community trials because of the economic importance. Simulation tests on a larger scale than the confirmatory test have been made using detergent-free natural sewage prepared from human excreta, kitchen waste, starch, etc. to which the surfactant under investigation was added. Another way of tackling the problem was to distribute free packets of detergent containing the new surfactant to small communities, e.g. caravan sites, institutions, connected to a sewage treatment plant treating waste water only from that community.

Finally, the new product can be sold in a large area, after withdrawing the old product, as in the 'Luton, UK, experiment' and monitoring the performance of the (Luton) sewage treatment works before, during and after the area had been saturated with the new product.

## 4.1.2   Ultimate biodegradability

The choice of which screening test for ready ultimate biodegradability to use for a given chemical depends largely on its physical properties — solubility, volatility, adsorptivity (Table 4.2). It should be noted that the OECD Guidelines

**Table 4.2** Application of test methods

| OECD Test No. | Analytical method | Suitability for compounds which are: | | |
|---|---|---|---|---|
| | | poorly soluble | volatile | adsorbing |
| DOC Die-Away (301A) | Dissolved organic carbon | − | − | ± |
| $CO_2$ Evolution (301B) | Respirometry: $CO_2$ evolution | + | − | + |
| MITI (I) (301C) | Respirometry: oxygen consumption | + | ± | + |
| Closed Bottle (301D) | Respirometry: dissolved oxygen | ± | + | + |
| Modified OECD Screening (301E) | Dissolved organic carbon | − | − | ± |
| Manometric Respirometry (301F) | Oxygen consumption | + | ± | + |

(OECD, 1981), and the EC methods listed in Table 4.3, indicate that the test methods are not immutable and can be modified, or even replaced, provided that the changes can be justified scientifically.

In order to avoid using too high an initial concentration of test substance, it is advisable to carry out a preliminary test for inhibition.

The first step in the OECD scheme (Figure 4.1) is to apply a suitable screening test for ready biodegradability (Level I). If the test substance is readily biodegradable (by a method in Table 4.3A) no further work is necessary, but if the test substance does not degrade by the required amount a test, at Level II (from Table 4.3B, or any other test), for inherent biodegradability is applied. At this level, a higher bacterial density is used, other organic compounds may be added and a much longer duration, up to 6 months, can be given thus greatly enhancing the chances that biodegradation will occur.

**Table 4.3** Tests for ultimate biodegradability — obligatory for chemicals not on EINECS list (1987) (EC Directives 67/758, 79/831, proposed 90/C33/03. See also EC (1986))

| Test No. | | | Test name | Conditions | Cell density (approx. no./ml) | Analysis |
|---|---|---|---|---|---|---|
| OECD | EC | ISO | | | | |
| *A. Screening for ready biodegradability* | | | | | | |
| 301A | $C_4$–A[a] | 7827 | DOC Die-Away | Shaken, open flask | $10^6$ (S/E) | DOC |
| 301B | $C_4$–C[a] | 9439 | Modified Sturm | Enclosed vessel: aerated with $CO_2$-free air | $10^6$ (S/E) | $CO_2$; DOC |
| 301C | $C_4$–F[a] | – | MITI I | Enclosed flask; air/medium | $10^6$ (special S) | $O_2$ uptake; DOC |
| 301D | $C_4$–E[a] | N.160 | Closed Bottle | Sealed bottle; completely filled; air-saturated | $10^6$ (E) | $O_2$ uptake |
| 301E | $C_4$–B[a] | 7827 | Modified OECD Screening | shaken, open flask | $10^{2-3}$ (E) | DOC |
| 301F | $C_4$–D[a] | 9408 | Manometric Respirometry | Enclosed flask; air/medium | $10^6$ (S/E) | $O_2$ uptake |
| *B. Inherent biodegradability* | | | | | | |
| 302A | X | 9887 | Semi-continuous activated sludge (SCAS) | Aerated sewage + sludge; open cylinders; daily fill and draw; 3–6 months | $10^8$ (S) | DOC/COD |
| 302B | X | 9888 | Zahn and Wellens | Aerated sludge; open cylinders 28 days, repeated | $3 \times 10^7$ (S) | DOC/COD |
| Any other test which can be substantiated | | | | | | |
| *C. Simulation tests* | | | | | | |
| 303A | X | N.140 | Activated sludge | Aerobic sewage treatment continuous operation | $10^8$ (S) | DOC/COD |
| Any other simulation which can be substantiated | | | | | | |

X, Off. J.E.C. (1988); S, activated sludge; E, effluent.
[a] Published in revised form with the seventh Amendment of 67/758(EC Directive 92/32).
Note: In nearly all tests the test substance may be determined by specific analysis.

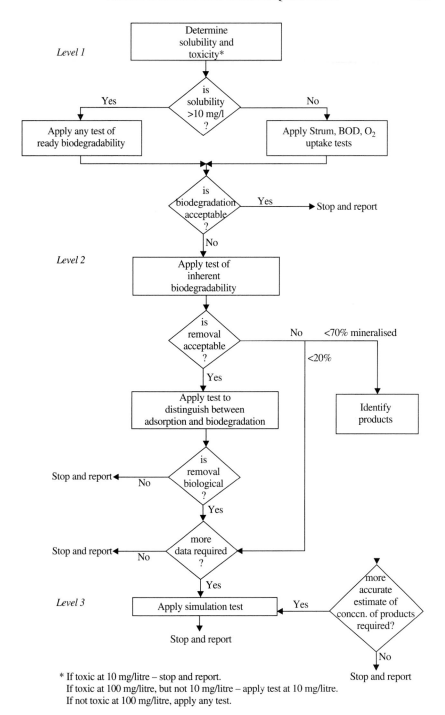

**Figure 4.1** OECD strategy for biodegradability screening.

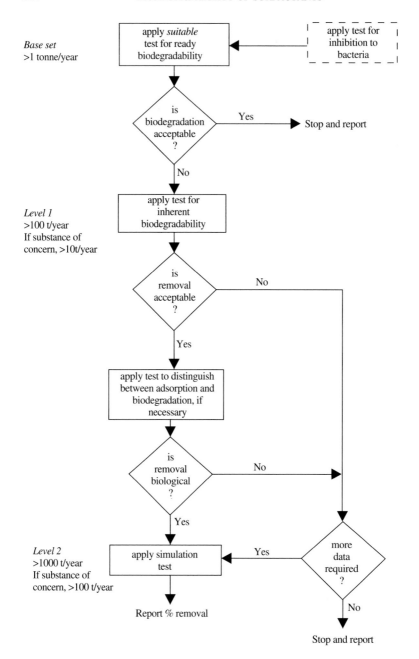

**Figure 4.2** EC strategy for testing.

A result of <20% degradation means that the test substance is probably not inherently biodegradable and further work is not called for. If >70% degradation occurs, the test substance is inherently biodegradable and if it is required to know if it is also treatable, a simulation test (Level III from Table 4.3C) is applied. This has to be ascertained because not all inherently biodegradable substances are removed in waste water treatment. The result of this test will give a firm prediction of the degree of removal on the large scale. Chemicals which are degraded by between 20 and 70% (DOC) (plateau values) in the Level II test or in the MITI test (Level I) are presumably only partially degradable and the identity of the intermediate metabolites (or recalcitrant metabolites) should be established for ecotoxicological purposes.

Whereas in the OECD scheme the progression from one level of testing to another is at the discretion of the experimenter, in the EC scheme (Figure 4.2) (EC, 1986) the main criterion triggering tests is the annual tonnage of the substance produced or its total tonnage. Testing normally begins at the Base Set ($\equiv$ level I, OECD) when the annual production reaches 1 tonne and a test for inherent biodegradability (level I $\equiv$ OECD level II) is normally required when production reaches 100 tonnes, or 500 tonnes total. Further testing (level II $\equiv$ OECD level III), such as a simulation test, is required when 1000 tonnes/year or 5000 tonnes total are reached. Chemicals of concern, e.g. those very toxic to fish, may have to be tested at a higher level even though the production indicated for that level has not been reached.

### 4.1.3   Other tests (not in Table 4.3)

As indicated earlier, any scientifically justifiable test method may be used and it is worth mentioning two, both of which measure $CO_2$ production.

#### 4.1.3.1   $^{14}C$-labelled substances.

Many investigations using $^{14}C$-labelled compounds, including surfactants (specific molecular species), have been made on a number of key surfactants, especially LABS, but there is no standard procedure. These tests have the advantage of being able to follow the removal of the test substance and ultimate biodegradation by measuring the $^{14}CO_2$ evolved at environmentally realistic concentrations. From the data, kinetic rate constants (and half-lives) of the first-order decay of the test substance and first-order production of $CO_2$ may be calculated (e.g. Larson, 1980, 1983) using the equation:

$$P_t = P_0 \left[1 - e^{k(t-c)}\right]$$

where $P_t$ is the %ThCO$_2$ at time $t$, $P_0$ is the maximum value (asymptote) of %ThCO$_2$, $k_1$ is the first-order constant, and $c$ is the lag time.

Larson (1983) showed that, where there was no inhibition at the higher concentrations of test substance, the $k_1$ values recorded in the normal Sturm test (301B) agreed with those obtained at the $\mu g/l$ level using $^{14}C$ labelling.

*4.1.3.2  Head-space $CO_2$ method.*  The Sturm method has some drawbacks; it is not applicable to volatile substances, $CO_2$ is not flushed out of solution as fast as it is produced (Weytjens *et al.*, 1994) and the apparatus is cumbersome. (In the EPA, USA, method smaller, shaken vessels are used and $CO_2$ is fairly rapidly flushed out; R.J. Larson, pers. comm.). Because of these drawbacks other methods have been sought for a ready biodegradability test. Struijs and Stoltenkamp (1990) have improved an earlier 'enclosed' method (Ennis *et al.*, 1978; Boatman *et al.*, 1986) by containing the OECD medium, activated sludge inoculum and the test substance in serum bottles of nominal volume 120 ml. The bottles are sealed with butyl rubber septa and the ratio of gas to liquid is 1:2. Incubation is carried out at 20°C in the dark on a shaker and bottles are sacrificed at frequent intervals for $CO_2$ analysis (and, if required, DOC). The bottle contents are acidified to pH less than 3 by injection of phosphoric acid through the septa and are shaken for 1 h to attain equilibrium. Samples (1 ml) of head-space gas are taken and directly injected in a standard way into the reaction chamber of a carbon analyser to obtain the $CO_2$ content of the gas.

In their version, Birch and Fletcher (1991) used larger vessels but did not acidify the test mixture preferring to analyse the head-space gas and also the liquid phase immediately after breaking the bottle seals. The sum of the gaseous $CO_2$-C and liquid DIC gives the total $CO_2$-C produced. In both methods, corrections were made for the $CO_2$ produced by inoculum controls.

Thus, these methods overcome the drawbacks of the Sturm procedure and deal successfully with poorly soluble and volatile chemicals. The Birch and Fletcher (1991) method is in draft SCA (UK) form and it is to be hoped that this and the Struijs–Stoltenkamp (1990) method will be discussed by the various organisations (OECD, EC, ISO) with a view to getting it accepted at least as an alternative to the Sturm method.

## 4.2  Validation and interpretation of results

### 4.2.1  Validity of results

Before a value for the percentage removal of a test substance can be accepted a number of conditions have to be fulfilled.

First, a test is considered valid if the difference between extremes of replicate values of the percentage removal at the plateau, at the end of the test or at the end of the 10-day window, as appropriate, is less than 20%. Next, the standard or reference chemical must be removed by certain values. In primary degradation tests, the soft surfactant standard must be removed by more than

90% in 7–10 days, while the hard standard should not be removed by more than 35%. In tests for ultimate biodegradability the minimum values to be attained by standard chemicals are 70% DOC, 60% ThOD and $ThCO_2$ by day 14, except in the MITI test in which at least 40% ThOD should be attained in 7 days and 65% in 14 days. Also, in tests involving DOC removal evidence that adsorption is not the significant mechanism must be obtained.

In the $CO_2$ evolution test the initial concentration of inorganic carbon should be less than 5% of the total carbon present. The $CO_2$ produced by the inoculum alone should be low; it is normally 20–40 mg/l of medium in 28 days but if it exceeds 70 mg/l the calculations, data and technique should be examined. Failing this a pre-conditioned inoculum or one from a different source should be used. Similarly, in manometric respirometric methods the inoculum blank should be 20–30 mg $O_2$/l in 28 days; if it exceeds 60 mg $O_2$/l the data and technique should be examined and, if necessary, a pre-conditioned inoculum or a new source should be used. In the Closed Bottle method the inoculum blank should be less than 1.5 mg $O_2$/l after 28 days; values greater than this call for an examination of techniques, leading perhaps to pre-conditioned or different inocula. The concentration of dissolved oxygen in the bottles should at no time be less than 0.5 mg/l and the method analysis must be capable of adequately determining dissolved oxygen at this level.

### 4.2.2   Pass levels

Rarely does the percentage removal of a surfactant or other test substance reach 100% within the period of a screening, static test, because the conditions are stringent mainly due to the very low bacterial density. Experience has shown that a surfactant which degrades by 80% (MBAS or BiAS) or more in the Screening test are subsequently removed by 90% or more in the Confirmatory (Husmann) test. Thus, it was decided that 80% should be the pass level in the Screening test for surfactants to be environmentally acceptable.

For ultimate biodegradability the OECD Chemical Group (OECD, 1980, 1981) agreed that, to be classified as readily biodegradable, chemicals should be removed in screening tests by more than 80% specific analysis, 70% DOC and 60% ThOD and $ThCO_2$. The values for respirometric methods are lower than for DOC since some of the carbon in the test substance in used to form new bacterial cells so that less is available to take up oxygen to form $CO_2$. If ThOD cannot be calculated, COD is used but %COD is less satisfactory than %ThOD since some chemicals are not fully oxidized in the COD test. If the formula or composition of the test substance is not available, the C content can be determined and used to calculate the $ThCO_2$. The values of 60% are contested since there are many examples, including two 'ring tests' (CEC, 1982, 1985), in which %DOC and %ThOD were determined simultaneously by the Manometric Respirometric method and although >90%DOC was removed, often only 45–50% ThOD was recorded.

These values have to be attained in a 10-day period ('window') within the 28 days of the test, the period starting when removal reaches 10% DOC, ThOD or ThCO$_2$. Chemicals reaching the pass level after 28 days, even though within a 10-day window, are not deemed to be readily biodegradable. A 14-day window is permitted in the Closed Bottle test if it is considered that the number of bottles required to evaluate the 10-day window causes the test to become unwieldy. The 10-day window concept does not apply to the MITI method.

### 4.2.3   Interpretation

*4.2.3.1   Primary degradation.*   Surfactants which are consistently degraded in the Screening test by more than 80%, with little difference between replicates are classed by the relevant Directive as environmentally acceptable (Table 4.1) and can be expected from past experience to degrade easily and completely in the aquatic environment. No further testing is required. Those which do not meet the pass level or show disparate values for replicates are not rejected as being insufficiently biodegradable, but should be subjected to the OECD Confirmatory test (OECD, 1976; Table 4.1). The value for percentage removal (over 21 days steady operation) in the latter test is taken as the percentage degraded on the large scale; the removal must be >90% MBAS or BiAS for the surfactant to be acceptable.

*4.2.3.2   Ultimate biodegradation.*   For chemicals which reach one or other of the respective pass levels in screening tests, other evidence is necessary to establish that the removal mechanism was biological. This evidence can be given by lack of removal in sterile, abiotic controls. The shape of the removal curve from inoculated test solutions can give additional evidence of the mechanism. A sigmoidal-shaped curve is good evidence of biodegradation, whereas a linear removal over all or most of the 28 days indicates abiotic removal, which can be confirmed from the sterile controls. If the only mechanism is adsorption, normally a rapid process, the curve would not be sigmoidal, but would soon reach a plateau. Adsorption would also be indicated in the separate control set up for the purpose.

Chemicals degraded by more than the set limits are deemed to be readily degradable in the environment. It has been found that, of such chemicals which have been subjected to an activated sludge simulation test, all have been removed by values equal to or greater than corresponding values in the screening tests. Of course, some chemicals not degraded well in a screening test have been more than adequately removed in a simulation test.

For chemicals which are not removed in the screening tests (Table 4.3A), one of the controls set up would have established whether inhibition was the cause; if so, another test should be performed at a lower initial concentration of the test substance. If the chemical does not degrade, it is subjected to a test for inherent biodegradability (in Table 4.3B, or any other test not listed) and if it still does not degrade (<20%) it is deemed to be non-biodegradable. Chemicals which degrade well (>70%) are then subjected to an activated sludge simulation test (Table 4.3C) since not all inherently degradable chemicals degrade at kinetic rates sufficiently high to be removed in sewage treatment plants.

Chemicals removed in tests for inherent biodegradability by only intermediate values (20–70% DOC) would appear to form recalcitrant metabolites, that is non-degradable substances. It is important that these intermediates be studied for their effects on the aquatic environment. This limit of 70% as a value above which recalcitrant metabolites are not formed was challenged by Gerike *et al.* (1984), who suggested that the limit is dependent on the number of C atoms in the molecule of the parent compound. On the unlikely basis of a one-C atom molecule being recalcitrant, and assuming only one metabolic pathway, they suggested, for example, that the limit for LABS (18 C atoms) should be 17/18 or 94% DOC. For a more realistic three-C atom recalcitrant molecule, the limit for LAS would be 15/18 or 83%. Gerike *et al.* (1984) devised a method based on the Husmann Simulation test operated in the coupled mode over a long period to determine whether residual organic C derived from a test substance accumulated; the identity of the organic matter was not identified. The test substance was added daily at 5 mg C/l to one vessel and nutrients were added in concentrated form to both vessels. Effluents were filtered and recycled daily back to their respective aeration vessels. Transinoculation was effected by transferring centrifuged sludge, the supernatant liquid being returned to the original vessel. Sampling was carried out sparingly, great care was taken with all operations and corrections were made in the calculations of percentage removal for all material transfer and removal. In the case of LABS, they concluded that no recalcitrant metabolites were formed.

## 4.3   Legal requirements

### 4.3.1   EEC

In dealing with surfactants in detergents, the 12 countries of the EEC are bound by a number of Directives (EC 73/404, 1973a, EC 73/405, 1973b, EC 82/242, 1982a, EC 82/243, 1982b). The objects of the Directives are to bring together (or approximate) the laws of the member states relating to methods of testing the biodegradability of surfactants and to ensure that member states put on the market only those surfactants which are adequately biodegradable. Additionally member states must not prohibit or restrict the placing on the market and use of detergents which comply with the provisions of the Directives.

These Directives lay down that all anionic and non-ionic surfactants present in washing, rinsing and cleaning products (i.e. detergents) must be tested, for primary biodegradability only, by methods described in the Directives. Although the original Directive (EEC 73/404) referred to four groups — anionic, non-ionic, cationic, ampholytic — the Commission has not yet dealt with the latter two types. The surfactants must degrade by at least 80% MBAS or BiAS in the screening tests or at least 90% in the Confirmatory test (Table 4.1) before they may be put on the market. There are some derogations (repeals, exemptions) for certain low-tonnage, low-foaming non-ionic surfactants when used in powders for dishwashing and industrial bottle-washing machines. They are of the blocked co-polymer ethoxy–propoxy type, degrade by very much less than 80%, and may be marketed until adequately degradable substitutes are found. However, these derogations were subject to examination every 3 years or so with the option to renew the derogation or to let it lapse. The decisions taken are reflected in Statutory Regulations (see below) in the UK (Directives were not necessarily issued). In 1984 the derogations were extended until 31 March 1986 (HMSO, 1984) and again, in 1986, until 31 December 1989 (HMSO, 1986). Since 1989 the derogations have not been extended, although it is not clear whether suitably biodegradable substitutes have been put on the market. If they have not, the existing products have presumably been sold illegally for the past 5 years or so.

These Directives do not require pre-notification of results to the EEC or to a member state. Member states are obliged to put into force necessary laws, regulations and administrative provisions to comply with the Directives and must send the Commission the texts of the main provisions of national law.

In the UK, the law takes the form of Statutory Regulations (HMSO, 1978a,b, 1984, 1986) which make it an offence to market a 'contravening detergent'. The Regulations instruct samples of detergents to be collected from shops, supermarkets and warehouses and sent to the Laboratory of the Government Chemist, where they are divided into three (equal?) parts. One sub-sample is sent to the manufacturer or importer, one is kept in reserve in case of a dispute and the third is analysed. The sub-sample is treated by the prescribed process (EC, 1982a,b) to extract the surfactant(s) which is then assessed for primary biodegradability by the OECD screening method (Table 4.1; Chapter 3: 3.3.1). The results are reported to the Department of the Environment. If more than 80% is degraded, no further action is needed, but if less than 80% is degraded the Confirmatory test (Table 4.1; Chapter 3: 3.3.2) is performed by the Water Research Centre, Medmenham, UK. If the removal value is still less than 80%, the manufacturer's and reserve samples should be tested by nominated laboratories. Values less than 80% confirm that an offence has been committed under the Regulations. On the international scale, in a dispute between two member states a reserve sample is tested by an accredited laboratory in a third member state and the results are considered by a committee established for the purpose.

However, in the UK Regulations, the original sample of detergent is a minimum of 900 g and the initial test is carried out on the extract from 300 g detergent. The Confirmatory test requires at least 10 g surfactant (24 l/day at 20 mg/l for 21 days) and at most 30 g (for 63 days). Thus, assuming a 100%

efficient extraction, the percentage surfactant in the detergent must be at least $(10/300 \times 100)$ 3.3% for a 3 week test or $(30/300 \times 100)$ 10% for a full 9 week test. This would have been adequate for detergent powders when the Directives were promulgated but over the years mixed surfactants and lower contents of surfactant have been used. Initially, 10–20% anionic surfactant was present in detergents but this has been reduced to 5–10% and the contents of non-ionic surfactant can be as low as 1–5%, and abrasive powders for cleaning surfaces usually contain only 1–2% anionic surfactants.

This situation should be rectified, although to date (1994) no sample has failed to be degraded by more than 80% (in UK) in screening tests; in fact, nearly all were degraded by more than 90%. No samples have had to be examined by the Confirmatory method.

### 4.3.2 Other countries

In other countries the results of testing surfactants are notified before the products are marketed. For example, in Switzerland, surfactants are tested and notified before the initial advertisement is published announcing their sale. The detergents must not contain non-biodegradable organic substances and the Swiss Government has the power to take samples from retail outlets for testing, as in the UK scheme.

### 4.3.3 USA

In the USA, there is no legislation on surfactants, although there is for the content of phosphate in detergents, but there is a voluntary arrangement between Government and industry on how surfactants should be tested before being marketed. The tests for primary biodegradation are those described by the Soap and Detergent Association (SDA, 1965), and are different from those in the EEC Directives. The inoculum used in the tests may be grown in the presence of 'soft' surfactants, not permitted in the EEC Directives, and in the screening test itself two sub-cultures are made giving further opportunity for adaptation. The initial concentration of surfactant is 30 mg/l, on the border of inhibitory concentrations, compared with only 5 mg/l in the Directive. A surfactant which is degraded by more than 90% in the screening test may be put on the market without further testing, but one which reaches between 80 and 90% must be subjected to the semi-continuous activated sludge (SCAS) test (not the continuous Husmann method as in the Directive). If the removal then reaches over 90% in the SCAS test, the surfactant may be put on the market. Surfactants which do not reach 80% in the screening test are rejected without further testing, unlike the strategy in the EEC Directives. Their rejection is presumably made because of the less stringent conditions in the SDA Die-Away test than in the OECD test and also because of the more lenient conditions in the SCAS method — 23 h retention compared with 3 h in the Confirmatory method.

In addition, the EPA has published a number of tests relevant to the possible effects of chemicals in the environment, as well as tests for ultimate biodegradability. The EPA does not dictate which tests are appropriate in a particular case; this is left to the manufacturer. When the test results are available — before marketing — a discussion ensues between the EPA and the manufacturer as to the acceptability of the surfactant (or other chemical) to the environment. This would include a consideration of the probable environmental concentration, the 'no effect concentration' and the existence and nature of stable intermediates and their properties. Further testing may be requested by the EPA after the discussion, so it may be prudent to negotiate with the EPA before any testing is done, although even then the need for further testing cannot be ruled out.

### 4.3.4    EEC Dangerous Chemicals Directives 67/548, 79/831, 90/C 33/03

In Section 4.3.1, it was assumed that the surfactants considered were on the European Inventory of Existing Chemical Substances (EINECS) list (EINECS, 1987), that is, they were registered before 18 September 1981. This implies that they were to be considered as 'existing' chemicals as defined in EEC Directive 67/548 (1987). 'New' surfactants — not on the list — have to be tested not only for primary biodegradability, as described above, but also for ultimate biodegradability as described in the 'new' or dangerous chemicals Directives 67/548, 79/831 and 92/69 (Off. J.E.C., 1984, 1988, 1992) (see Section 3.4, Chapter 3). The other important difference in testing requirements is that 'new' chemicals have to be tested for ecotoxicity (on fish, invertebrates, algae, etc.) whereas 'existing' chemicals do not. Which tests have to be applied depends largely on the amount of surfactant marketed each year (Fig. 4.2), as well as on the route to the environment and other properties of the substance.

However, it can be said that whether tests are compulsory or not, a wide variety of tests for biodegradability and ecotoxicity, including studies at very low concentrations using $^{14}$C-labelled surfactants, have been applied to most existing surfactants by the manufacturers, because of their concern that their products should not harm the environment.

### 4.3.5    The future

The present schemes for surfactants have been in place for about 20 years and in general have worked well. There are, however, pressures to change and extend the schemes. For example, The Netherlands government has proposed that surfactants should be assessed for ultimate biodegradability and that the development of methods which couple biodegradation and toxicity testing should be encouraged (Van Leeuwen, 1990). The effluent from a biodegradation test unit could, for example, be used as the influent for an aquatic toxicity test. Also, the behaviour of surfactants in some of the revised OECD test methods for ready biodegradability has been investigated (Struijs, 1990).

In February 1993, DG III of the European Commission initiated discussions on the possible extension of the Detergent Directives, not only to cover the primary biodegradability of cationic and amphoteric surfactants, but also the ultimate biodegradability of all four types of surfactant. It is to be expected that, before the amendments to the Directives are promulgated, the test methods proposed, especially the simulation test, will have been thoroughly substantiated by the use of calibration exercises and exchanges of experience.

The discussions on the proposals to extend the Detergent Directives to cover ultimate biodegradation (or mineralization) have been delayed due to administrative difficulties.

More importantly the OECD and EEC are currently extending the more comprehensive programme of testing, including toxicological and ecotoxicological testing, under the 'new' chemicals Directive (EC 67/548, EC 92/32) to the testing of 'existing' chemicals, including surfactants. (EC Regulation on Existing Substances, 793/93, 1993). The main object of the collection of data on chemicals is to carry out risk evaluations of the chemicals on various parts of the environment. A risk evaluation is made by calculating how people, animals and plants may be exposed to the substance under study and estimating the likely effects as a result of the exposure.

Producers or importers of chemicals are to send to the Commission what data on the products they already have or can glean from the literature. Chemicals produced or imported at more than 1000 tonnes per year are to be dealt with first and have to be reported within 1 (or in some cases 2) years after the promulgation of Regulation 793/93. If the substance is listed in Annex I of the Regulation, the information must have been sent in by 3 June 1994; if not, by 3 June 1995. Chemicals produced at over 10 tonnes need to have small packages of information reported within 5 years, i.e. by 3 June 1998.

The original data on the chemicals will be screened and as a result some chemicals will be assigned to a priority list. Full details of the Priority Setting Scheme are due to be published by the EC some time in 1994. If insufficient information of a reliable quality is available to enable a risk evaluation to be made, more information will have to be obtained under conditions of Good Laboratory Practice by or on behalf of the producer/importer. Non-priority chemicals will be dealt with in a less urgent manner.

Although under Regulation 793/93 the exposure element of the evaluation requires a test for ultimate biodegradability, it will presumably still be necessary to determine primary biodegradation of the substance, including surfactants, since data on toxicity to fish, etc. are expressed in concentrations of the parent molecule. This and other facets are covered in the Guidance Document on how to carry out risk evaluations, published by the Commission in May, 1994.

A useful necessary guide to the problems of if, when and how data should be reported has been published by HMSO, entitled 'How to Report Data on Existing Chemical Substances' (1994). The sub-title reveals more: 'a guidance to companies on how to comply with EC Regulation 793/93 on existing substances'. It includes the text of the Regulation as well as two Annexes detailing chemicals on the EINECS (1987) list which are produced or imported at more than 1000 tonnes per year and those which are exempt from some of the Articles of the Regulation.

## References

Birch, R.R. and Fletcher, R.J. (1991) The application of dissolved inorganic carbon measurements to the study of aerobic biodegradability. *Chemosphere* **23**, 507–524.

Boatman, R., Cunningham, S.L. and Ziegler, D.A. (1986) A method for measuring the bio-degradation of organic chemicals. *Environ. Toxicol. Chem.* **5**, 233–243.

CEC (1982) Commission of the European Communities. First Ring Test: Respirometry method, Degradation Accumulation sub-group, Ring Test Programme, June 1982.

CEC (1985) Commission of the European Communities. Second Ring Test: Respirometry method, Degradation Accumulation sub-group, Ring Test Programme, 1983–1984.

EC Directive 67/548 (1967) Dangerous substances — classification, packaging and labelling. *Off. J.E.C.* **196**, 1–67.

EC Directive 73/404 (1973a) On the approximation of the laws of the Member States relating to detergents. *Off. J.E.C.* **L347**, 51–52.

EC Directive 73/405 (1973b) Relating to methods of testing the biodegradability of anionic surfactants. *Off. J.E.C.* **L347**, 53–54.

EC Directive 79/831 (1979) Amending 67/548 EC for the sixth time — dangerous substances — classification, packaging and labelling. *Off. J.E.C.* **L259**, 10–28.

EC Directive 82/242 (1982a) Relating to methods of testing the biodegradability of non-ionic surfactants. 31 March 1982.

EC Directive 82/243 (1982b) Relating to the methods of testing the biodegradability of anionic surfactants. 31 March 1982.

EC (1986). Guiding principles for a strategy for biodegradability testing. DG X1/841/86, 9 pp. (being revised).

EC Directive 92/32. (1992) 7th Amendment to Directive 67/548. *Off. J.E.C.* **L154** Vol. 35, 30 April 1992.

EC Directive 92/69. (1992) Annex. Compendium of test methods. *Off. J.E.C.* **L383A** Vol. 35, 29 Dec 1992.

EC Regulation on Existing Substances 793/93. (1993) *Off. J.E.C.* **L84**/1, 5 April 1993.

EC (1994) Technical Guidance Document: Risk Assessment of New Notified Substances, May, 1994.

EINECS (1987) European Inventory of Existing Commercial Chemical Substances, Advance Edition. Office for Official Publications of the European Communities, Luxembourg, 1987.

Ennis, D.M., Kramer, A., Jameson, C.W., Mazzochi, P.H. and Bailey, W.J. (1978) Structural factors influencing the biodegradation of imides. *Appl. Environ. Microbiol.* **35**, 51–53.

Gerike, P., Holtmann, W. and Jasiak, W. (1984) A test for detecting recalcitrant metabolites. *Chemosphere* **13**, 121–141.

HMSO (Her Majesty's Stationery Office) (1978a) Statutory Instruments, No. 564. The Detergents (Composition) Regulations, 1978, London.

HMSO (1978b) Statutory Instruments, The Detergents (Composition) (Amendment) Regulations, 1978, London.

HMSO (1984) Statutory Instruments No. 1369. The Detergents (Composition) (Amendment) Regulations, 1984, London.

HMSO Statutory Instruments No. 560. The Detergents (Composition) (Amendment) Regulations 1986, London.

HMSO (1994) How to report data on existing chemical substances. London.

Larson, R.J. (1980) Role of biodegradation kinetics in predicting environmental fate, In *Biotransformation and Fate of Chemicals in the Aquatic Environment*, eds. A.W. Maki, K.J. Dickson and J. Cairns, Am. Soc. Microbiol., Washington, DC, pp. 67–86.

Larson, R.J. (1983) Comparison of biodegradation rates in laboratory screening studies with rates in natural waters. *Residue Rev.* **85**, 159–171.

OECD (1976) (Organization for Economic Cooperation and Development). Determination of the biodegradability of surfactants used in synthetic detergents. Paris, 1976.

OECD (1980) Chemicals Group. Report on chemicals testing programme, degradation, accumulation — Final report. Berlin and Tokyo. Paris, 1980.

OECD (1981) Chemicals group. Guidelines for testing chemicals. Paris, 1981.

Official Journal of the European Communities (1984) Tests for Degradation. Office for Official Publications. L2985, Luxembourg. L **251**, pp. 160–211.

Official Journal of the European Communities (1988) Methods for the determination of ecotoxicology. Office for Official Publications L2985, Luxembourg. L **133**, pp. 88–127.

SDA (1965) (Soap and Detergent Association). A procedure and standard for the determination of the biodegradability of alkyl benzene sulfonate and linear alkylate sulfonate. *J. Am. Oil Chem. Soc.* **42**, 986–993.

Struijs, J. (1990) A proposal to screen surfactants for ultimate biodegradability. Paper presented at meeting on Ultimate Biodegradability, Expert Group on Detergents and the Environment, Zeist, The Netherlands, 9 November, pp. 61–96.

Struijs, J. and Stoltenkamp, J. (1990) Headspace determination of evolved carbon dioxide in a biodegradability screening test. *Ecotoxicol. Environ. Safety,* **19**, 204–211.

Van Leeuwen, C.J. (1990) Intended objectives, scope and environmental benefits of ultimate biodegradability of detergent surfactants. Paper presented at meeting on Ultimate Biodegradability, Expert Group on Detergents and the Environment, Zeist, The Netherlands, 9 November, pp. 37–46.

Weytjens, D., Van Ginneken, I. and Painter, H.A. (1994) The recovery of carbon dioxide in the Sturm text for ready biodegradability, *Chemosphere,* **28**, 801–812.

# 5 Biodegradability of anionic surfactants

## J. STEBER and H. BERGER

### 5.1 General characteristics of anionic surfactants

Like all surfactants, anionic surfactants ('anionics') are surface-active compounds being composed of a hydrophobic alkyl chain which is connected to one or two hydrophilic groups. In aqueous solutions, anionics dissociate into a negatively charged anion and the (positively charged) counterion. The anion is the carrier of the surfactant properties and its negative charge is due to the presence of a sulfonate, sulfate, phosphate or carboxyl group in the molecule. Generally, most of the commercially available surfactants represent mixtures of homologues having a different alkyl chain length in the range $C_8$–$C_{18}$. The selection of specific homologue mixtures and, in several cases additionally, the existence of a multitude of isomers influence the specific properties and performance characteristics of the respective surfactant.

Soaps, sulfonates and sulfates represent the technically most important anionic surfactant groups used in the household and personal care sector and applied for industrial purposes, e.g. for industrial cleaners, as auxiliaries for process steps in the textile and other industries, in pharmaceutical products, in plant protecting agents, etc. Apart from soap, by volume, the most important groups of anionic surfactants are linear alkylbenzene sulfonates, fatty alcohol ether sulfates, alkane sulfonates and fatty alcohol sulfates; in addition, $\alpha$-olefine sulfonates, sulfosuccinates, $\alpha$-sulfo fatty acid esters and phosphate esters play a significant role. The anionics represent the dominating class of surfactants on the market having an estimated consumption of almost 1.1 million tons in Western Europe in 1990 corresponding to about 55% of the total (excluding soap) (Richtler and Knaut, 1991).

### 5.2 Application of anionic surfactants and their environmental relevance

By volume, the most prominent application field for anionic surfactants is the use in consumer products, e.g. detergents, dishwashing and cleaning agents and personal care products. After being used, all these products will be passed into the sewerage system and then, eventually, into the environment. Hence, the use of those products is closely connected with the issue of their environmental compatibility. This is also the reason why the history of surfactant applications in the detergent field is also a history of searching an optimum of technical performance and, at the same time, of reduction of their environmental impact.

### 5.2.1   Synthetic anionic surfactants and the 'detergent problem'

Until the end of the Second World War, detergents were usually formulated with soap as the main surfactant. Although synthetic anionics surfactants, e.g. alkyl sulfates, had already been used in the 1930s in detergents for 'easy care fabrics' (Felletschin *et al.*, 1981), the broad replacement of soap in detergents started in the USA in the 1940s and somewhat later in Europe and Japan. The first generation of synthetic anionic surfactants was represented by alkylbenzene sulfonates of the tetrapropylenebenzene sulfonate type (TPS). It was the most important synthetic anionic until the middle sixties, having excellent performance properties and being easily producible and, hence, cheap. In comparison with soap, synthetic anionics like TPS have a much lower sensitivity to water hardness.

A disadvantage of TPS became obvious in Germany and other European countries during the drought-like summers of 1959 and 1960 when large mountains of foam could be observed on rivers and sewage treatment plants, particularly at sites where mechanical agitation and turbulence favoured this process, e.g. at weirs and in aerators of sewage works. The reason for this unexpected and unacceptable foam problem was the deficient biodegradability of TPS. This shortcoming of TPS was essentially connected with the structural features of this surfactant, the highly branched chain structure of its alkyl group (cf. 5.3.2). Fortunately, the so-called 'detergent problem' of the late 1950s was more an aesthetic one than a real environmental disaster — no fish deaths or other significant detrimental effects towards aquatic life in rivers were observed — but far-reaching consequences were drawn resulting in governmental regulations or recommendations concerning the use of surfactants in detergents.

The experience with TPS pinpoints clearly the environmental relevance of biodegradability of all compounds which enter the aqueous environment after their use. Microbial biodegradation is the main mechanism for the irreversible removal of these substances from the aquatic and soil environment preventing a continuous increase of these materials up to dangerous concentrations. It became clear from the experiences with TBS that the biodegradation behaviour of a chemical must be known before it is broadly used, and that test methods must be available enabling a reliable prognosis of the biodegradation in the environment.

### 5.2.2   Legal requirements on biodegradability of anionic surfactants

In West Germany a law was passed by the government in 1961 enabling them to set minimum standards of biodegradability for synthetic detergents. This law was followed in 1962 by a statutory order which specified this standard and suitable laboratory test procedures for determining the biodegradability of anionic surfactants. The two test procedures specified by the detergent legislation have been called — after their adaptation by the Organization for Economic Cooperation and Development (OECD) — OECD Screening test and OECD Confirmatory test (OECD, 1971), respectively. At least in one of these tests the anionic surfactant had to be biodegradable at a minimum of 80% removal of the parent compound (primary biodegradation). This limit was not passed by TPS, but by the second generation of alkylbenzene sulfonate surfactants, the linear alkylbenzene sulfonates (LABS). The German statutory order became legally enforceable in 1964. Other European countries (Switzerland, France, Italy) followed the German example and introduced legal requirements forcing the use of LABS and prohibiting TPS in detergent formulations.

In the late 1960s most Western European countries had limited the TPS use in domestic detergents and in 1968 the Council of Europe endorsed an agreement requiring all types of surfactants to be biodegradable. Based on this agreement which came into force in 1971, the European Economic Community (EEC) released two directives on the biodegradability of surfactants in 1973. The first EEC Directive (73/404/EEC) stipulated an average biodegradability of ≥90% of all types of surfactants in detergent products (EEC, 1973a). The second Directive (73/405/EEC) described the biodegradability test procedures to be applied for anionic surfactants, and specified a minimum of 80% degradation in a single test (EEC, 1973b).

For the determination of the biodegradation degree of anionic surfactants according to the detergent legislation, a semi-specific analysis of the original surfactant is used, i.e. the methylene blue analytical procedure. The method is based on the fact that sulfonate and sulfate substituted anionics, representing the mostly used groups, form adducts with the cationic dye, methylene blue, which can be extracted in a chloroform phase and quantified colorimetrically in a photometer. So, the corresponding degradation tests follow the disappearance of the original surfactant with its primary attributes, e.g. surface-active properties such as foaming power. For that reason, the loss of methylene blue activity — expressed as MBAS (methylene blue active substance) removal — is also called 'primary degradation' of the respective anionic surfactant. The measurement of MBAS removal does not identify whether organic intermediates remain in the test solution or that the surfactant decays to its inorganic constituents. Nevertheless, the primary degradation of a surfactant represents the first stage in the sequence of degradation steps leading, in the end, to their ultimate degradation. Due to the primary degradation, the surface-active properties of the anionic are lost and, hence as a rule, also the environmentally most undesirable property, i.e. the high aquatic toxicity of the parent compound. Therefore, primary biodegradation has maintained its relevance in the context of environmental compatibility assessment of surfactants in spite of the fact that ultimate biodegradability is the essential prerequisite to exclude long-term effects in the environment.

### 5.2.3   Anionic surfactants in surface waters

The environmental consequences resulting from the application of synthetic anionic surfactants of the first generation (TPS) and from the following continuous efforts to improve their biodegradability can be impressively followed by observing the results of a systematic monitoring program conducted in West Germany since 1958 (Fischer and Winkler, 1976). The first systematic examinations of surfactant concentrations in West-German rivers started at that time after the MBAS analytical method was available. Figure 5.1 shows the MBAS loads and concentrations in the river Rhine at Düsseldorf–Himmelgeist sampling site in the course of three decades (1958–1989) (Gerike *et al.*, 1989a, 1991; Steber, 1991). Several clearly distinguishable periods can be observed. The first phase between 1958 and 1964, covers the period when TPS was used in detergents. This resulted in a continuous and dramatic increase of the MBAS loads and concentrations. The average MBAS concentration in 1958 was 0.11 mg/l and 0.47 mg/l in 1964; the maximum value was determined in autumn 1964 with 0.72 mg MBAS/l. In tributaries of the Rhine river, like Neckar, Main and Ruhr, even considerably higher MBAS concentrations up to >1 mg MBAS/l were observed (Fischer and Winkler, 1976).

After implementation of the Germany statutory order to the detergent law in 1964, the replacement of TPS by LABS led to a continuous decrease of the MBAS concentrations. The annual average values ranged between 0.24 and 0.32 mg/l, corresponding to a reduction by up to 50% of the previous concentrations (Fischer, 1980). Also the MBAS loads displayed a continuous decrease by 26%

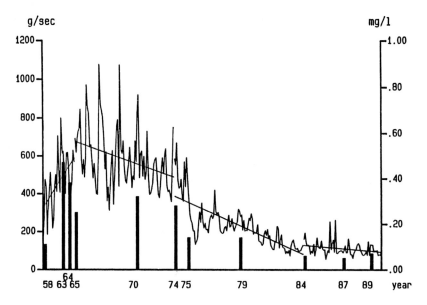

**Figure 5.1** MBAS loads and concentrations in the Rhine river at Düsseldorf.

between 1965 and 1974 (Gerike *et al.*, 1989a) in spite of the fact that the annual anionic surfactant consumption had increased by 70% during this period (Fischer, 1980).

Beginning in 1975, an additional significant drop of MBAS concentrations and loads was noted at the Düsseldorf–Himmelgeist site of the Rhine river; at the same time a considerable dumping of the MBAS load amplitudes was noticed. These striking changes were connected with the start of operation of the new sewage treatment works at Düsseldorf-Süd. The development of the MBAS loads in the following years showed a further continuous decrease of the anionic surfactants, but since the middle of the 1980s at a lower rate. Overall, the MBAS load fell by about 90% between 1965 and 1989. Based on the statistical treatment of the data from the weekly Rhine river analyses (applying a refined MBAS analytical method which reduced the below mentioned 'background' problems) it could be calculated that the real load of linear alkylbenzene sulfonate at the Düsseldorf–Himmelgeist sampling point corresponds to less than 2% of the theoretical value expected if degradation of this surfactant had not taken place (Gerike *et al.*, 1989a). For comparison, in 1963 the anionic surfactant consumption was only about one-third of the current value; nevertheless, about 85% of the TPS consumed in detergents was recovered as non-degraded residual anionic surfactant material (Fischer and Winkler, 1976).

The process of reduction of MBAS loads in surface waters is still continuing, but nowadays at considerably lower rates. This is understandable since the MBAS concentrations measured today are generally $\leq 70$ $\mu g/l$ (Gerike *et al.*, 1989b) being lower than the values reported for 1958. These pristine MBAS values were mainly attributed to biogenic materials in the rivers (Fischer and Winkler, 1976). The more the current MBAS concentrations approach these 'natural' background values, the less reliable is the information with respect to the real anionic surfactant load of river waters. Thus, for obtaining realistic data on anionic surfactant concentrations in rivers, substance-specific analytical methods have to be applied. At present, such a specific analytical method is only available for LABS (Matthijs and de Henau, 1987), which has been applied in the Rhine river monitoring program since 1989. The measurements in 1990 showed that the real LABS concentrations were in the range of about 10 $\mu g/l$ whereas MBAS concentrations remained on average at 70 $\mu g/l$; as LABS is the most important anionic surfactant by volume, this discrepancy is mainly due to the biogenic MBAS background concentration (Steber, 1991).

## 5.3   Particular structure and application features of anionic surfactants

Generally, the starting materials of surfactants are based either on vegetable and animal fats and oils (native raw materials) or on crude oil and natural gas (petrochemical raw materials). Between both raw material families, combinations and overlappings are possible. Petrochemical raw materials are the dominant feedstock for alkylbenzene sulfonates, olefine sulfonates and alkane sulfonates. On the other hand, alkyl sulfates and alkyl ether sulfates are mainly produced from native fats and oils but can, in principle, also originate from petrochemical material sources as well. As previously mentioned, sulfonates and sulfates are the most important groups of anionic surfactants. The sulfonation (C–S coupling) and the sulfation (C–O–S coupling) reactions are technologically realized by various processes. For production of sulfonates the processes of sulfoxidation, sulfochlorination, and the sulfonation are most important. Sulfation may be executed by use of oleum, chlorosulfonic acid or sulfur trioxide as reagents (Biermann *et al.*, 1987).

In the following the particular structural and application features of the most important anionic surfactant groups are discussed in more detail.

## 5.3.1 Soaps

Soaps are the alkali salts of fatty acids.

$$R\!-\!CH_2\!-\!C\!\!\underset{O^-Na^+}{\overset{O}{\lessgtr}} \qquad (R = C_{10-16})$$

Large-scale production of sodium soap is based on neutral oils (triglycerides) or fat- and oil-derived fatty acids obtained in a saponification process with NaOH. The mostly used feedstocks for soap production are tallow fat ($C_{16-18}$), palm kernel oil ($C_{8-14}$) and coconut oil ($C_{12-16}$). Toiletry soaps are prepared by addition of specific ingredients, e.g. dyestuffs, fragrances and other additives (Biermann *et al.*, 1987).

The surface-active properties of soap are dependent on the carbon chain length of the fatty acids with an optimum at 12–18 carbon atoms. Soap is the oldest surfactant being already in use for cleaning purposes for more than four millennia. Whereas soap was the only surfactant used in detergents at levels up to 40% in the product until the Second World War, it has been substituted in modern products by so-called synthetic surfactants exhibiting a higher efficiency, i.e. better performance at lower concentration levels (Coons *et al.*, 1987).

The main reason for the replacement of soaps in modern detergents and cleaning agents is their sensitivity to water hardness. Hard water containing alkaline-earth ions like Ca- and Mg-ions, forms precipitates of insoluble soaps (lime soap) and deposits on fabrics and washing machine components. These lime soaps tend to accumulate on textiles resulting in discoloring of fabrics and in developing a disagreable odour. Due to the deactivation of the detergency of soap by water hardness, a detergent formulation based on soap requires two to three times more surface-active material than a product based on synthetic surfactants (Richtler and Knaut, 1988). Nowadays, the relevance of soap in modern detergents is restricted to their role as foam regulator.

For 1990 a ratio of 56% for soap and 44% for all synthetic surfactants was estimated at a total world production of 15.9 million tons of surfactants (Richtler and Knaut, 1991). The major part of the 8.9 million tons of soap worldwide produced is used as bar soap (66%) and toilet soap (30%), i.e. as washing and cleaning agent. The main producers of soap are — with the exception of USA — countries having a lower gross national product like India, (former) USSR, China, Brazil, Egypt, etc. In Western Europe, the USA and Japan, about 1.6 million tons of soaps are consumed which is equivalent to less than 18% of the total world soap production (Richtler and Knaut, 1991). In Western Germany the soap usage is in the range of 100 000 tons per year (Schöbert *et al.*, 1988).

### 5.3.2  Alkylbenzene sulfonates

Alkylbenzene sulfonates are quantitatively the most prominent group of anionic surfactants being mainly used in heavy-duty laundry powders, light-duty liquid dish detergents, heavy-duty laundry liquids and speciality cleansers. Favourable solubility, detergency and foaming properties, the easy control of its sensitivity to water hardness as well as economics have made linear alkyl benzene sulfonates the most important anionic for detergents today. Apart from its use in household and industrial cleaning agents, LABS exhibits a broad spectrum of applications, e.g. in the textile and fibre industry (e.g. pretreatment of natural fibres, wetting in the bleach process, auxiliary in the dying process, etc.), in the plastics industry, in the manufacture of coating materials, paints, binding materials, etc. and in the leather and photographic industry (Piorr, 1987).

As previously discussed (5.2) the branched alkyl chain tetrapropylene benzene sulfonate (TPS) represented the first generation of alkylbenzene sulfonates replacing soaps in detergent formulations. TPS has the following chemical structure:

$$H_3C-\overset{\overset{\displaystyle CH_3}{|}}{C}H-CH_2-\overset{\overset{\displaystyle CH_3}{|}}{\underset{\underset{\displaystyle SO_3^-Na^+}{\bigcirc}}{C}}-CH_2-\overset{\overset{\displaystyle CH_3}{|}}{C}H-\overset{\overset{\displaystyle CH_3}{|}}{C}H-CH_3$$

TPS is manufactured by tetramerizing propylene to a highly branched α-dodecylene which is subsequently substituted by benzene and sulfonated (Biermann *et al.*, 1987). Due to the insufficient biodegradability of the alkyl chain and the resulting environmental problems TBS has been replaced in Europe, USA, Japan and many other countries by the linear alkylbenzene sulfonate (LABS). However, there are still many countries using TPS, e.g. in the Middle and Far East, Central and South America and Africa (Richtler and Knaut, 1988).

LABS has the following structure:

$$H_3C-(CH_2)_x-CH-CH_2-(CH_2)_y-CH_3$$

$$(x + y = 6 - 9)$$

$$SO_3^-Na^+$$

The starting material for the technical production of LABS are linear olefines ($C_{10-13}$) obtained from petroleum fractions which are then reacted with benzene in the presence of a catalyst. The technical sulfonation of alkylbenzene is accomplished mainly with gaseous sulfur trioxide in specific reactors. The sulfonate group is introduced into the benzene ring primarily in the *para* position (Biermann *et al.*, 1987).

LABS consists of a mixture of more than 20 different homologues and isomers covering an alkyl chain length range of $C_{10}$–$C_{13}$ (average chain length: $C_{11.6}$) and an isomer distribution (sulfophenyl substitution at all carbons except for carbon-1) which is influenced by the synthesis process and the catalyst used (Coons *et al.*, 1987). Also the amounts of technical side-products, non-linear alkyl benzene sulfonates and cyclic alkylates (dialkyl-tetralin sulfonates and dialkyl-indane sulfonates) are dependent on the production process used (Painter, 1992). Recently, a modified synthesis process for LABS has been described with a high yield (98%) of linear components (Cavalli *et al.*, 1992).

The annual world production of LABS is approximately 1.8 million tons (Berth and Jeschke, 1989). Its consumption in Western Europe, Japan and USA — an area which covers 75% of the total worldwide surfactant demand (without soap) — amounted to about 1 million tons in 1990 (Richtler and Knaut, 1991). The annual LABS usage in Western Europe was almost 500 000 tons in 1987 (Berth and Jeschke, 1989); the 60 000 tons of LABS consumed in West-German detergents represent more than 50% of the anionic surfactant usage in this application field (Malz, 1991).

### 5.3.3  Alkane sulfonates

Among alkane sulfonates only the secondary alkane sulfonates (SAS) with the following chemical structure

$$R_1-CH-R_2$$

$$(R_1 + R_2 = C_{11-17})$$

$$SO_3^-Na^+$$

have technical importance. They are manufactured from linear paraffins either by sulfochlorination or sulfoxidation (Biermann *et al.*, 1987). Technically produced SAS contain 85–87% mono-sulfonate, 7–9% di- and polysulfonates and 1% unreacted paraffins (Painter, 1992).

Alkane sulfonates are water-soluble products which are — due to the C–S bond — fully resistant to hydrolysis even under extreme pH conditions. The detergency properties of SAS, e.g. soil removal, foaming ability and sensitivity towards water hardness, are very similar to LABS. SAS are primarily used in light-duty liquid dishwashing detergents and to a lesser extent in laundry detergents (Coons *et al.*, 1987). Their application field also covers industrial cleaning and aid for process steps in the textile and fibre, photographic, metal processing and leather industries (Piorr, 1987).

The usage of SAS in Western Europe is in the range of 50 000 tons/year (Painter, 1992). The annual consumption of SAS in detergent products was 19 000 tons in Western Germany in 1989 corresponding to almost 17% of the total anionics usage (except soap) (Malz, 1991).

### 5.3.4   α-Olefine sulfonates (AOS)

α-Olefine sulfonates are petrochemicals-based anionic surfactants representing a mixture of alkene sulfonates (about 60–65%) with the formula

$$H_3C-(CH_2)_m-CH=CH-(CH_2)_n-SO_3^-Na^+ \qquad (n=0, 1, 2, \ldots; m=1, 2, 3, \ldots;$$
$$n+m=9-15)$$

and hydroxyalkane sulfonates (about 35–40%) having the structure

$$R-CH_2-CH(OH)-(CH_2)_x-SO_3^-Na^+ \qquad (R=C_{7-13}, x=1, 2, 3)$$

(Piorr, 1987; Painter, 1992).

For the technical production of AOS, α-olefines with 12–18 carbons obtained by ethylene oligomerization (Ziegler process) are used; crack olefines are not suitable as raw materials. For the sulfonation of α-olefines, the reaction with gaseous sulfur trioxide plays the major role. Since a number of side reactions are possible and influence the quality of the reaction product, several multi-step process methods have been developed which require further optimization (Biermann *et al.*, 1987).

AOS are mainly used in Japan and the USA. The most important application fields are laundry powders, light-duty liquid dishwashing detergents and shampoos. Compared with LABS and SAS, AOS with a chain length of $C_{14-18}$ are even less sensitive to water hardness. Due to their higher foam stability AOS may cause foam problems in drum-type washing machines which can only be overcome by application of special foam regulators but not very well by calcium soaps (Coons *et al.*, 1987)

The use of AOS in the European detergent field is relatively low. The consumption in Western Europe was around 5000 tons/year in 1976 (Painter, 1992). The annual consumption in Western Germany was less than 1000 tons out of a total anionics usage (except soap) of about 116 000 tons/year in 1989 (Malz, 1991).

### 5.3.5  α-Sulfo fatty acid esters (methyl ester sulfonates, FES)

The technically most important group of ester sulfonates are the α-sulfo fatty acid esters with the chemical structure

$$R{-}CH{-}C\overset{\displaystyle /\!\!O}{\underset{\displaystyle \diagdown O\,CH_3}{}} \qquad (R = C_{10-16})$$
$$\underset{SO_3^-Na^+}{|}$$

The starting material for the ester sulfonate production are fatty acid methyl esters which are sulfonated with sulfur trioxide yielding a mixture of sulfo ester acids, anhydrides and free fatty acids. In the consecutive neutralization process with aqueous sodium hydroxide the ester sulfonic acids are converted to the sodium salts, whereas disodium salts of the respective α-sulfo fatty acids (disalts) are formed from the anhydrides (Biermann et al., 1987).

The ester linkage of FES is stabilized by the adjacent sulfonate group so that ester sulfonates are very stable against hydrolysis (Stein and Baumann, 1975). Methyl ester sulfonates have good detergency, and lime soap dispersing properties. If the hydrophilic sulfonate group is located closer to the centre of the surfactant molecule, e.g. in ester sulfonates having an alcohol moiety with a longer chain length, the detergency decreases but those molecules exhibit good wetting properties (Biermann et al., 1987).

Methyl ester sulfonates are used in detergents presumably in Japan due to their good detergency properties, their hydrolytic stability and their relatively low sensitivity to water hardness. In the plastics industry the polymerization of vinyl monomers can be carried out by use of FES as emulsifier; also in the textile industry FES play a certain role (Piorr, 1987).

Compared with other sulfonate-substituted anionics like LABS and SAS, the usage figures of FES are relatively low. For instance, the annual consumption figures of this surfactant in the detergent field in Western Germany were below 1000 tons out of 116 000 tons of anionics in 1989 (Malz, 1991).

### 5.3.6  Fatty alcohol sulfates (AS)

Alkyl sulfates are predominantly manufactured from primary fatty alcohols, i.e. on the basis of native raw materials. Their general structure formula is

$$R{-}CH_2{-}O{-}SO_3^-Na^+ \qquad (R = C_{11-17})$$

The large-scale production of primary fatty alcohols is based on the high-pressure hydrogenation of fatty acid methylesters. For converting primary fatty alcohols into AS, the sulfation with sulfur trioxide/air mixtures in special reactors is the dominant industrial process which is followed by neutralization in a continuous process (Biermann *et al.*, 1987).

AS exhibiting excellent detergency and cleaning properties have been used already for a long time in the USA and Japan in heavy-duty detergents in combination with LABS. In Europe, AS are applied to the greatest extent in speciality detergents. Their sensitivity to water hardness has to be compensated for by their use in combination with sequestrants or ion exchangers in detergent formulations (Coons *et al.*, 1987). On the other hand, AS plays a major role among surfactants for cosmetic purposes due to their mildness and their dermatological and toxicological compatibility. Furthermore, their application field also covers industrial use, e.g. as cleaners, emulsifiers, dispergators, synergists in the pharmaceutical field, as auxiliaries in the textile and fibre production, as well as in the plastics, paints, leather, photographic and metal industries (Piorr, 1987).

The annual alcohol sulfate consumption in Western Europe was 56 000 tons in 1987, compared with 51 000 tons in Japan and 129 000 tons in the USA (Richtler and Knaut, 1988). The annual AS consumption in the detergent area was 10 000 tons in Western Germany in 1989 corresponding to almost 10% of the anionic surfactant consumption in this application field (Malz, 1991).

### 5.3.7   Alcohol ether sulfates (AES)

Also for AES, primary fatty alcohols with a carbon chain length of $C_{12-14}$ or $C_{12-18}$ form the most important feedstock; to some extent also synthetic alcohols such as Ziegler- and oxo-alcohols are used for AES manufacture. The respective alcohol ethoxylates are obtained by ethoxylation and subsequently sulfated with gaseous sulfur trioxide following neutralization with sodium hydroxide, ammonium or alkanolamines (Biermann *et al.*, 1987).

AES have the following chemical structure(s):

$$R—CH—CH_2—O—(CH_2—CH_2—O)_n—SO_3^-Na^+$$
$$\overset{|}{R'}$$

(fatty alcohol ether sulfates: $R' = H$, $R = C_{10-12}$; oxoalcohol ether sulfates: $R' = H$, $C_1$, $C_2$; $R + R' = C_{11-13}$; $n = 1-4$).

AES are highly water soluble and have a low sensitivity to water hardness, a high stability in the alkaline region, particularly good foaming properties and an excellent skin compatibility. They are used in Japan and USA mainly in heavy-duty detergents and in Europe especially in light-duty detergents and personal care products like foam baths and shampoos. In addition, AES play an important role as constituents of industrial cleaners and as auxiliaries in some industrial process steps (e.g. production of paints, coating materials and dyestuffs, leather processing) (Piorr, 1987).

The annual AES consumption in Western Europe amounted to 150 000 tons in 1987; in USA 160 000 tons were used and in Japan 40 000 tons (Richtler and Knaut, 1988). With reference to their use in detergents, AES represent the second most important anionic surfactant group in Western Germany with an annual consumption of 22 000 tons in 1989 corresponding to 19% of the total anionics consumption (without soap) (Malz, 1991).

### 5.3.8 Sulfosuccinates

Sulfo succinic alkyl esters having the structure formula

$$Na^+{}^-O_3S—CH—CO—O—R_1 \qquad R_1 = H \ (monoester)$$
$$| \qquad\qquad\qquad\qquad alkyl\ group\ (diest\epsilon$$
$$H_2C—CO—O—R_2 \qquad R_2 = alkyl\ group$$

are mainly manufactured from maleic anhydride in a catalysed reaction with an excess of the respective alcohol in the presence of azeotropic agents to remove the reaction water from the reaction mixture. After neutralization of the reaction mixture the excess alcohol and the azeotropic solvent are removed and the sulfonated product is obtained after reaction with $NaHSO_3$ (Biermann et al., 1987).

Sulfosuccinates can be based on linear or branched alcohols. For manufacture of diesters $C_{5-8}$-alcohols or fatty alcohol ethanolamides are preferred. Monoesters are mainly prepared from fatty alcohols, fatty alcohol ethoxylates or fatty acid alkanolamides.

Dialkyl sulfosuccinates based on alcohols with a total of less than nine carbons are water soluble; branching of the alkyl group favours water solubility. The 2-ethylhexanol-derived dialkylester has very good wetting properties and is mainly used in the textile industry. Monoalkyl sulfosuccinates are mainly based on linear alcohols and are soluble in water to only a limited extent so that dilute solutions are strongly turbid. Sulfosuccinic half-esters based on alcohol ethoxylates are clearly soluble. The monoalkyl succinates have good detergency and foam properties and lime soap dispersion capacity as well (Piorr, 1987).

Apart from their use in special detergent formulations and their application in the textile, plastics, photography and leather industries, a main application field is the cosmetic sector, i.e. their use in personal care products. Compared with other groups of anionic surfactants, the consumption figures are relatively low. In Western Germany the annual consumption of these materials in the detergent field was 2000 tons in 1989 being less than 2% of the total anionics usage for this application (Malz, 1991).

### 5.3.9   *Alkylphosphates and alkyl ether phosphates*

This group of anionic surfactants contains the phosphate group as the hydrophilic structure component. Alkylphosphates represent mixtures of dialkyl and monoalkyl phosphoric esters

$$\underset{\underset{OH}{|}}{\overset{\overset{O}{\|}}{R-O-P-OH}} \quad + \quad \underset{\underset{O-R}{|}}{\overset{\overset{O}{\|}}{R-O-P-OH}} \qquad R = alkyl$$

which are obtained after addition of phosphorus pentoxide to fatty alcohols or fatty alcohol ethoxylates. Water soluble products are formed by neutralization of the acids (Biermann *et al.*, 1987).

Dialkyl esters have good wetting properties whereas monoalkyl phosphates have an inhibiting effect on the foam generation by other anionic and non-ionic surfactants. Their application field is limited to special purposes, e.g. in the textile industry (pretreatment of natural fibres) and the leather industry (emulsifiers) (Piorr, 1987). The usage of alkylphosphates in the detergent field is very low as indicated by an annual consumption of <100 tons in Western Germany in 1989 (Malz, 1991).

## 5.4   Biodegradation of anionic surfactants

### 5.4.1   *Soap*

*5.4.1.1   Primary and ultimate biodegradation.*   Soaps representing the alkali salts of mainly $C_{12}$–$C_{18}$ fatty acids are readily accessible to biodegradation (Table 5.1). Although the poor solubility of calcium or magnesium soaps influences the biodegradation rate — e.g. a comparison of the $CO_2$ evolution in the Sturm test yielded mineralization rates of 80–90% for the sodium salt but only 63% for the calcium salt (Schöberl *et al.*, 1988) — the published biodegradation data prove the general assumption that salts of fatty acids are ultimately biodegradable within a short period of time. $BOD_5$ values between 50 and 60% have been reported for sodium soaps in the $C_{12-18}$ chain range (Swisher, 1987). $BOD_5$ measurements in a manometric respirometry test system (sapromat) yielded values of 85% for sodium-laurate and sodium-palm kernel soap ($C_8$–$C_{14}$), approximately 75% for sodium-oleate and sodium-tallow soap ($C_{16/18}$), 55% for sodium-stearate and 43% for the $C_{22}$-homologue (behenate); the respective values after 10 days incubation were 100% BOD/COD for the sodium soaps based on lauric, oleic and tallow fatty acids, ≥90% for sodium soap from palm kernel oil, almost 90% for stearic acid and almost 80% for behenate (Mix-Spagl, 1990). In standardized screening test systems like the

**Table 5.1** Biodegradation test data of fatty acids and soaps

| Test substance | Test method | Analysis | % Biodegradation | Type of biodegrad. | Reference |
|---|---|---|---|---|---|
| Sodium soap ($C_{12-22}$) | Sturm | $CO_2$ | 80–90% | Ultimate/ready | Schöberl et al., 1988 |
| Calcium soap ($C_{12-22}$) | Sturm | $CO_2$ | 63% | Ultimate | Schöberl et al., 1988 |
| Sodium soaps ($C_{12-18}$) | Respirometric | $BOD_5$/COD | ca. 50–60% | Ultimate | Swisher, 1987 |
| Sodium-laurate ($C_{12}$) | Respirometric | $BOD_{5/10}$/COD | 85%/100% | Ultimate | Mix-Spagl, 1990 |
| Sodium-palm kernel ($C_{8-14}$) s. | Respirometric | $BOD_{5/10}$/COD | 85%/≥90% | Ultimate | Mix-Spagl, 1990 |
| Sodium-oleate ($C_{18}$) | Respirometric | $BOD_{5/10}$/COD | 75%/100% | Ultimate | Mix-Spagl, 1990 |
| Sodium-tallow soap ($C_{16-18}$) | Respirometric | $BOD_{5/10}$/COD | 75%/100% | Ultimate | Mix-Spagl, 1990 |
| Sodium-stearate ($C_{18}$) | Respirometric | $BOD_{5/10}$/COD | 55%/>85% | Ultimate | Mix-Spagl, 1990 |
| Sodium-behenate ($C_{22}$) | Respirometric | $BOD_{5/10}$/COD | 43%/>75% | Ultimate | Mix-Spagl, 1990 |
| Sodium-stearate ($C_{18}$) | Closed Bottle | $BOD_{30}$/COD | 85–100% | Ultimate/ready | Schöberl et al., 1988 |
| $C_{8-18}$ fatty acids | Modif. Blok[a] | $BOD_{28}$/COD | 100% | Ultimate | Gerike, 1984 |
| $C_{16}$ fatty acid | Modif. Blok[a] | $BOD_{28}$/COD | 84% | Ultimate | Gerike, 1984 |
| $C_{18}$ fatty acid | Modif. Blok[a] | $BOD_{28}$/COD | 79% | Ultimate | Gerike, 1984 |
| $C_{22}$ fatty acid | Modif. Blok[a] | $BOD_{28}$/COD | 69% | Ultimate | Gerike, 1984 |
| Sodium-stearate ($C_{18}$) | Modif. Blok[a] | $BOD_{28}$/COD | 100% | Ultimate | Gerike, 1984 |

[a] See text.

Closed Bottle test, 85–100% $BOD_{30}/COD$ has been reported for sodium-stearate (Schöberl et al., 1988). Gerike (1984) investigated a number of poorly soluble fatty acids in a modification of the Closed Bottle test especially suited for testing poorly soluble compounds (Modified Blok test) and found $BOD_{28}$ values of 100% for a $C_{8-18}$ mixture, 84% for $C_{16}$, 79% for $C_{18}$ and 69% for $C_{22}$; the better soluble sodium salt of the $C_{18}$-acid (sodium-stearate) exhibited a degradation extent of virtually 100% stressing again an influence of the water solubility on the degradation result of these compounds.

The high ultimate biodegradability of soaps — substantiated by a multitude of additional test data contained in the compilation of Swisher (1987) — presupposes that also a very fast and complete primary biodegradation of these molecules has taken place. However, concrete data are not readily available. The MBAS analytical method generally used for monitoring the primary degradation of anionic surfactants is not applicable to soaps since the carboxylate anion does not form a complex with the dye cation under the acidic conditions used in this method (Swisher, 1987). Mix-Spagl (1990) reported on the primary biodegradation of sodium-oleate in the OECD Screening test measuring the removal of the dissolved fraction of this soap by gas chromatographic analysis: within 10 days incubation a 96% primary degradation was observed, thus confirming a quick and virtually complete transformation of these substances to degradation intermediates being, in the end, ultimately biodegraded.

*5.4.1.2 Biodegradation pathway.* The main pathway for degradation of soaps is $\beta$-oxidation, being the usual degradation process of fatty acids in living cells. The $\beta$-oxidation process represents a series of enzymatic reactions involving coenzyme A, a complex sulfur-containing compound (Figure 5.2). In the first step the carboxyl group of the fatty acid is esterified with the thiol group of the coenzyme; an $\alpha$-, $\beta$-unsaturated compound is formed after enzymatic removal of two hydrogens. This intermediate is hydrated to the corresponding $\beta$-hydroxy derivative and subsequently dehydrogenated yielding the $\beta$-keto derivative.

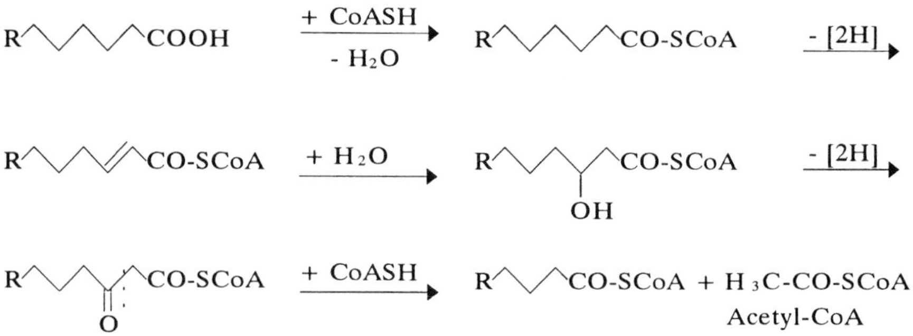

**Figure 5.2** Schemes of $\beta$-oxidation (CoASH = coenzyme A).

Ultimately, a further coenzyme A (CoA) molecule is necessary to split off a $C_2$ fragment (acetyl-CoA) and leaving a fatty acid-CoA ester with an alkyl chain shortened by a $C_2$-unit. This sequence of reactions is repeated until the whole alkyl chain is transformed into acetyl-CoA fragments. These fragments are subsequently used in living cells for energy production (via the citric acid cycle) or biomass formation. More detailed reviews of the $\beta$-oxidative degradation process of fatty acids and its interconnection with degradative and assimilative pathways in microbial cells are provided by most biochemistry textbooks.

*5.4.1.3 Anaerobic degradation.* Although the degradation of fatty acids by $\beta$-oxidation is oxidative, the removal of hydrogen in this process does not require molecular oxygen. Thus, dissimilation of fatty acids by anaerobic organisms can also proceed via $\beta$-oxidation allowing one to anticipate that soaps are anaerobically as well degradable as under aerobic conditions. The number of published specific investigations into the anaerobic degradability of soaps or the respective fatty acids is not very large; however, the available data confirm unanimously the expected anaerobic biodegradability of these compounds. For a $C_{16}$-fatty acids which was [14]C-radiolabelled at carbon-1, carbon-16 or uniformly at all carbons, respectively, an ultimate degradation extent of 92–97% was determined in a digester model system after incubation for 28 days and measurement of the evolved [14]C-methane and [14]CO$_2$ (Steber and Wierich, 1987). Also in a more stringent anaerobic degradation screening test, the ECETOC test (Birch *et al.*, 1989), this positive evaluation could be confirmed: based on the measurement of gas (i.e. $CO_2$ + methane) and DIC (i.e. dissolved inorganic carbon) formation, a degradation extent of 79–94% was found for sodium-palmitate within a 3- to 4-week period (Birch *et al.*, 1989). Finally, in a systematic investigation by Mix-Spagl (1990) a series of sodium and calcium soaps were tested in an anaerobic fermentor simulating a sewage works digester. Semi-continuous addition of sodium soaps to the fermentor yielded gas productions equivalent to 95% degradation of laurate, 70% of oleate and palm kernel-based soap, 60% of tallow-based soap and only 14% of behenate. In the same test system a single addition of calcium soaps and subsequent 10–13 days incubation resulted in degradation rates of ≥90% for laurate, behenate and stearate. The reason for this considerably improved degradation behaviour was unclear showing, nevertheless, that even very poorly soluble long-chain soaps are easily accessible to anaerobic degradation.

### 5.4.2 Alkyl benzene sulfonates

*5.4.2.1 Data on primary degradation.* The available large database for the biodegradability evaluation of TPS and LABS confirms that the branched-chain TPS is considerably less biodegradable than the linear ABS molecules (Swisher, 1987). Due to the multiple branching of TPS even the primary biodegradation

**Table 5.2** Primary and ultimate biodegradation of TPS and LABS in standard tests

| Test substance | Test method | Analysis | % Biodegradation | Type of biodegrad. | Reference |
|---|---|---|---|---|---|
| TPS | OECD Screening | MBAS | 18–50% | Primary | Swisher, 1987 |
| | OECD Confirmatory | MBAS | 20–45% | Primary/STP model | Swisher, 1987 |
| | Closed Bottle | BOD/COD | <10% | Ultimate | Swisher, 1987 |
| | Coupled Units (CAS) | DOC | 41% | Ultimate/STP model | Fischer, 1975 |
| LABS | OECD Screening | MBAS | 95% | Primary | Schöberl et al., 1988 |
| | OECD Confirmatory | MBAS | 93–97% | Primary/STP model | Schöberl et al., 1988 |
| | Closed Bottle | BOD/COD | 55–65% | Ultimate/ready | Schöberl et al., 1988 |
| | Modif. OECD Screening | DOC | 73–84% | Ultimate/ready | Schöberl et al., 1988 |
| | Sturm | $CO_2$ | 50–75% | Ultimate/ready | Schöberl et al., 1988 |
| | Zahn–Wellens | DOC | 95% | Ultimate/inherent | Gerike and Fischer, 1979 |
| | SCAS | DOC | 92–93% | Ultimate/inherent | Larson, 1979 |
| | EPA Activ. Sludge | DOC | 98% | Ultimate/inherent | Gerike and Fischer, 1981 |
| | Coupled Units (CAS) | DOC | 73–82% | Ultimate/STP model | Gerike and Fischer, 1979 |
| | CAS (uncoupled) | DOC | 93% | Ultimate/STP model | Schöberl et al., 1988 |
| | 'Metabolite test' | DOC | 94.9 ± 1.2% | Ultimate | Gerike and Jasiak, 1986 |

rates in die-away and continuous activated sludge tests are, in most cases, below 80% MBAS removal (Table 5.2) which is the required (primary) biodegradability limit in standardized tests like the OECD Screening and the OECD Confirmatory tests. In the 19-day OECD Screening test degradation values of 18–50% MBAS removal have been reported; in the continuous activated sludge (CAS) tests the MBAS removals were mainly in the range of 20–45%. Data from a number of sewage treatment processes tested in labs and pilot plants confirm this evaluation: in oxidation ponds, trickling filters and activated sludge plants the average MBAS removal of TBS was below 50% (Swisher, 1987).

In contrast to TPS, commercial LABS (with an average chain length of approximately $C_{12}$) shows MBAS removals of more than 90–95% in these standard tests (Table 5.2). In a recent compilation of ecological data of the major surfactant groups (Schöberl et al., 1988) primary degradation rates of LABS were reported with 95% MBAS removals in the OECD Screening test and 93–97% in the OECD Confirmatory test. This high extent of primary biodegradation of LABS was confirmed by the outcome of a number of pilot sewage treatment plant investigations and field studies (Swisher, 1987). Painter and Zabel (1989) compiled the published data from real sewage treatment plant studies and reported on MBAS removal rates of 85–90% and specific LABS removals of 95–99.5% measured in activated sludge plants; the respective figures for biological filters were 85–96% MBAS removal and 73–91% LABS removal.

There exist also primary biodegradation data of LABS from kinetic studies. In river water a half-life time of about 3 days was determined (Larson and Payne, 1981); in sea water the half-life was considerably longer with 6–9 days (Vives-Rego et al., 1987).

It has to be recognized that the primary degradation figures reported on commercial LABS represent composite values resulting from the different degradation rates of the individual homologues and isomers which LABS is composed of. It is discussed below (cf. 5.4.2.3) that the rate and extent of primary biodegradation of the individual LABS molecules depends on the alkyl chain length and the position of the sulfophenyl substituent within the molecule.

5.4.2.2    *Data on ultimate degradation.*    Whereas poor primary biodegradation, as in the case of TBS, will indicate poor ultimate biodegradability, good primary biodegradation does not necessarily indicate good ultimate biodegradation. However, for linear alkyl benzene sulfonate (LABS) the evidence is abundant that the primary biodegradation of the surfactant is followed by further degradation processes resulting, ultimately, in formation of mineralization products and bacterial biomass (ultimate degradation).

Evidence for ultimate biodegradation of a chemical in lab tests can be based on different analytical parameters like oxygen uptake (biological oxygen demand compared with chemical oxygen demand: BOD/COD), $CO_2$ evolution and — although with a certain reservation — organic carbon removal (Table

5.2). From Swisher's (1987) vast data list of LABS biodegradation results it is evident that the majority of oxygen uptake data obtained in die-away tests exceed a threshold of 40–50% of the theoretical oxygen demand (ThOD). In a German report on surfactants' ecological data (Schöberl *et al.*, 1988) the results from the Closed Bottle test range between 55% and 65% of ThOD and BOD/COD, respectively. According to the criteria of OECD (1981), a compound is regarded as 'readily biodegradable' (i.e. readily accessible to ultimate degradation) if it passes the limit of 60% of ThOD or BOD/COD within 28 days. Thus it can be concluded that the LABS degradation data obtained in the Closed Bottle test are in the borderland of 'ready biodegradability'. However, it has to be recognized that the Closed Bottle test represents the most stringent type of the OECD ready biodegradability tests (EEC, 1986). Compared with LABS, the BOD-based ultimate degradation rates of TPS are extremely low being in most cases below 10% of the theoretical oxygen demand (Swisher, 1987).

Provided the test compound and its degradation intermediates are water-soluble and non-volatile under the employed test conditions, organic carbon removal rates (e.g. DOC removal) determined in screening tests also constitute a reliable parameter for measuring ultimate biodegradability of a compound. The majority of DOC removal data of LABS from biodegradation screening tests indicate a high extent of ultimate degradation of this surfactant (Swisher, 1987). In the Modified OECD Screening test, being one of the OECD tests for ready biodegradability, DOC removals of 73–84% have been reported (Schöberl *et al.*, 1988). Since the OECD limit for obtaining the attribute 'readily biodegradable' is 70% DOC removal (OECD, 1981), LABS meets this requirement. It is not surprising that in less stringent degradation tests like the OECD tests for inherent biodegradability (EEC, 1986) even more favourable DOC removal rates were obtained, e.g. 92–93% in the SCAS test (Larson, 1979), 95% and 98% in the Zahn–Wellens test and in the EPA activated sludge test, respectively (Gerike and Fischer, 1979, 1981).

DOC removal rates of LABS in model sewage treatment plants, e.g. continuous activated sludge tests (CAS), are in the range between 70 and 90% (Painter and Zabel, 1989). In the Coupled Units test with a 3 h retention time, values between 73 and 82% DOC removal were obtained (Gerike and Fischer, 1979); in the uncoupled version of this test (without sludge exchange) a 93% carbon removal was reported (Schöberl *et al.*, 1988). For comparison, the COD removal for TBS was 41 ± 9% in the Coupled Units test (Fischer, 1975). When organic carbon removal rates are measured in test systems with a high sludge load like CAS tests or tests for inherent biodegradability, it has to be considered that the carbon removals can also be influenced by adsorption processes onto sludge. However, based on the knowledge about the oxidative degradation steps of LABS and the concomitant formation of polar intermediates with decreasing adsorption behaviour (cf. 5.4.3.3) it can be concluded that the high DOC removals measured in these tests are mainly due to ultimate degradation.

This conclusion is supported by the degradation results based on carbon dioxide formation from LABS in screening and simulation tests. Measurement of $CO_2$ evolution provides unequivocal information about the mineralization extent of a chemical. Generally, in screening tests like the $CO_2$ evolution test (Sturm test), about 50–75% of the total carbon of LABS was recovered as carbon dioxide after a test duration of approximately 4 weeks (Swisher, 1987; Schöberl et al., 1988); a $CO_2$ evolution of $\geq 60\%$ in the Sturm test indicates 'ready biodegradability' according to the OECD (1981) criteria. It has to be considered that the total extent of ultimate biodegradation will be higher than the mineralization results indicate since these values do not reflect the portion of ultimate degradation which is due to the transformation of the surfactant's carbon into bacterial biomass. This was clearly shown in a number of experiments with [ring-[14]C]radiolabelled LABS (Swisher, 1987; Steber, 1979) in screening and CAS tests: in those cases where a balance of the [14]C distribution was given, the $CO_2$ evolution extent of 50–80% paralleled with a biomass production of 10–20% yielding a total ultimate degradation rate of 70–>80%. This figure is congruent with the organic carbon removal rates discussed previously confirming once again that the DOC removals reflected real ultimate degradation of LABS.

Finally, also the data from a special test, the 'Test for detecting recalcitrant metabolites' (Gerike and Jasiak, 1986), provide evidence that no poorly degradable intermediates are formed in the degradation process of LABS. In this modification of the Coupled Units test, the biodegradation potential of the underlying CAS system is considerably increased by recycling the daily effluent back to the CAS unit and replenishing it with a concentrate of fresh food and test compound. After running this test for more than 50 cycles, a carbon removal rate of $85 \pm 5\%$ was measured (Gerike and Jasiak, 1986); in the final stage of the test only synthetic food was replenished daily giving the LABS-derived compounds an additional opportunity for degradation. At the end of the test a $94.9 \pm 1.2\%$ DOC removal was measured allowing one to exclude the formation of even the smallest theoretically possible metabolite, a $C_1$-molecule, during the degradation process of LABS. The difference between the obtained result and the value indicative for complete degradation, i.e. 100% DOC removal, is assumed to be due to technical impurities of LABS (Gerike and Jasiak, 1986).

*5.4.2.3   Biodegradation pathways.*   The structure of the LABS molecule offers three potential areas where the initial enzymatic attack could take place: the alkyl chain, the aromatic ring and the sulfonate substituent. As a matter of fact, there exist a number of studies into the biodegradation routes of LABS which indicate that the degradation of this surfactant is effected by simultaneous as well as sequential enzymatic reactions of various bacterial groups. Based on degradation experiments with pure bacterial cultures, Cain et al. (1972) discussed five different possibilities for starting the microbial degradation of

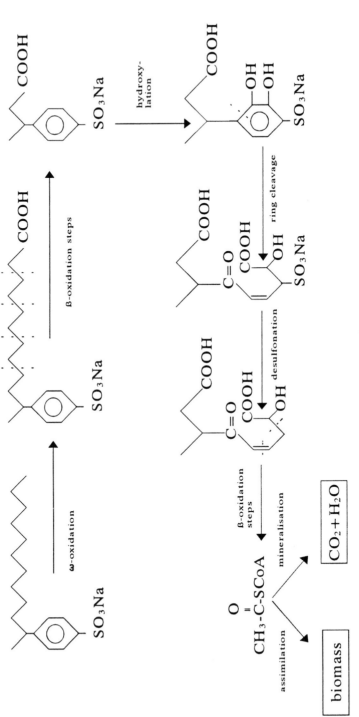

**Figure 5.3** Scheme of the LABS biodegradation pathway.

LABS; however, it became clear that the full pathway to transform the parent compound into final degradation products, i.e. mineralization and assimilation products, is brought about by the actions of mixed bacterial cultures. According to the present knowledge, the microbial degradation route of LABS in microbial biocenoses consists of three main steps (Figure 5.3): (i) oxidation of the terminal methyl groups of the alkyl chain and subsequent degradation of the chain yielding short-chain sulfophenyl alkanoic acids, (ii) oxidative fission of the aromatic ring leaving sulfonate-substituted dicarboxylic acids, (iii) desulfonation of the ring degradation products and further catabolism by general metabolic routes.

*5.4.2.3.1 Oxidation of the alkyl chain.* The investigations by Huddleston and Allred (1963), Willets (1973) and other authors discussed in Swisher (1987) provided evidence that the initial microbial attack of the LABS molecule takes place at the terminal methyl groups of the alkyl chain according to mechanisms described for the oxidation of aliphatic hydrocarbons (McKenna and Kallio, 1965). Accordingly, this terminal attack, called $\omega$-oxidation, is initiated by the addition of molecular oxygen to the methyl group catalysed by an oxygenase enzyme. The resulting primary hydroperoxide is then reduced to the primary alcohol which can be oxidized by dehydrogenase enzymes to the aldehyde and subsequently to the corresponding acid (Swisher, 1987). The first detectable LABS oxidation product is the $\omega$-carboxylated derivative as shown by Huddleston and Allred (1963) who isolated sulfophenyldodecanoic acid after incubation of 2-phenyldodecane sulfonate. The occurrence of LABS degradation intermediates carboxylated at both ends of the alkyl chain has also been reported (Willets, 1973).

The following steps of the oxidative degradation of the alkyl chain are accomplished by the $\beta$-oxidation mechanism, i.e. the successive shortening of sulfophenyl alkanoic acids by removing $C_2$-units (acetyl-CoA residues) at a time as described for the degradation pathway of fatty acids (cf. 5.4.1.2). Huddleston and Allred (1963), Swisher (1963) and several other authors (quoted in Swisher, 1987) made contributions to the relatively consistent picture of the $\beta$-oxidative degradation of the LABS alkyl chain which continues until the alkyl chain only has 4–7 carbon atoms (Schöberl and Kunkel, 1977; Divo and Cardini, 1980). The observed transient accumulation of short-chain sulfophenyl carboxylic acids in the test media can be explained by steric hindrance of the continuation of the $\beta$-oxidation route due to the sulfophenyl substituent. Investigations by Divo and Cardini (1980) showed that a pure culture of *Pseudomonas* sp. was unable to proceed beyond the degradation of the chain. This resulted in DOC concentrations becoming constant after a 9-day incubation of LABS but yielded exactly the products expected from the $\omega$-/$\beta$-degradation pathway: even-numbered sulfophenylcarboxylic acids were formed from LABS molecules with an even-numbered carbon chain and odd-numbered intermediates were obtained from odd-numbered LABS homologues.

Although $\beta$-oxidation represents the main mechanism of the alkyl chain degradation of LABS, an oxidative removal of a $C_1$-unit by cleavage of the terminal carboxy group ($\alpha$-oxidation) may also be involved to a minor extent, as indicated by the findings of Swisher (1963) and Baggi et al. (1972).

Already early investigations into the biodegradation of LABS (Swisher, 1963; Huddleston and Allred, 1963; Allred et al., 1964) have shown that the disappearance kinetics of individual LABS homologues in a degradation test depend on the chain length: the degradation rate (determined by the MBAS analytical procedure) increases with the chain length. In addition, more specific analyses by Huddleston and Allred (1963) showed that the isomers of LABS mixtures having their sulfophenyl substituent near the end of the chain are removed quicker than those having the aromatic ring in a central position. This relationship — primary degradation rate is positively correlated to the distance between the sulfophenyl group and the most remote methyl group of the alkyl chain — is called 'distance principle' and has been verified by many investigators (Swisher, 1987). The distance principle was demonstrated nicely by the work of Wickbold (1975) who examined the effluent of a continuous model sewage treatment plant (OECD Confirmatory test) measuring the percent MBAS removals of each individual homologue and isomer contained in the technical LABS ($C_{10}$–$C_{13}$, Marlon A). It was observed for instance, that the 2-sulfophenylalkanes were removed at the quickest rate (95–96% MBAS removal) whereas the 5-sulfophenylalkane isomer was degraded least (52% MBAS removal). Nevertheless, short-chain homologues with a central aromatic ring position are also well degradable after a sufficient adaptation period of the bacteria involved (Schöberl, 1979, 1989). Schöberl (1989) suggested that the molecular explanation for the distance principle is based on the steric hindrance of the enzymatic oxidation of the parent LABS molecule by the sulfophenyl group. As discussed, the primary LABS degradation intermediates were identified as sulfophenylcarboxylic acids all having a shortened alkyl chain; thus, it is obvious that the steric hindrance of the $\beta$-oxidation process by the sulfophenyl substituent is not the molecular basis of the distance principle. Consequently, the hindrance must occur at an earlier stage. Schöberl (1989) concluded that the first oxidation step, the hydroxylation of the methyl group is critical. Based on Wickbold's data (1975) it was proposed that the ring-free alkyl chain site of the LABS molecule must have at least seven carbons in order to ensure unhindered hydroxylation. If fewer carbons are present, the alkyl chain terminus is able to reach the active site in the interior of the enzyme only in a very special spatial arrangement due to the hindrance by the sulfophenyl substituent so that a decrease of the hydroxylation rate results.

Although the distance principle of primary LABS degradation has been verified many times, more recent investigations by Larson (1990) conducted with radiolabelled LABS at lower concentrations (10–100 $\mu$g/l) revealed only little variation of the $^{14}CO_2$ evolution rates and extent between the individual homologues and isomers: half-lifes for mineralization of the benzene ring of

$C_{10}$–$C_{14}$ homologues were $20 \pm 2$ h in river water and $15 \pm 2$ h in the river sediment. This indicates that the $\omega$-oxidation step is not the rate-determining reaction of the ultimate biodegradation process of LABS.

*5.4.2.3.2 Aromatic ring oxidation and desulfonation.* Although not directly explored by investigations on LABS but on short-chain and 1-sulfophenyl-substituted alkylbenzene sulfonates, there exists a sound view of the further degradation of short-chain sulfophenylalkanoates. The fact that the sulfophenyl carboxylic acids constitute the main fraction of LABS degradation intermediates (Schöberl and Kunkel, 1977; Steber, 1979) and that sometimes also a pause of further degradation of these metabolites was observed (Swisher, 1987) is indicative that the ring cleavage is the rate-limiting step in LABS degradation.

Basically, several enzymatic mechanisms for aromatic ring cleavage exist. All types require the introduction of hydroxy groups catalysed by an oxygenase enzyme in the presence of molecular oxygen. Schöberl (1989) discussed the so-called *meta*-cleavage as the most appropriate mechanism for the ring opening of LABS intermediates. Accordingly, the first step is the incorporation of molecular oxygen into the ring by the action of a hydroxylase enzyme forming a 2,3-dihydroxy derivative; the latter undergoes subsequently *meta*-fission of the ring yielding aliphatic sulfo-keto unsaturated dicarboxylic acids. Although alternative ring cleavage mechanisms might also be possible requiring, for instance, premature desulfonation of the ring (Swisher, 1987; Schöberl, 1989), the sulfate liberation prior to the ring opening seems to play a minor role in real microbial biocenoses. The detection of sulfonated aliphatic LABS catabolites in the effluents of CAS systems (Krüger, 1964; Schöberl and Kunkel, 1977; Steber, 1979) supports the pathway outlined in the scheme (Figure 5.3).

The next stage of the further biodegradation of the dicarboxylated and sulfonate-substituted aliphatic intermediates includes the desulfonation step yielding the keto-unsaturated dicarboxylic acids plus inorganic sulfate. These aliphatic intermediates are further decomposed by mechanisms which may include $\beta$-oxidation and other enzymatic reactions further integrating the products into the general microbial metabolic pathways like the citrate and glyoxylate pathways. However, these final steps of LABS ultimate degradation have not been investigated in detail.

*5.4.2.4 Anaerobic degradation.* Having the basic degradation steps of LABS under aerobic conditions in mind, it is quite obvious that the degradation of this compound in the anaerobic environment either must be based on alternative mechanisms or it is not possible. $\omega$-Oxidation as the initial step of the bacterial alkyl chain attack requires molecular oxygen which is not sufficiently present in anaerobic environmental systems like sludge digesters, septic tanks, anoxic sediments and soils. Similarly, the mechanisms proposed for the oxidative cleavage of the benzene ring require molecular oxygen for the formation of a dihydroxylated aromatic ring.

Although only a limited number of investigations into the anaerobic biodegradability of LABS have been published, there is no indication for the existence of alternative degradation routes by-passing the discussed dependence of basic LABS degradation steps on the presence of molecular oxygen. The few data in Swisher's (1987) list referring to LABS biodegradability under anaerobic conditions suggest a poor primary biodegradation (MBAS removal). This underlines the assumption that the primary degradation step of LABS in aerobic systems, i.e. the oxidation of the terminal alkyl chain methyl group(s) does not proceed anaerobically. Although published experimental data are lacking, it can be expected that the cleavage of the LABS aromatic ring system will also not take place in anaerobic environments. Furthermore, no indication of degradability was obtained in a model sludge digester when a $^{14}$C-LABS-derived fraction of sulfonated intermediates having an open aromatic ring was incubated anaerobically (Steber and Wierich, 1989). Thus, the desulfonation step also seems to be dependent on the presence of molecular oxygen; this conclusion could be substantiated by studies on several other sulfonated compounds (Steber and Wierich, 1989). Further support of this reasoning comes from the findings by Itoh *et al.* (1987) who compared the anaerobic biodegradabilities of several surfactants: LABS and $\alpha$-olefine sulfonate yielded the poorest results.

As a matter of fact, there is a lot of evidence that LABS is anaerobically not biodegraded in practice. Bruce *et al.* (1966) reported that 'hard' (i.e. TPS) as well as 'soft' (i.e. LABS) alkylbenzene sulfonates are equally resistant to degradation under anaerobic conditions. These early conclusions are supported by the findings that in anaerobically treated, i.e. digester, sludges relatively high concentrations of LABS can be found (Painter and Zabel, 1989). Activated sludges and air-dried digester sludges, i.e. aerobically treated sludges, only contain low LABS concentrations (0.1–0.8 g LABS/kg dry matter). However, primary sludges from the primary settlement tanks of a sewage treatment plant and digester sludges as well exhibit considerably higher LABS concentrations (4–30 g LABS/kg dry matter). Since approximately 20% of the LABS load entering sewage treatment plants is eliminated by adsorption onto primary sludge (Painter and Zabel, 1989) which is subsequently treated in digesters, these data show that no significant reduction of the LABS concentrations takes place during digestion of sewage sludge (primary and secondary sludges), i.e. LABS is not degraded under these circumstances. If such sludges are spread on agricultural land for fertilizing purposes, the prevailing aerobic conditions will allow the continuation of the degradation of LABS.

Ultimately, the anaerobic recalcitrance of LABS was clearly demonstrated in a recent investigation by Federle and Schwab (1992) who examined sediments of a wastewater pond which had been exposed to the discharge of waste water from a local laundromat for more than 25 years. The anaerobic microbial communities of this pond sediment and of a control sediment (not exposed to surfactants previously) as well were not able to mineralize $^{14}$C-radiolabelled LABS whereas compounds known to be anaerobically degradable were rapidly degraded by the sediment microorganisms from either pond.

## 5.4.3   Secondary alkane sulfonates (SAS)

*5.4.3.1   Data on primary degradation.*   Since alkane sulfonates react as methylene blue active substances (MBAS) their primary biodegradation can be followed easily using this analytical parameter (Table 5.3). The available data basis (Swisher, 1987) shows clearly that SAS undergo a rapid primary biodegradation with MBAS removal of >90% in die-away tests within a few days. The examination of the $C_{13}$–$C_{18}$ homologues of commercial-type SAS by several authors, as compiled by Swisher (1987), gave no indications of differences in the primary biodegradation rate between the individual homologues after an incubation time of 4–9 days. In the 19-day OECD Screening test a 96% MBAS decrease was reported (Schöberl *et al.*, 1988).

The same positive conclusion on primary biodegradation of SAS can be drawn when looking at the test data obtained in continuous model sewage treatment plants. In CAS systems like the OECD Confirmatory test (3 h retention time) 97–98% MBAS removals have been reported for commercial SAS having a chain length of $C_{13}$–$C_{18}$ (Schöberl *et al.*, 1988). In addition, Painter (1992) quotes 97–99% MBAS removals in small-scale trickling filters. The data list for SAS (Swisher, 1987) contains considerably more primary degradation results confirming this range of a high primary degradation extent.

*5.4.3.2   Data on ultimate biodegradation.*   Secondary alkane sulfonates perform very well in the OECD tests for ready biodegradability indicating a rapid ultimate biodegradability (Table 5.3). In a compilation of ecological data of important surfactant groups (Schöberl *et al.*, 1988) an 88–96% DOC removal in the Modified OECD Screening test, 63–95% BOD/COD in the Closed Bottle test and a $CO_2$ evolution of 56–91% in the Modified Sturm test have been reported. The majority of these results surpass the OECD limits for 'ready biodegradability' and suggests that the quick primary biodegradation of this surfactant is followed immediately by degradation processes leading, in the end, to ultimate degradation. The already mentioned systematic examination of individual SAS homologues from $C_{13}$ to $C_{19}$ (reported in Swisher, 1987) showed little difference in the degradation extent determined as COD removal after 21 days. Biodegradation studies by Lötzsch *et al.* (1979) using a uniformly labelled $C_{17}$-SAS in a discontinuous screening test system yielded a 53–61% $^{14}CO_2$ evolution within the 12-day test period. The same labelled test material was also used in a study under CAS test conditions (Neufahrt *et al.*, 1980). After addition of $^{14}$C-SAS for 1 day to the acclimated test system and subsequent dosing of unlabelled SAS for another 3 days, the total extent of $^{14}CO_2$ evolution was 47% of the total $^{14}$C fed; 25% was found in the activated sludge and was attributable to $^{14}$C-biomass; thus, the extent of ultimate degradation was around 70%. In CAS test systems, like the Coupled Unit test, considerably higher carbon removal rates were obtained ranging between 83 and 96% (Schöberl *et al.*, 1988).

**Table 5.3** Primary and ultimate biodegradation of SAS and AOS in standard tests

| Test substance | Test method | Analysis | % Biodegradation | Type of biodegrad. | Reference |
|---|---|---|---|---|---|
| SAS (C$_{13-18}$) | OECD Screening | MBAS | 96% | Primary | Schöberl et al., 1988 |
| | OECD Confirmatory | MBAS | 97–98% | Primary | Schöberl et al., 1988 |
| | Lab-scale trickling filter | MBAS | 97–99% | Primary | Painter, 1992 |
| | Modif. OECD Screening | DOC | 88–96% | Ultimate/ready | Schöberl et al., 1988 |
| | Closed Bottle | BOD/COD | 63–95% | Ultimate/ready | Schöberl et al., 1988 |
| | Sturm | CO$_2$ | 56–91% | Ultimate/ready | Schöberl et al., 1988 |
| | Coupled Units (CAS) | DOC | 83–96% | Ultimate/STP model | Schöberl et al., 1988 |
| | 'Metabolite test' | DOC | 97.4 ± 0.5% | Ultimate | Gerike and Jasiak, 1986 |
| AOS (C$_{14-18}$) | OECD Screening | MBAS | 99% | Primary | Schöberl et al., 1988 |
| | OECD Confirmatory | MBAS | 98% | Primary/STP model | Schöberl et al., 1988 |
| | Modif. OECD Screening | DOC | 85% | Ultimate/ready | Schöberl et al., 1988 |
| | Closed Bottle | BOD/COD | 85% | Ultimate/ready | Schöberl et al., 1988 |
| | Sturm | CO$_2$ | 65–80% | Ultimate/ready | Schöberl et al., 1988 |
| (C$_{16/18}$) | Coupled Units (CAS) | DOC | 70–78% | Ultimate/STP model | Schöberl et al., 1988 |

The anticipation that in the course of SAS degradation no poorly degradable intermediates are formed was confirmed by the outcome of the 'Test for detecting recalcitrant metabolites' (Gerike and Jasiak, 1986). The DOC removal of $97.4 \pm 0.5\%$ in this test excludes the formation of the smallest theoretically possible metabolite, a $C_1$-molecule, but suggests the presence of a small percentage (2.1–3.1%) of recalcitrant material which might be attributable to technical impurities like di- and polysulfonates.

*5.4.3.3  Biodegradation pathways.*  Little information is available on the biodegradation mechanisms of secondary alkane sulfonates. Based on enzymatic studies on *n*-dodecane-2-sulfonate and linear alkane sulfonates by Thysse and Wanders (1974) it is proposed that the first step is a biodesulfonation reaction in which a monooxygenase enzyme reaction requiring molecular oxygen might be involved (Swisher, 1987). The hydroxysulfonate (keto-bisulfite) formed may then be hydrolysed yielding 2-dodecanone and sulfate; the sulfur-free intermediate may be subsequently oxidized to an ester by sub-terminal oxidation. Ester cleavage yields acetic acid and an alcohol which can enter the β-oxidation pathway (Figure 5.4). However, Swisher (1987) also discusses the possibility of an alkane sulfonate degradation starting with the oxidation of the terminal methyl group of the alkyl chain (ω-oxidation), continuing with β-oxidation and, finally, cleaving the sulfonate group; this pathway has not been proved yet but some reported degradation results would be compatible with such a mechanism having some resemblance to the LABS degradation pathway.

*5.4.3.4  Anaerobic degradation.*  Few direct data referring to the anaerobic biodegradability of alkane sulfonates are available. According to Bruce *et al.* (1966), SAS is not degraded anaerobically. Assuming that the previously discussed proposals of the degradation mechanism of alkane sulfonate are valid in practice, it can be expected that this surfactant is poorly biodegradable in the absence of oxygen. Both proposed mechanisms for the initial microbial attack of alkane sulfonate, i.e. oxidative desulfonation or ω-oxidation, require molecular oxygen. Thus, in analogy to LABS whose anaerobic recalcitrance and the underlying biochemical reasons have been discussed previously (5.4.2.4), SAS is also expected to be recalcitrant under anaerobic conditions.

$$C_9\text{—}CH_2\text{—}CH(CH_3)\text{—}SO_3^- \longrightarrow C_9\text{—}CH_2\text{—}CHOH(CH_3)\text{—}SO_3^- \longrightarrow C_9\text{—}CH_2\text{—}CO\text{—}CH_3 (+ HSO_3^-)$$
$$\longrightarrow C_9\text{—}CH_2\text{—}O\text{—}CO\text{—}CH_3 \longrightarrow C_9\text{—}CH_2OH + H_3C\text{—}CO_2H$$

**Figure 5.4** Proposed scheme of the primary steps of SAS degradation (acc. to Swisher, 1987).

### 5.4.4  Alpha olefine sulfonates (AOS)

*5.4.4.1  Data on primary degradation.*  AOS representing a mixture of about equal portions of linear alkene sulfonates and linear hydroxyalkane sulfonates reacts with the cationic dye, methylene blue, thus allowing the tracing of their primary biodegradation with the MBAS analytical procedure (Table 5.3). The data available for primary biodegradation of AOS in discontinuous tests (Swisher,1987) indicate unequivocally a rapid and virtually quantitative removal of the parent surfactant in those systems. Kravetz *et al.* (1982) reported MBAS removals between 96 and 100% within 3 days for defined alkyl chain length homologues ($C_{12}$–$C_{18}$) and mixtures as well. Correspondingly, for commercial $C_{14-18}$-AOS a 99% MBAS removal was found in the OECD Screening test; in the CAS-type OECD Confirmatory test the corresponding result was 98% (Schöberl *et al.*, 1988). Based on investigations by Sekiguchi *et al.* (1975) analysing test samples of individuals of AOS homologues ($C_{15}$–$C_{18}$) which were approximately 50% degraded, it could be concluded that the primary biodegradation of the alkene sulfonate is faster than the disappearance of the hydroxy-alkanes.

*5.4.4.2  Data on ultimate biodegradation.*  The data on the AOS biodegradation in OECD tests for ready biodegradability characterize commercial AOS as a surfactant group being readily accessible to ultimate biodegradation. In the Modified OECD Screening test an 85% DOC removal, in the Closed Bottle test 85% BOD/COD and in the Sturm test a 65–80% $CO_2$ evolution were reported (Schöberl *et al.*, 1988). There seems to be only a slight or no significant difference of the mineralization extent between the individual $C_{14-18}$ homologues, as indicated by the results from Tuvell *et al.* (1978) and Kravetz *et al.* (1982) in a $CO_2$ evolution test. No difference was found either between the mineralization rates of $C_{14-18}$ alkane sulfonates and the corresponding hyroxyalkane sulfonates (Tuvell *et al.*, 1978).

In the continuous sewage model plant test system, the Coupled Units test, DOC removals for $C_{16-18}$-AOS were $70 \pm 5\%$ and $78 \pm 3\%$, respectively, being somewhat lower than the rates reported for SAS (Schöberl *et al.*, 1988).

*5.4.4.3  Biodegradation pathways.*  Similarly to SAS, very little knowledge exists about the degradation mechanisms of AOS. In the investigation into the enzymatic biodesulfonation of primary alkane sulfonates, AOS-type model compounds were also tested (Thysse and Wanders, 1974; Swisher, 1987). The results might indicate that biodesulfonation is also the initial degradation step in the case of AOS catalysed by an alkane sulfonate-$\alpha$-hydroxylase yielding, in the end, a desulfonated ketene which could be hydrolysed to the corresponding acid (Swisher, 1987) (Figure 5.5). As yet, no concrete studies are known to prove or disprove this hypothesis.

$$C_6C=C-SO_3^- \longrightarrow C_6C=COH-SO_3^- \longrightarrow C_6C=C=O(+ HSO_3^-) \longrightarrow C_6C-CO_2H$$

**Figure 5.5** Proposed scheme of the initial degradation steps of primary alkane sulfonate (acc. to Swisher, 1987).

*5.4.4.4 Anaerobic degradation.* The assumption that AOS degradation may proceed according to a mechanism similar to SAS and, hence, anaerobic biodegradability should be poor, is supported by the findings of Itoh *et al.* (1987) who compared the anaerobic biodegradability of several anionic and non-ionic surfactants on the basis of digester gas formation: AOS and LABS constituted the compounds with the poorest degradation results. Thus, these results may support the discussed reasoning for anaerobic recalcitrance of SAS and AOS.

### 5.4.5 α-Sulfo fatty acid esters/methyl estersulfonates (FES)

*5.4.5.1 Data on primary biodegradation.* The primary degradation of α-sulfo fatty acids ('disalts') and their esters (mainly methyl) can be followed by the MBAS analytical procedure and proceeds rapidly (Table 5.4). The data list (Swisher, 1987) quotes a number of references proving MBAS removals of >90% in discontinuous tests within a few days. This agrees with the observation (Gode *et al.*, 1987) of a 99% MBAS removal in the OECD Screening test within 5 days when tallow-based ($C_{16/18}$) estersulfonate was tested. Also the primary degradation rates in the CAS-type OECD Confirmatory test were high exhibiting a 95% MBAS removal for $C_{16/18}$-FES. A similar degradation extent (98%) was obtained for the $C_{12/14}$ derivative in this model sewage plant test system; the results for the ester and the disalt were virtually the same (Stein and Baumann, 1975).

*5.4.5.2 Data on ultimate biodegradation.* Ultimate biodegradability of methyl ester sulfonates was proved in a number of investigations (Table 5.4). In the Closed Botle test 76% BOD/COD was found (Gode *et al.*, 1987) classifying the surfactant as readily biodegradable according to OECD criteria. On the other hand, the data obtained in the Sturm test (42–57% $CO_2$ evolution) (Schöberl *et al.*, 1988) are below this limit. Degradation tests with [14]C-radiolabelled α-sulfo palmitic acid, methyl ester (Steber and Wierich, 1989) in a discontinuous shake flask system similar to the OECD Screening test showed a test concentration-dependent mineralization extent: after 28 days the [14]$CO_2$ evolution was 62–67%, 62% and 55% at concentrations of 0.1, 1 and 5 mg/l, respectively. Since in the Sturm test considerably higher test concentrations are used compared with the Closed Bottle test, this might be the explanation for the mentioned divergent results in ready biodegradability tests. Maurer *et al.* (1977) investigated a series of methyl estersulfonates with individual chain lengths between $C_9$ and $C_{18}$. The

**Table 5.4** Primary and ultimate biodegradation of FES in standard tests

| Test substance | Test method | Analysis | % Biodegradation | Type of biodegrad. | Reference |
|---|---|---|---|---|---|
| FES ($C_{16/18}$) | OECD Screening | MBAS | 99% | Primary | Gode et al., 1987 |
| | OECD Confirmatory | MBAS | 95% | Primary/STP-model | Gode et al., 1987 |
| ($C_{12/14}$) | OECD Confirmatory | MBAS | 98% | Primary/STP-model | Stein and Baumann, 1975 |
| ($C_{16/18}$) | Closed Bottle | BOD/COD | 76% | Ultimate/ready | Gode et al., 1987 |
| | Sturm | $CO_2$ | 42–57% | Ultimate/ready | Schöberl et al., 1988 |
| | Coupled Units (CAS) | DOC | 98% | Ultimate/STP-model | Gode et al., 1987 |
| | 'Metabolite test' | DOC | $99.6 \pm 1.6\%$ | Ultimate | Gode et al., 1987 |

DOC removal in the 4-week die-Away test increased from 54 to 83% upon increasing the fatty acid chain length.

The data obtained from testing $C_{16/18}$-FES in a continuous sewage treatment model plant, the Coupled Units test, characterize this surfactant as excellently biodegradable under practical conditions: the DOC removal was 98 ± 6% (Gode et al., 1987), i.e. only a very low percentage of the FES-derived carbon was contained in the plant effluent. Additional investigations with $^{14}$C-labelled $C_{16}$-FES as model substance tested in a miniature CAS unit (Gode et al., 1987; Steber and Wierich, 1989) revealed that the mineralization rate of the surfactant is around 60% under continuous sewage plant conditions (3 h retention time); taking the radiolabelled moiety of the activated sludge into account having been analysed as bacterial biomass, an ultimate degradation extent of approximately 90% was obtained which confirms that the high Coupled Unit test result is due to ultimate degradation of this surfactant.

Finally, this conclusion was further substantiated by the result of the 'Test for detecting recalcitrant metabolites' (Gerike and Jasiak, 1986): with 99.6 ± 1.6% C removal the formation of any recalcitrant degradation intermediate can be excluded (Gode et al., 1987).

*5.4.5.3   Biodegradation pathways.*   Based on comparisons of the MBAS and DOC removals and sulfate formation rates of a sulfostearate isopropylester, Swisher (1987) concluded that microbial biodegradation of long-chain sulfo fatty acid derivatives does not start with a desulfonation step. This conclusion was confirmed by the studies of Steber and Wierich (1989) into the primary and ultimate degradation kinetics of $\alpha$-sulfo [U-$^{14}$C]palmitic acid in a discontinuous screening test. The disappearance of the parent compound was complete within 5 days; at this time the maximum of a new fraction of $^{14}$C-labelled compounds emerged which decreased in the course of the following weeks while $^{14}$CO$_2$ formation increased. The closer examination of this transient $^{14}$C intermediate fraction isolated from the effluent of a model CAS unit revealed that the major part of these compounds represented carboxyalkyl derivatives of the parent surfactant ('carboxy-FES'). About 20% of the degradation intermediate fraction consisted of acidic unsulfonated compounds.

Based on these results a degradation pathway for FES was proposed (Steber and Wierich, 1989) starting with $\omega$-oxidation of the terminal alkyl group and continuing with $\beta$-oxidation (Figure 5.6) leaving short-chain carboxyestersulfonates which are subsequently desulfonated. The latter step is assumed to take place prior to the scission of the methyl ester bond since this bond is stabilized by the $\alpha$-sulfo-substituent.

*5.4.5.4   Anaerobic degradation.*   Taking account of the fact that the degradation mechanism of $\alpha$-sulfo fatty acids and the corresponding methyl esters may have some similarity with the LABS degradation route, i.e. starting with $\omega$-oxidation, one can expect that also $\alpha$-sulfo fatty acids and their esters are

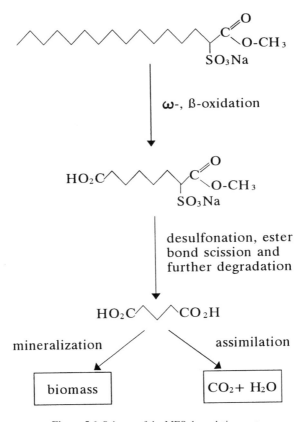

**Figure 5.6** Scheme of the MES degradation route.

poorly degradable under anaerobic conditions. This anticipation was supported by the previous findings of Maurer *et al.* (1965) who could not find any MBAS removal when testing $C_{18}$-$\alpha$-estersulfonates in a continuous anaerobic digester within a 30-day test duration. More detailed investigations into the anaerobic biodegradation behaviour of a [14]C-chain labelled FES in a model digester system confirmed this view (Steber and Wierich, 1989); after a 4-week digestion period the formation of [14]$CO_2$ and [14]$CH_4$ from the $\alpha$-sulfo palmitic acid ('disalt') and the corresponding methyl ester (FES) was less than 5%. The anaerobic incubation of the main (aerobic) degradation intermediate fraction, carboxy-MES, did not yield a significantly higher [14]C gas production confirming again that the sulfonate substituent of a molecule cannot be split off anaerobically. Thus, the data obtained in this FES study fit very well in the view of structural prerequisites for anaerobic biodegradability of surfactants as outlined in the LABS chapter (cf. 5.4.2.4). Also the fate of FES being present in digested sludges will be similar to LABS when the sludge is spread on agricultural land.

Investigations with radiolabelled FES have shown that a fast ultimate degradation of this surfactant happens in aerated soils (Steber and Wierich, 1989).

### 5.4.6 Fatty alcohol sulfates (AS)

*5.4.6.1 Data on primary biodegradation.* The linear primary alkyl sulfates mainly derived from fatty alcohols biodegrade very rapidly in terms of primary degradation with removal rates of almost 100% within 1 or a few days when tested in discontinuous systems (Swisher, 1987). The number of reported test data is abundant (Swisher, 1987) showing that the chain length of the AS molecules has no significant influence on the MBAS removal of this surfactant group. In the standardized OECD Screening test a 99% MBAS removal was reported for fatty alcohol sulfates (Table 5.5) with a $C_{12-18}$ chain length and for $C_{12/15}$-oxoalcohol sulfate (Schöberl *et al.*, 1988). For tallow-based AS the extent of primary degradation was already 95–98% within 5 days incubation (Steber *et al.*, 1988).

Correspondingly, also very high MBAS removals have been found for this surfactant class under model sewage treatment plant conditions, i.e. in CAS systems: the MBAS removals in the OECD Confirmatory test were in the range of 98–99% for linear $C_{12-18}$ alcohol and $C_{12-15}$-oxo-alcohol sulfates (Schöberl, 1988; Steber *et al.*, 1988). Also in sea water a very fast primary biodegradation of AS was measured (Vives-Rego *et al.*, 1987): the half-life time of sodium dodecyl sulfate was in the range of 0.25–0.34 days, whereas a period of 6–9 days was found for LABS. The data compilation by Swisher (1987) shows very clearly, on the other hand, that multiple branchings of the alkyl chain reduce the rate and the extent of primary biodegradation of alcohol sulfates dramatically.

*5.4.6.2 Data on ultimate biodegradation.* Linear or single branched-chain alcohol sulfates are excellently degradable not only in terms of primary biodegradation but also regarding their ultimate biodegradability (Table 5.5). All types of OECD tests for 'ready biodegradability' confirm unequivocally that this surfactant group ($C_{12-18}$-AS, $C_{12-15}$ oxo alcohol sulfate) exceeds the OECD criteria easily: 88–96% DOC removal in the Modified OECD Screening test, 63–95% BOD/COD in the Closed Bottle test and 64–96% $CO_2$ evolution in the Sturm test (Schöberl *et al.*, 1988). The comparison of the degradation rates of coco-based ($C_{12/14}$) and tallow-based ($C_{16/18}$) AS in the Modified OECD Screening test and in the Closed Bottle test indicated a slightly higher degradation extent of the shorter chain homologues which may be explained by the lower water solubility of the $C_{16/18}$-AS; however, in both cases the BOD/COD was more than 50% within 5 days incubation in the Closed Bottle test underlining the very rapid ultimate biodegradation of AS in this relatively stringent test system (Steber *et al.*, 1988). Of course, there exist a vast number of further published data on ultimate biodegradation of AS (Swisher, 1987) but the

**Table 5.5** Primary and ultimate biodegradation of AS in standard tests

| Test substance | Test method | Analysis | % Biodegradation | Type of biodegrad. | Reference |
|---|---|---|---|---|---|
| AS ($C_{12-18}$) | OECD Screening | MBAS | 99% | Primary | Schöberl et al., 1988 |
| ($C_{16/18}$) | OECD Screening | MBAS | 99% | Primary | Steber et al., 1988 |
| ($C_{16/18}$) | OECD Confirmatory | MBAS | 98–99% | Primary/STP model | Steber et al., 1988 |
| ($C_{12-18}$, OXO-$C_{12-15}$) | OECD Confirmatory | MBAS | 98–99% | Primary/STP model | Schöberl et al., 1988 |
| ($C_{12-18}$, OXO-$C_{12-15}$) | Modif. OECD Screening | DOC | 88–96% | Ultimate/ready | Schöberl et al., 1988 |
| ($C_{12-18}$, OXO-$C_{12-15}$) | Closed Bottle | BOD/COD | 63–95% | Ultimate/ready | Schoberl et al., 1988 |
| ($C_{12-18}$, OXO-$C_{12-15}$) | Sturm | $CO_2$ | 64–96% | Ultimate/ready | Schöberl et al., 1988 |
| ($C_{12/14}$) | Modif. OECD Screening | DOC | 91% | Ultimate/ready | Steber et al., 1988 |
| ($C_{12/14}$) | Closed Bottle | BOD/COD | 90–94% | Ultimate/ready | Steber et al., 1988 |
| ($C_{16/18}$) | Modif. OECD Screening | DOC | 85–88% | Ultimate/ready | Steber et al., 1988 |
| ($C_{16/18}$) | Closed Bottle | BOD/COD | 77% | Ultimate/ready | Steber et al., 1988 |
| ($C_{12/14}$) | Coupled Units (CAS) | DOC | 97% | Ultimate/STP model | Steber et al., 1988 |
| ($C_{16/18}$) | Coupled Units (CAS) | DOC | 96% | Ultimate/STP model | Steber et al., 1988 |
| ($C_{16/18}$) | CAS (uncoupled) | DOC | 94–96% | Ultimate/STP model | Schöberl et al., 1988 |
| ($C_{12/14}$) | 'Metabolite test' | DOC | $99.9 \pm 1.6\%$ | Ultimate | Steber et al., 1988 |

discussed results seem sufficiently representative for the characterization of this surfactant group.

The results from testing linear or single branched alcohol sulfates in continuous sewage model plants are in full agreement with the conclusions drawn from screening tests. In the Coupled Units-CAS test $97 \pm 7\%$ and $96 \pm 2\%$ carbon removals were found for coco- and tallow-based AS (Steber et al., 1988). In a CAS test without sludge exchange the carbon removal rates were in the range of 94–99% for $C_{16-18}$-AS (Schöberl et al., 1988). Additional studies were carried out with the poorly soluble stearyl fatty alcohol sulfate which was uniformly $^{14}C$-labelled in the alkyl chain (Steber et al., 1988). After continuous dosage of this material to a miniature CAS unit, a 60% $^{14}CO_2$ evolution rate at steady-state conditions was determined while just 10% of the added radioactivity ($A_0$) was recovered as dissolved radioactive material in the effluent. Only 0.3% of $A_0$ was found in this model plant effluent to be due to residual surfactants proving a primary degradation rate of >99% and a carbon removal of >90%. About 30% of $A_0$ found its way into the activated sludge; more than 90% of this was attributable to bacterial biomass. This high degree of incorporation of radiocarbon into biomass and the high mineralization rate confirm that the substantial carbon elimination rates measured in CAS systems reflect real ultimate degradation even in the case of a poorly soluble AS homologue. On account of these results it is not surprising that the residue-free ultimate biodegradability of fatty alcohol sulfates could also be verified in the 'Test for detecting recalcitrant metabolites' (Gerike and Jasiak, 1986). In this test the result of $99.9 \pm 1.6\%$ carbon removal for the water-soluble coco fatty alcohol sulfate ruled out the formation of even the smallest theoretically possible stable metabolite, a $C_1$-compound (Steber et al., 1988).

*5.4.6.3  Biodegradation pathways.*    Since linear alcohol sulfates exceed all other surfactants in speed of primary and ultimate biodegradation (Swisher, 1987) it can be assumed that the processes involved in the biodegradation of AS are effective in a broad range of microbial species. There are several in-depth studies dealing with the primary degradation steps of linear alcohol sulfates (Swisher, 1987). According to these results, the initial attack of these molecules is effected by the enzymatic cleavage of the sulfate ester bond leaving inorganic sulfate and a fatty alcohol (Figure 5.7). It is understandable that this hydrolytic step catalysed by alkylsulfatases will lead to the immediate loss of the ability to form a complex with the methylene blue dye explaining the rapid primary biodegradation of AS. It is noteworthy that the alkylsulfatase activity of microbial cells is in many cases due to the presence of several different enzymes having individual specificities with regard to the position of the sulfate group (primary/secondary AS) and chain length of the alkylsulfate (Swisher, 1987). In many of the bacterial strains used for closer examination of the alkyl sulfatase these enzymes seemed to be adaptive, i.e. their synthesis in the bacterial cell was induced by the presence of primary fatty alcohol sulfate. Correspondingly,

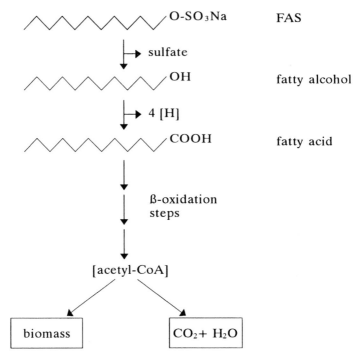

**Figure 5.7** Scheme of the FAS degradation route.

an investigation of the occurrence of sulfatases in environmental bacteria by White *et al.* (1985) has shown that bacterial strains having a constitutive sulfatase are considerably more prevalent in polluted than in clean river sites.

The next step in the degradation route of alkylsulfates is the enzymatic oxidation of the fatty alcohol by dehydrogenases yielding at first the corresponding aldehyde and subsequently the fatty acid. The dehydrogenase enzymes active on linear primary alcohols from $C_2$ to $C_{12}$ seem to be constitutive (Swisher, 1987), i.e. permanently present in the degrading organisms, whereas the capability of oxidation of the higher alcohols may be readily induced when the constitutive dehydrogenase is synthesized in larger amounts. The further degradation of the fatty alcohol sulfate intermediates, the fatty acids, proceeds via β-oxidation enabling the utilization of the organic moiety of the molecule for energy and biomass production. This was nicely demonstrated in a study by Thomas and White (1989) on degradation of [14]C-labelled sodium dodecyl sulfate by *Pseudomonas* C12 B strain: 70% of the label was released as $CO_2$; the residual radioactivity was incorporated almost quantitatively into cells. Analysis of the extractable lipids established the sequential formation of 1-dodecanol, dodecanal and dodecanoic acid. At this point the pathway diverged leading either to the formation of carbon dioxide via β-oxidation or to elongation of fatty acyl residues by $C_2$ fragments (originating from $\beta$-oxidation, too) with rapid incorporation into lipid fractions.

In a few publications on the biochemical pathways of linear alkylsulfates the possible existence of an alternative degradation route has been discussed, i.e. via $\omega$- and subsequent $\beta$-oxidation leaving carboxylated alkylsulfate intermediates. However, the critical evaluation of these findings by Swisher (1987) led to the conclusion that none of these experiments gave unequivocal evidence for the existence of this pathway.

*5.4.6.4 Anaerobic degradation.* Whereas information on anaerobic biodegradability of most surfactants is relatively scarce, more literature reporting on the anaerobic biodegradation behaviour of fatty alcohol sulfates is available. Already from the knowledge on the degradation pathway of AS in aerobic systems it can be anticipated that the mechanisms involved should also work in the absence of molecular oxygen. As a matter of fact, MBAS removals of $\geq 90\%$ have been reported by several authors after incubation of $C_{12-18}$ fatty alcohol sulfates in anaerobic test systems (Swisher, 1987). The work of Oba *et al.* (1967) suggested that anaerobic degradation of these surfactants is a simple hydrolysis without further degradation of the fatty alcohol formed. This is in contrast to more recent findings by Steber *et al.* (1988) who studied the anaerobic degradation of radiolabelled $C_{12}$- and $C_{18}$ fatty alcohol sulfates in a model digester. After a 4-week test period more than 90% of the surfactant radioactivity was recovered in the gaseous final degradation products, i.e. carbon dioxide and methane. Also in another more stringent anaerobic test system, the ECETOC Screening test (Birch *et al.*, 1989) the excellent anaerobic biodegradability of stearyl sulfate was confirmed by the result of a gas production equivalent to 88% of the surfactant's organic carbon. Ultimately, also the test data from Wagener and Schink (1987) confirmed the anaerobic degradation of AS to $CO_2$ and methane and showed the conversion of sulfate to sulfide.

## 5.4.7  Alcohol ether sulfates (AES)

*5.4.7.1 Data on primary biodegradation.* Alcohol ether sulfates with a linear or single branched alkyl chain show an excellent primary biodegradation (Table 5.6) which is comparable with the MBAS removals found for fatty alcohol sulfates (Swisher, 1987). Even for linear AES with higher ethoxylation numbers, e.g. 8.5 EO units, MBAS removals of 96% in the Closed Bottle test after 30 days incubation have been reported (Fischer, 1982). However, multiple branchings of the alkyl chain reduce the primary biodegradation extent considerably (Swisher, 1987). For $C_{12/14}$-fatty alcohol + 2 EO sulfate and $C_{12/15}$-oxoalcohol + 3 EO sulfate MBAS removals of 98–99% were obtained in the OECD Screening test (Schöberl *et al.*, 1988). A similar high extent of primary degradation, i.e. 96% MBAS removal, was reported for these surfactants when tested in the

**Table 5.6** Primary and ultimate biodegradation of AES in standard tests

| Test substance | Test method | Analysis | % Biodegradation | Type of biodegrad. | Reference |
|---|---|---|---|---|---|
| AES | | | | | |
| $(C_{12/14} + 2\ EO)$ | $(C_{12-18} + 8.5\ EO)$ | Closed Bottle | MBAS | 96% (30 days) | Primary Fischer, 1982 |
| and $(oxo\text{-}C_{12-15} + 3\ EO)$ | OECD Screening | MBAS | 98–99% | Primary | Schöberl et al., 1988 |
| | OECD Confirmatory | MBAS | 96% | Primary/STP model | Schöberl et al., 1988 |
| $(C_{12-18} + 8.5\ EO)$ | OECD Confirmatory | MBAS | 100% | Primary/STP model | Fischer, 1982 |
| $(C_{12/14} + 2\ EO)$ | Modif. OECD Screening | DOC | 96–100% | Ultimate/ready | Schöberl et al., 1988 |
| and $(oxo\text{-}C_{12-15} + 3\ EO)$ | Closed Bottle | BOD/COD | 58–100% | Ultimate/ready | Schöberl et al., 1988 |
| | Sturm | $CO_2$ | 65–83% | Ultimate/ready | Schöberl et al., 1988 |
| | Coupled Units (CAS) | DOC | 67–99% | Ultimate/STP model | Schöberl et al., 1988 |
| $(C_{12-18} + 8.5\ EO)$ | Closed Bottle | BOD/COD | 100% | Ultimate/ready | Fischer, 1982 |
| $(C_{12/14} + 2\ EO)$ | 'Metabolite test' | DOC | $102.1 \pm 3.9\%$ | Ultimate | Gerike and Jasiak, 1986 |

CAS-type OECD Confirmatory test (Schöberl *et al.*, 1988). Even a $C_{12-18}$-fatty alcohol + 8.5 EO sulfate showed an almost 100% MBAS removal in this model sewage treatment plant (Fischer, 1982) suggesting again that the primary biodegradation of linear AES is excellent and is largely unaffected by the alkyl chain length and the degree of ethoxylation.

*5.4.7.2  Data on ultimate biodegradation.*  The overwhelming portion of data reported on the ultimate biodegradability of linear AES form a sound basis for the conclusion that this surfactant group is readily and completely biodegradable (Swisher, 1987). In standardized tests (Table 5.6) like the OECD tests for ready (ultimate) biodegradability, degradation rates of 96–100% carbon removal in the Modified OECD Screening test, 58–100% BOD/COD in the Closed Bottle test and 65–83% $CO_2$ evolution in the Sturm test have been reported for $C_{12/14}$-fatty alcohol + 2 EO sulfate and $C_{12/15}$-oxoalcohol + 3 EO sulfate; these figures are very similar to those measured for AS (Schöberl *et al.*, 1988). Even long-chain homologues like $C_{12/18}$-fatty alcohol + 8.5 EO sulfate, exhibited a degradation rate of almost 100% BOD/COD in the Closed Bottle test (Fischer, 1982) under-lining that ultimate biodegradation of linear AES is not very strongly affected by the chain lengths of the alkyl and the polyoxyethylene portions of the molecule. However, in the case of linear secondary AES, multiple branched oxoalcoho-lether sulfates or AES based on alcohol propoxylates, the extent of ultimate degradation deteriorates considerably (Swisher, 1987).

The data on ultimate biodegradation of alcohol ether sulfates obtained in CAS test systems exhibit a broader range of DOC removal rates than reported for fatty alcohol sulfates: the carbon removals for $C_{12/14}$-fatty alcohol + 2 EO sulfate and $C_{12/15}$-oxoalcohol + 3 EO sulfate were between 67 and 99% in the Coupled Units test (Schöberl *et al.*, 1988). It might be suggested that the presence of the polyoxyethylene group in the molecule and, as Gerike and Jakob (1988) showed on alcohol ethoxylates, the strong dependence of their biodegradability on the retention time in the CAS system may account for these wavering results. Nevertheless, in the case of AES, a complete ultimate biodegradability without formation of any poorly degradable intermediates was proved by testing $C_{12/14}$-fatty alcohol + 2 EO sulfate in the 'Test for detecting recalcitrant metabolites' (Gerike and Jasiak, 1986): the result of $102.1 \pm 3.9\%$ carbon removal is sufficient to exclude any theoretically possible 'biologically hard' metabolite.

*5.4.7.3  Biodegradation pathways.*  Based on the knowledge about basic degradation routes of sulfated and sulfonated anionics and alcoholethoxylates, three starting points for the microbial degradation of AES might be possible: (i) $\omega$-/$\beta$-oxidation of the alkyl chain, (ii) cleavage of the sulfate substituent and subsequent degradation of the residue according to mechanisms discussed for alcohol ethoxylates (discussed elsewhere), (iii) cleavage of an ether bond in the molecule, e.g. between the alkyl chain and the polyethylene glycol sulfate moiety or between two EO-units. It seems that all of these possible mechanisms are realized (Swisher, 1987).

A detailed investigation (Hales *et al.*, 1982) of dodecylalcohol + 3 EO sulfate (SDTES) biodegradation by four bacterial isolates showed that the *Pseudomonas* strains studied used the ether cleavage pathway predominantly and the desulfation route only to a minor extent. In two of the isolates the involved etherases were not specific, thus, cleaving each of the three ether bonds of the molecule in reproducible proportions. In two other strains the specificity of the etherase enzyme was much higher and cleavage occurred almost entirely at the alkyl–ether linkage. However, the $\omega$-/$\beta$-oxidation pathway also appears to be followed to a minor extent by some test strains and mixed cultures (Yoshimura and Masuda, 1982; Hales *et al.*, 1986; White and Russell, 1988). To sum up, the ether cleavage pathway (Figure 5.8) seems to be the predominant one in AES degradation because of several advantages over the alternative routes (Hales *et al.*, 1986): (i) although sulfatase cleavage is established in the degradation of alkylsulfates, desulfonation of alkyl ether sulfates produces alkylethoxylates which still must undergo ether cleavage to render the alkyl chain accessible to further degradation/assimilation, (ii) $\beta$-oxidation becomes progressively

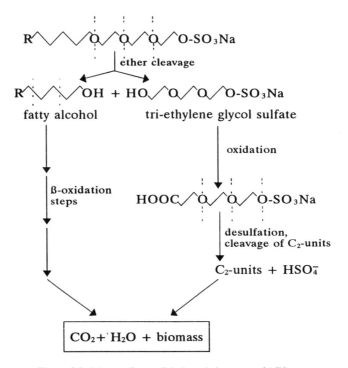

**Figure 5.8** Scheme of a possible degradation route of AES.

inefficient when the site of cleavage approaches the hydrophilic moiety; in contrast, ether cleavage at the alkyl–ether linkage liberates an alkyl, the rest being fully accessible to further utilization.

Although the glycol ether sulfate residues resulting from ether cleavage of SDTES were only slowly oxidized to the corresponding carboxylates but were not further degraded by a *Pseudomonas* isolate (Hales *et al.*, 1982, 1986), these intermediates were fully degraded to inorganic sulfate and carbon dioxide in mixed cultures (Griffiths *et al.*, 1986; White and Russell, 1988). Hales (1981) proposed a stepwise pathway for the polyglycol sulfate degradation similar to the mechanisms for PEG (Kawai *et al.*, 1978). Accordingly, the polyglycol sulfate is oxidized to a carboxylated fragment followed by hydrolysis of the terminal $C_2$ unit to yield the next shorter polyglycol sulfate, and repeating this cycle again. It was shown by Hales (1981) that in mixed cultures the desulfation step yielding inorganic sulfate could take place at any stage of the polyglycol sulfate degradation route.

*5.4.7.4 Anaerobic degradation.*   Although there is very little published about the anaerobic biodegradation behaviour of alcohol ether sulfates, nevertheless, the information on existing aerobic degradation pathways allows at least general considerations on possible routes in the absence of oxygen. For linear alcohol ethoxylates, anaerobic biodegradation via central fission (forming the free alcohol and a hydrophilic PEG moiety) or by stepwise cleavage of $C_2$ units from the OH-terminus of the molecule has been established (Swisher, 1987; Steber and Wierich, 1987). Thus, both the central ether cleavage pathway as shown for aerobic degradation and the desulfation mechanism followed by stepwise ether cleavage reactions are possible routes for anaerobic degradation of AES. Indications for the real existence of anaerobic degradation mechanisms of AES were provided by Oba *et al.* (1967) who recorded a significant MBAS removal when $C_{12/14}$-alcohol + 3 EO sulfate was incubated anaerobically; however, a $C_{16}$-alcohol substituted by one butoxy sulfate or one propoxy sulfate group was degraded to a considerably less extent in an anaerobic river water test system (Maurer *et al.*, 1971). In addition, Painter (1992) reported data from long-term studies showing a high MBAS removal after a 6–8 months incubation of $C_{14-18}$-alcohol + 3 EO-sulfate in a lab septic tank system and an anaerobic digester, respectively. Ultimately, the experimental comparison of several types of anionic surfactants under anaerobic test conditions (Itoh *et al.*, 1987) revealed $CO_2$ and methane production from AES, thus proving ultimate biodegradability, at a less rapid rate, however, than found for alkylsulfates.

### 5.4.8  Sulfosuccinates

*5.4.8.1 Data on primary degradation.*   Biodegradation of sulfosuccinates, i.e. the diesters of sulfosuccinic acid, is strongly influenced by the structural properties

of the alkyl groups. Biodegradation tests of several sulfosuccinate types in a
river water die-away system (Hammerton, 1956) yielded MBAS removal rates
equivalent to a half-life time of 4.5 days for *n*-octyl sulfosuccinate and 7.5 days
for the 2-ethylhexyl isomer. Other substituents like 3.5.5-trimethylhexanol or
isobutanol decreased the speed of primary degradation even more. Thus it can
be concluded that branching of the alkyl moiety results in a drop of the MBAS
removal rate. Nevertheless, the data of diethylhexyl sulfosuccinates from
discontinuous degradation tests are in the range of >90% MBAS removal within
2–3 weeks (Swisher, 1987). In the standardized OECD Screening test lasting 19
days a 97% MBAS decrease was reported for dioctyl sulfosuccinate (Table 5.7);
correspondingly, the evaluation of the same compound in the CAS-type OECD
Confirmatory test yielded a 95% MBAS removal (Schöberl *et al.*, 1988).

*5.4.8.2 Data on ultimate degradation.* The basis of published data for
evaluating the ultimate biodegradability of sulfosuccinaes is relatively small.
Cordon *et al.* (1970) found an 83% DOC removal for di-2-ethylhexyl sulfosuccinate in a 3-week screening test inoculated with small amounts of activated
sludge; however, the release of inorganic sulfate was very poor in this experiment. Since sulfosuccinic acid itself was not degraded under these test conditions, it could be concluded that the high carbon removal was only due to the
ultimate degradation of the two alkyl residues. Dioctyl sulfosuccinate was
degraded in the Closed Bottle test only to an extent of 50% BOD/COD within
the 4-week test duration (Table 5.6) so that this surfactant does not seem to be
'readily biodegradable' according to OECD criteria (Schöberl *et al.*, 1988). Also
in the continuous model sewage treatment plant, the Coupled Units test, the
DOC removal was moderate with a DOC removal of 49 ± 13% (Schöberl *et al.*,
1988). According to the discussed influence of the alkyl structure on the
(primary) biodegradation rate of sulfosuccinates it is anticipated that representatives of this surfactant group with linear alkyl residues will exhibit higher results
in the OECD tests for ready biodegradability.

*5.4.8.3 Biodegradation pathways.* As the scarce database indicates, relatively little work has been done so far to elucidate the biodegradation routes of this
surfactant group. The large MBAS removal of a number of primary sulfosuccinates (cf. 5.4.8.1) and the findings by Cordon *et al.* (1970), i.e. high carbon
removal but no release of inorganic sulfate (cf. 5.4.8.2), may indicate that degradation starts with the scission of the ester bonds and is followed by $\beta$-oxidation
of the carboxylates formed from the alcohol residues. The poor degradation
result reported for sulfosuccinic acid may indicate that this core element of the
sulfosuccinate surfactants is not readily biodegraded by environmental microorganisms. Nevertheless, it can be anticipated that this molecule is inherently
biodegradable since no principal obstacles of an oxidative desulfonation seem to
exist according to mechanisms known from other sulfonated surfactants.

**Table 5.7** Primary and ultimate biodegradation data of sulfosuccinates in standard tests

| Test substance | Test method | Analysis | % Biodegradation | Type of biodegrad. | Reference |
|---|---|---|---|---|---|
| Dioctyl sulfosuccinate | OECD Screening | MBAS | 97% | Primary | Schöbert et al., 1988 |
| | OECD Confirmatory | MBAS | 95% | Primary/STP model | Schöberl et al., 1988 |
| | Closed Bottle | BOD/COD | 50% | Ultimate/ready | Schöberl et al., 1988 |
| | Coupled Units (CAS) | DOC | 49% | Ultimate/STP model | Schöberl et al., 1988 |

*5.4.8.4  Anaerobic biodegradation.*  No information exists in the literature on the anaerobic biodegradability of sulfosuccinic acid diesters. Due to the presence of the ester bonds in the molecule it can be expected, however, that primary biodegradation by ester cleavage will take place followed by oxidation of the alcohol residues to the corresponding acids which are subsequently degraded via $\beta$-oxidation. The water soluble sulfosuccinate is not expected to undergo anaerobic biodegradation since cleavage of the C–S bond has been found to occur only in the presence of molecular oxygen. Thus, in spite of the deficiency of concrete data, it can be assumed that a primary but not an ultimate degradation of sulfosuccinates takes place in the anaerobic environment.

## 5.4.9  Phosphate ester surfactants

*5.4.9.1  Data on biodegradation.*  In line with their minor technical importance, the information on the biodegradation behaviour of phosphate ester surfactants is very scarce. Phosphate-substituted surfactants do not react with the methylene blue dye so that this analytical tool for tracing primary biodegradation is not applicable. Cooper and Urfer (1964) reported on a complete loss of surface tension after a 10-day incubation of didecylphosphate in a river water die-away system. Based on measurements of foamability and surface tension, Ropuszynski *et al.* (1979) determined the primary biodegradation of a number of phosphate esters in discontinuous tests. Different salts of $C_{12}$-alcohol + 4–10 EO phosphates and nonylphenol + 3–14 EO phosphates lost their primary surfactant properties almost completely within an incubation time of 3–15 days; based on surface tension measurements, the primary biodegradation of octyl- and nonylphenol phosphates was complete within 1–3 days. When measuring the degradation extent of these surfactants by the determination of the loss of phosphate esters linkages, values between 20–90% were obtained. In the case of $C_{12}$-alcoholethoxylate and nonylphenolethoxylate derivatives this decrease of ester bonds corresponded well with the loss of surface tension; in the case of octyl- and nonylphenol phosphates the organic ester removal took considerably more time than the decrease of foamability and surface tension.

Even less data are available on the ultimate degradation of phosphate ester surfactants. Winter (1962) reported a 66% oxygen uptake when testing the triethanolamine salt of a primary alkylphosphate in the Warburg respirometer for 1 day. Experimental results from sewage treatment simulation tests with phosphate esters are completely lacking.

*5.4.9.2  Biodegradation pathways.*  No particular information on the biodegradation route of phosphate ester surfactants has been described in the literature. However, it seems sensible to assume that the multitude of phosphate ester compounds being present in living organisms (e.g. nucleic acids, phospholipids, etc.) parallel with the ubiquitous presence of phosphatases, i.e. enzymes splitting phosphate ester bonds. Thus, it can be expected that some of these phosphatases are also able to attack the ester linkages of phosphate ester surfactants. Also the possible step sequence of the degradation process is speculative. However, the relatively good comparability of the rates determined for the loss of surface tension and the ester linkage removal as found by Ropuszynski *et al.* (1979) for alkylethoxylate and nonylphenolethoxylate phosphates (cf. 5.4.9.1), may indicate that the ester cleavage is the primary degradation step. The discrepancy of these rates observed in the case of octyl- and nonyl-phenol phosphates suggests, however, that it is not yet possible to propose a consistent view of the degradation route of this surfactant class.

*5.4.9.3 Anaerobic degradation.* As one can expect from the scarcity of aerobic degradation data, no information on anaerobic biodegradability is available. Assuming that ester cleavage is the primary degradation step followed by oxidative degradation of the released alkyl residues, no principal obstacles of the anaerobic biodegradation of these surfactants are recognizable. However it must be emphasized that too little knowledge exists as yet to regard this as more than a speculation.

# References

Allred, R.C., Setzkorn, E.A. and Huddleston, R.L. (1964) Detergent biodegradability as shown by various analytical techniques. *J. Am. Oil Chem. Soc.*, **41**, 13–17.

Baggi, G., Catelani, D., Galli, E. and Treccani, V. (1972) The microbial degradation of phenylalkanes: 2-phenylbutane, 3-phenylpentane and 4-phenylheptane. *Biochem. J.* **126**, 1091–1097.

Berth, P. and Jeschke P. (1989) Consumption and fields of application of LABS. *Tenside Deterg.* **26**, 75–79.

Biermann, M., Lange, F. Piorr, R., Ploog, U., Rutzen, H., Schindler, J. and Schmid, R. (1987) Synthesis of surfactants, in *Surfactants in Consumer Products. Theory, Technology and Application*, ed. J. Falbe, Springer-Verlag, Berlin, Heidelberg, New York, pp. 23–132.

Birch, R.R., Biver, C., Campagna, R., Gledhill, W.E., Pagga, U., Steber, J., Reust, H. and Bontinck, W.J. (1989) Screening of chemicals for anaerobic biodegradability. *Chemosphere*, **19**, 1527–1550.

Bruce, A.M., Swanwick, J.D. and Ownsworth, R.A. (1966) Synthetic detergents and sludge digestion: Some plant observations. *J. Proc. Inst. Sewage Purif. Pt.* **5**, 427–447.

Cain, R.B., Willetts, A.J. and Bird, J.A. (1972) Surfactant biodegradation: Metabolism and enzymology. in *Biodeterioration of Materials* (eds. A.H. Walters and Hueck van der Plas), Vol. 2. 2nd Int. Biodet. Symp. Lunteren, 1971, Wiley, New York, pp. 136–144.

Cavalli, L., Divo, C., Giufridda, G., Pellizon, T., Radici, P., Valorta, L. and Zatta, A. (1992) Linear alkyl benzene (LAB) from linear olefins with $AlCl_3$ catalyst. In: 3rd CESIO International Surfactant Congress & Exhibition — A world market. 1–5 June, 1992. Proceedings Section–Plenary Lecture — A and B, pp. 105–114.

Coons, D., Dankowski, M., Diehl, M., Jakobi, G., Kuzel, P., Sung, E. and Trabitzsch, U. (1987) Performance in detergents, cleaning agents and personal care products. In: *Surfactants in Consumer Products. Theory, Technology and Application*, ed. J.Falbe, Springer-Verlag Berlin, Heidelberg, New York, pp. 197–398.

Cooper, R.S. and Urfer, A.D. (1964) Sodium dialkyl phosphates: Surfactant properties and use in heavy duty detergents. *J. Am. Oil Chem. Soc.* **41**, 337–340.

Cordon, T.C., Maurer, E.W. and Stirton, A.J. (1970) The course of biodegradation of anionic detergents by analyses for carbon, MBAS and sulfate ion. *J. Am. Oil Chem. Soc.* **47**, 203–206.

Divo, C. and Cardini, G. (1980) Primary and total biodegradation of alkylbenzene-sulfonates. *Tenside Deterg.* **17**, 30–36.

EEC (European Economic Community) (1973a) Council Directive of 22 November 1973 on the approximation of the laws of the member states relating to detergents (73/404/EEC). *Off. J. E. C.* No. L **347**/51 (17.12.73)

EEC (European Economic Community) (1973b) Council Directive of 22 November 1973 on the approximation of the laws of the member states relating to methods of testing biodegradability of anionic surfactants (73/405/EEC). *Off. J. E. C.*, No. L **347**/53 (17.12.73)

EEC (European Economic Community) (1986) Guiding principles for a strategy for biodegradability testing. EEC XI/841/86 Brussels.

Federle, T.W. and Schwab, B.S. (1992) Mineralization of surfactants in anaerobic sediments of a laundromat waste water pond. *Water Res.* **26**, 113–127.

Felletschin, G., Knaut, J. and Schöne M. (1981) Deutsche Hydrierwerke (DEHYDAG) Stationen ihrer Geschichte. Schriften des Werksarchivs der Henkel KGaA Düsseldorf, p. 53. Henkel KGaA, Düsseldorf.

Fischer, W.K. (1975) Biodegradability — An important criterion for the environmental compatibility of surfactants and other product compounds. *Riv. Ital. Sostanze Grasse* **LII**, 373–376.

Fischer, W.K. (1980) Entwicklung der Tensidkonzentration in den deutschen Gewässern 1960–1980. *Tenside Deterg.* **17**, 250–261.

Fischer, W.K. (1982) Important aspects of the ecological evaluation of fatty alcohols and their derivatives, in *Fatty Alcohols — Raw Material Methods and Uses* Henkel, Düsseldorf, pp. 187–222.

Fischer, W.K. and Winkler, K. (1976) Detergentienuntersuchungen im Stromgebiet des Rheins 1958–1975. *Vom Wasser* **47**, 81–129.

Gerike, P. (1984) The biodegradability testing of poorly water soluble compounds. *Chemosphere* **13**, 169–190.

Gerike, P. and Fischer, W.K. (1979) A correlation study of biodegradability determinations with various chemicals in various tests. *Ecotoxicol. Environ. Safety* **3**, 157–173.

Gerike, P. and Fischer, W.K. (1981) A correlation study of biodegradability determinations with various chemicals in various tests. II. Additional results and conclusions. *Ecotoxicol. Environ. Safety* **5**, 45–55.

Gerike, P. and Jasiak, W. (1986) How completely are surfactants biodegraded? *Tenside Deterg.* **23**, 300–304.

Gerike, P and Jakob, W. (1988) Nonionic surfactants in the Coupled Units Test. *Tenside Deterg.* **24**, 166–168.

Gerike, P., Winkler, K., Schneider, W. and Jakob, W. (1989a). Zur Wasserqualität des Rheins bei Düsseldorf. *Tenside Deterg.* **26**, 136–140.

Gerike, P., Winkler, K. and Jakob, W. (1989b). Gewässeruntersuchungen im Stromgebiet des Rheins und ökologische Folgerungen. *Tenside Deterg.* **26**, 270–275.

Gerike, P., Winkler, K., Schneider, W., Jakob, W. and Steber, J. (1991) Mengenbilanzen von Wasch- und Reinigungsmittel-Inhaltsstoffen mit Auswirkungen auf die Gewässer. *Tenside Deterg.* **28**, 86–89.

Gode, P., Guhl, W. and Steber, J. (1987) Ökologische Bewertung von $\alpha$-Sulfofettsäuremethylestern. *Fat Sci. Technol.*, **89**, 548–552.

Griffiths, E.T., Hales, S.G., Russell, N.J., Watson, G.K. and White, G.F. (1986) Metabolite production during the biodegradation of the surfactant sodium dodecyltriethoxy sulphate under mixed-culture die-away conditions. *J. Gen. Microbiol.* **132**, 963–972.

Hales, S.G. (1981) Microbial degradation of linear ethoxylate sulfates. Ph.D. Thesis, Univ. Wales, Cardiff (quoted in Swisher, 1987).

Hales, S.G., Dodgson, K.S., White, G.F., Jones, N. and Watson, G.K. (1982) Initial stages in the biodegradation of the surfactant sodium dodecyltriethoxy sulfate by *Pseudomonas* sp. Strain DES 1. *Appl. Envir. Microbiol.* **44**, 790–800.

Hales, G.S., Watson, G.K., Dodgson, K.S. and White, G.F. (1986) A comparative study of the biodegradation of the surfactant sodium dodecyltriethoxy sulfate by four detergent-degrading bacteria. *J. Gen. Microbiol.* **132**, 953–961.

Hammerton, C. (1956) Synthetic detergents and water supplies. *Proc. Soc. Water Treat. Exam.*, **5**, 145–174.

Huddleston, R.L. and Allred, R.C. (1963) Microbial oxidation of sulfonated alkylbenzenes. *Dev.Int. Microbiol.* **4**, 24–38.

Itoh, S., Naito, S., and Unemoto, T. (1987) Comparative studies on anaerobic biodegradation of anionic and nonionic surfactants. *Eisei Kagaku* **33**, 415–422.

Kawai, F., Kimura, T., Fukaya, M., Tani, Y., Ogata, K., Ueno, T. and Fukami, H. (1978) Bacterial oxidation of polyethylene glycol. *Appl. Environ. Microbiol.* **35**, 679–684.

Kravetz, L., Chung, H. and Rapean, L.C. (1982) Ultimate biodegradation studies of AOS. *J. Am. Oil Chem. Soc.* **59**, 206–210.

Krüger, R. (1964) Recent studies on alkylbenzene sulfonates. *Fette Seifen Anstrichm.* **66**, 217–221.

Larson, R.J. (1979) Evaluation of biodegradation potential of xenobiotic organic chemicals. *Appl. Environ. Microbiol.* **38**, 1153–1161.

Larson, R. (1990) Structure-activity relationships for biodegradation of linear alkylbenzenesulfonates. *Environ. Sci. Technol.* **24**, 1241–1246.

Larson, R.J. and Payne, A.G. (1981) Fate of the benzene ring of linear alkyl-benzene sulfonate in natural waters. *Appl. Environ. Microbiol..* **41**, 621–627.

Lötzsch, K., Neufahrt, A. and Täuber, G. (1979) Comparative tests on the biodegradation of secondary alkane sulfonates using [14]C-labelled preparations. *Tenside Deterg.* **16**, 150–155.

Malz, F. (1991) Neue Ergebnisse aus dem Arbeitsfeld `Tenside und Gewässerschutz'. *Tenside Deterg.* **28**, 482–486.

Matthijs, E. and de Henau, H. (1987) Determination of linear alkyl-benzenesulfonates in aqueous samples, sludges and soils using HPLC. *Tenside Deterg.* **24**, 193–199.

Maurer, E.W., Cordon, T.C., Weil J.K., Nuñez-Ponzoa, M.W., Ault, W.C. and Stirton, A.J. (1965) The effect of tallow based detergents on anaerobic digestion. *J. Am. Oil Chem. Soc.*, **42**, 189–192.

Maurer, E.W., Cordon, T.C. and Stirton, A.J. (1971) Microaerophilic biodegradation of tallow-based anionic detergents in river water. *J. Am. Oil Chem. Soc.* **48**, 163–165.

Maurer, E.W., Weil, J.K. and Linfield, W.M. The biodegradation of esters of α-sulfo fatty acids. *J. Am. Oil Chem. Soc.* **54**, 582–584.

McKenna, E.J. and Kallio, R.E. (1965) The biology of hydrocarbons. *Annu. Rev. Microbiol.* **19**, 183–208.

Mix-Spagl, K. (1990) Untersuchungen zum Umweltverhalten von Seifen. In: Umweltverträglichkeit von Wasch- und Reinigungsmitteln. *Muenchener Beitr. Abwasser Fisch. Flussbiol.* **44**, 153–171.

Neufahrt, A., Lötzsch, K. and Weimer, K. (1980) Radiometric studies on the biodegradation of secondary alkane sulfonates in a model activated sludge system. XI. *Jornadas Comm. Español Deterg.* 105–118.

Oba, K., Yoshida, Y. and Tomiyama, S. (1967) Studies on biodegradation of synthetic detergents. I. Biodegradation of anionic surfactants under aerobic and anaerobic conditions. *Yukagaku* **16**, 517–523.

OECD (Organization for Economic Cooperation and Development) (1971) Pollution by detergents. Determinaton of the biodegradability of anionic surface active agents. OECD, Paris, France.

OECD (Organization for Economic Cooperation and Development) (1981) OECD Guidelines for Testing of Chemicals. Section III. Degradation and Accumulation, pp. 1–11. OECD, Paris, France.

Painter, H.A. (1992) Anionic surfactants, in *The Handbook of Environmental Chemistry,* ed. O. Hutzinger, Vol. 3, Part F: *Anthropogenic Compounds. Detergents,* ed. N.T. de Oude, Springer-Verlag, Berlin, Heidelberg, New York, pp. 1–88.

Painter, H.A. and Zabel, T. (1989) The behaviour of LABS in sewage treatment. *Tenside Deterg.* **26**, 108–115.

Piorr, R. Structure and application of surfactants, in *Surfactants in Consumer Products. Theory, Technology and Application,* ed. J. Falbe, Springer-Verlag, Berlin, Heidelberg, New York, pp. 5–22.

Richtler, H.J. and Knaut, J. (1988) World prospects for surfactants, in *Proc. 2nd World Surfactants Congress,* Paris, 24–27 May 1988. Vol. I, Plenary Lecture: *Surfactants Economics, Raw materials for surfactants,* ASPA, Paris, pp. 3–58.

Richtler, H.J. and Knaut, J. (1991) Surfactants in the nineties. *Seifen Öle Fatte Wachse* **117**, 545–553.

Ropuszynski, S. Perka, J., Mularczyk, E. and Rutkowska, K. (1979) Biodegradation of selected organic phosphate surfactants. *Pollena-TSPK* **23**, 33–36 (data reported in Swisher, 1987).

Schöberl, P. (1979) Biodegradation of 4-phenyldecane sulfonate. *Tenside Deterg.* **16** 146–149.

Schöberl, P. (1989) Basic principles of LAS biodegradation. *Tenside Deterg.* **26**, 86–94.

Schöberl P. and Kunkel, E. (1977) Fish compatibility of the residual surfactants and intermediates from the biodegradation of LAS. *Tenside Deterg.* **14**, 293–296.

Schöberl, P., Bock, K.J. and Huber, L. (1988) Ökologisch relevante Daten von Tensiden in Wasch- und Reinigungsmitteln. *Tenside Deterg.* **25**, 86–98.

Sekiguchi, H., Oba, K., Masuda, M. and Sugiyama, T. (1975) Studies on relative biodegradation rates of alkene vs. hydroxyalkane sulfonates in AOS. *Yukagaku* **24**, 675–679.

Steber, J. (1979) Untersuchungen zum biologischen Abbau von $^{14}$C-ringmarkiertem linearen Alkylbenzolsulfonat in Oberflächenwasser- und Kläranlagenmodellen. *Tenside Deterg.* **16**, 140–145.

Steber, J. (1991) Long term monitoring of residual surfactant concentrations in German river waters. *Expert Meeting on Ultimate Biodegradability.* Bilthoven (NL) 9 November 1990. Proceedings, Overleggroep Deskundigen Wasmiddelen-Milieu, Zeist, pp. 23–36.

Steber, J. and Wierich, P. (1987) The anaerobic degradation of detergent range fatty alcohol ethoxylates. Studies with $^{14}$C-labelled model surfactants. *Water Res.* **21**, 661–667.

Steber, J. and Wierich, P. (1989) The environmental fate of fatty acid α-sulfomethyl esters. Biodegradation studies with a $^{14}$C-labelled model surfactant. *Tenside Deterg.* **26**, 406–411.

Steber, J., Gode, P. and Guhl, W. (1988) Fatty alcohol sulfates. The ecological evaluation of a group of important detergent surfactants. *Soap Cosmet. Chem. Spec.* **64**, 44–50.

Stein, W. and Baumann, H. (1975) α-Sulfonated fatty esters: manufacturing process, properties and applications. *J. Am. Oil Chem. Soc.* **52**, 323–329.

Swisher, R.D. (1963) Transient intermediates in the biodegradation of LAS. *J. Water Pollut. Control Fed.* **35**, 877–892.

Swisher, R.D. (1987) *Surfactant Biodegradation* 2nd edition, revised and expanded. Marcel Dekker, New York and Basel.

Thomas, O.R.T. and White, G.F. (1989) Metabolic pathway for the biodegradation of sodium dodecylsulfate by *Pseudomonas* spec. C 12 B. *Biotechnol. Appl. Biochem.* **11**, 318–327.

Thysse, G.J.E. and Wanders, T.H. (1974) Initial steps in the degradation of *n*-alkane-1-sulfonates by *Pseudomonas. Antonie van Leeuwenhoek* **40**, 25–37.

Tuvell, M.E., Kuehnhanss, G.O., Heidebrecht, G.D., Hu, P.C. and Zielinski, A.D. (1978) AOS–An anionic surfactant system: Its manufacture composition, properties and potential applications. *J. Am. Oil Chem. Soc.* **55**, 70–80.

Vives-Rego, J., Vaqué, M.D., Sanchez, L.J. and Parra J. (1987) Surfactant biodegradation in sea water. *Tenside Deterg.* **24**, 20–22.

Wagener, S. and Schink B. (1987) Anaerobic degradation of nonionic and anionic surfactants in enrichment cultures and fixed-bed reactors. *Water Res.* **21**, 615–622.

White, G.F. and Russell, N.J. (1988) Mechanisms of bacterial biodegradation of alkyl sulphate and alkyl-polyethoxy sulphate surfactants, in *7th Int. Biodeterioration Symp.*, Cambridge, eds. D.L. Houghton, R.N. Smith, and H.O.W. Eggins. Elsevier, Barking, UK, pp. 325–332.

White, G.F., Russell, N.J. and Day, M.J. (1985) A survey of SDS-resistance and alkyl-sulfatase production in bacteria from clean and polluted river sites. *Environ. Pollut.* **A37**, 1–11.

Wickbold, R. (1975) Biodegradation of ABS. *Tenside Deterg.* **12**, 25–27.

Willets, A.J. (1973) Microbial aspects of the biodegradation of synthetic detergents: A review. *Int. Biodetn. Bull.* **9**, 3–10.

Winter, W. (1962) Biodegradation of detergents in sewage treatment. *Wasserwirtsch. Wassertech.* **12**, 265–271.

Yoshimura, K. and Masuda, F. (1982) Biodegradation of sodium alkyl poly(oxyalkylene) sulfates. *J. Am. Oil Chem. Soc.* **59**, 328–332.

# 6 Biodegradability of cationic surfactants

C.G. VAN GINKEL

## 6.1 Introduction

Cationic surfactants refer to molecules with at least one hydrophobic hydrocarbon tail attached to a hydrophilic head-group carrying a positive charge. Of the cationic surfactants especially quaternary ammonium salts are of commercial significance. These compounds consist of one or more hydrophobic alkyl chain(s), benzyl, hydroxyethyl, polyethylene glycol and/or methyl groups linked to a positively charged nitrogen atom (Figure 6.1). Primary, secondary and tertiary long-chain amine salts may be positively charged and can therefore be regarded as cationic surfactants. Most cationic surfactants have straight alkyl chain(s) with lengths between 8 and 24 carbon atoms. These compounds are primarily produced on the basis of natural fats and oils such as tallow fat, coconut oil and palm oil, resulting in mixed alkyl chain lengths in most of the products.

Four properties of the cationic surfactants are the basis for their widespread use: surface activity, adsorption onto negatively charged solids, biocidal activity and their reaction with anionics. Because of these properties the cationic surfactants are used in large amounts in a number of applications such as fabric softeners, disinfectants, demulsifiers, emulsifiers, wetting agents and processing aids. The estimated world production amounts to 350 000 tonnes per year. The discrete positive charge on the quaternary ammonium salts promotes strong adsorption onto negatively charged fabrics. Hence certain derivatives of this class of surfactants are used as fabric softeners accounting for more than 50% of the total production. Use of quaternary ammonium salts as fabric softeners results in a concomitant discharge of quaternary ammonium salts into the environment. Once the cationic surfactants get into the environment they are widely distributed. In the environment quaternary ammonium salts strongly adsorb on negatively charged sludge and sediments.

This distribution and adsorption necessitate extensive knowledge on the characteristics of processes on which the decomposition of the quaternary ammonium salts depends. Degradation should prevent the accumulation of cationic surfactants in the environment because these compounds can pose aquatic hazards. Cationic surfactants may be hazardous to aquatic ecosystems since most of these chemicals are toxic at relatively low concentrations to algae, daphnids and fish (Kappeler, 1982; Lewis and Suprenant, 1983; Lewis and Wee, 1983; Lewis and Hamm, 1986; Lewis et al., 1986; van Leeuwen et al., 1990).

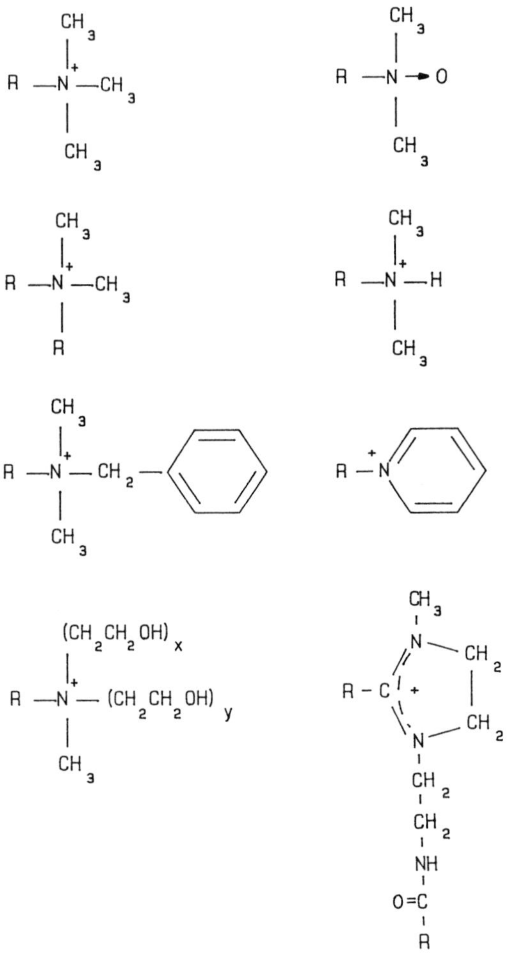

**Figure 6.1** Chemical structures of various cationic surfactants. The R represents a long alkyl chain.

Neufarth and Pleschke (1982) showed that ultraviolet light is able to bring about partial oxidation of quaternary ammonium salts. The resulting products are biodegradable. On the other hand, photodegradation by sunlight results in the formation of 'recalcitrant' products (Ruiz Cruz, 1981). Although the outcome with respect to the biodegradability of the resulting products is contra-dictory, it is very likely that light degrades quaternary ammonium salts to some extent. Recently, Valls *et al.* (1990) assessed the photooxidation of long-chain alkyl amines in pure water and sea water. Biodegradation, however, is by far the most important process to prevent accumulation of cationic surfactants in the environment because photodegradation is slow compared with biological degradation (Krzeminski *et al.*, 1973).

This review contains results obtained with standard biodegradability tests, die-away tests at low concentrations using radiolabelled material and studies on the behaviour of cationic surfactants in biological waste water purification plants. Furthermore, the review deals with the biodegradation route(s) of quaternary ammonium salts and fatty amine salts and provides statements on generalizations for their biodegradability.

## 6.2.  Biodegradability of cationic surfactants in OECD Screening tests

### 6.2.1  OECD screening test results

Test methods to determine the biodegradation primarily include non-specific measurements such as the biological oxygen demand, carbon dioxide production and decrease of dissolved organic carbon. In some cases 'specific' analytical methods are used to determine the concentration of quaternary ammonium salts. A simple method is based on complex formation with a coloured anionic. This complex is extracted from the water phase into a suitable solvent layer. Subsequently the colour of the complex in the solvent layer is measured photometrically (Waters and Kupfer, 1976). A more selective analytical procedure for the quantification of cationic surfactants by coupling HPLC with conductivity detection has been described by Wee and Kennedy (1982).

All standard tests using dissolved organic carbon or the parent molecule concentration as determinant are ignored in this section since disappearance of the test substance by adsorption should not be equated with biodegradation. Semi-continuous activated sludge tests and confirmatory tests using these determinants are primarily simulation tests of activated sludge plants and will therefore be discussed in the next section.

Closed bottle tests, MITI tests and Sturm tests are ready biodegradability tests described in both OECD and EEC test guidelines. In these tests the biodegradability of organic compounds is determined by measuring the oxygen consumption or the production of carbon dioxide. Biodegradation in the ready biodegradability tests is expressed as the ratio of oxygen uptake or carbon dioxide production to the theoretical oxygen demand or theoretical carbon dioxide production. Therefore, a result of 100% biodegradation is impossible because part of the organic carbon is used for the formation of biomass. Experience suggests that a biodegradation percentage over 60 in a screening test may be consistent with the total degradation of the organic compound.

Table 6.1 summarizes the test results of various cationic surfactants in closed bottle tests (CBT), MITI tests and Sturm tests. When comparing the data in Table 6.1, several discrepancies between biodegradability percentages of the alkyltrimethylammonium salts with long alkyl chains attract attention. These discrepancies are caused by inhibitory effects of the quaternary ammonium salts. This phenomenon is discussed in the next paragraph. Despite these toxic effects

**Table 6.1** Biodegradation percentages of quaternary ammonium salts in standard screening tests selected from Swisher (1987) and van Ginkel and Kolvenbach (1991)

| Test substance | Test | Duration (days) | Biodegra-dation (%) | Reference |
|---|---|---|---|---|
| Octyltrimethylammonium chloride | MITI | 10 | 73 | Masuda et al., 1976 |
| Decyltrimethylammonium chloride | MITI | 10 | 63 | Masuda et al., 1976 |
| Dodecyltrimethylammonium chloride | MITI | 10 | 59 | Masuda et al., 1976 |
| Tetradecyltrimethylammonium chloride | MITI | 10 | 35 | Masuda et al., 1976 |
| Hexadecyltrimethylammonium chloride | MITI | 10 | 0 | Masuda et al., 1976 |
| Octadecyltrimethylammonium chloride | MITI | 10 | 0 | Masuda et al., 1976 |
| Decyltrimethylammonium chloride | CBT | 8 | + | Dean-Raymond and Alexander, 1977 |
| Hexadecyltrimethylammonium chloride | CBT | 17 | + | Dean-Raymond and Alexander, 1977 |
| Hexadecyltrimethylammonium chloride | Sturm | 13 | 84 | Larson and Vashon, 1983 |
| Octadecyltrimethylammonium chloride | Sturm | 25 | 0 | Larson and Vashon, 1983 |
| Octadecyltrimethylammonium chloride | Sturm | 25 | 81[a] | Games et al., 1982 |
| Octadecyltrimethylammonium chloride | Sturm | 57 | 57[a] | Itoh and Naito 1982 |
| Didecyldimethylammonium chloride | MITI | 10 | 50 | Masuda et al., 1976 |
| Didodecyldimethylammonium chloride | MITI | 10 | 0 | Masuda et al., 1976 |
| Ditetradecyldimethylammonium chloride | MITI | 10 | 0 | Masuda et al., 1976 |
| Dihexadecyldimethylammonium chloride | MITI | 10 | 0 | Masuda et al., 1976 |
| Dioctadecyldimethylammonium chloride | MITI | 10 | 0 | Masuda et al., 1976 |
| Ditallowdimethylammonium chloride | Sturm | 10 | 0 | Itoh and Naito, 1982 |
| Dioctadecyldimethylammonium chloride | Sturm | 33 | 4 | Itoh and Naito, 1982 |
| Ditallowdimethylammonium chloride | CBT | 283 | 68 | van Ginkel and Kolvenbach, 1991 |
| Benzyloctyldimethylammonium chloride | MITI | 10 | 79 | Masuda et al., 1976 |
| Benzyldecyldimethylammonium chloride | MITI | 10 | 95 | Masuda et al., 1976 |
| Benzyldodecyldimethylammonium chloride | MITI | 10 | 89 | Masuda et al., 1976 |
| Benzyltetradecyldimethylammonium chloride | MITI | 10 | 83 | Masuda et al., 1976 |
| Benzylhexadecyldimethylammonium chloride | MITI | 10 | 5 | Masuda et al., 1976 |
| Benzyloctadecyldimethylammonium chloride | MITI | 10 | 0 | Masuda et al., 1976 |

[a] Tests performed in the presence of equimolar amounts of an anionic surfactant.
+, Growth of microorganisms observed.

it is clear that alkyltrimethylammonium salts and benzylalkyldimethylammonium salts are better degradable than dialkyldimethylammonium salts. The recalcitrance of quaternary ammonium salts in the screening tests increases with increasing alkyl chain lengths of both benzylalkyldimethylammonium and alkyltrimethylammonium salts. This effect is especially evident from test results of Masuda *et al.* (1976). Results of Miura *et al.* (1979) confirm this structure–biodegradability relationship. Recently, quaternary ammonium salts containing ester groups have been developed. The primary event in the degradation of an ester group containing quaternary ammonium salts is probably an abiotic hydrolysis giving rise to the formation of, e.g., fatty acids and diol quaternary ammonium salts. The fatty acids and the diol quaternary ammonium salts are readily biodegradable (Waters *et al.*, 1991; Simms *et al.*, 1992; Puchta *et al.*, 1993).

Long-chain fatty amines are included in this review because they act as cationic surfactants in acidic environments. Indeed, salts of fatty amines are applied as cationic surfactants. One should realize, however, that these environments do not prevail in nature. Using the MITI test various amines have been evaluated for their biodegradability and generalizations have been formulated by Yoshimura *et al.* (1980) as follows. The primary and secondary long-chain amines are readily biodegradable. Tertiary amines containing two methyl groups and one long alkyl chain are biodegradable but compared with primary and secondary amines biodegradation takes place at a slower rate. Tri(long-chain)alkyl amines are 'recalcitrant'. The biodegradation of fatty amines found by Yoshimura *et al.* (1980) in their MITI test does not always reach 60% which suggests partial mineralization of the fatty amines. However, the biodegradation curves of various fatty amines obtained in closed bottle tests demonstrate that these compounds are totally mineralized within 28 days (Akzo results). The biodegradation percentages at day 28 permit classification of the fatty amines as readily biodegradable (Figure 6.2). Finally, these closed bottle test results confirm more or less the generalizations formulated by Yoshimura *et al.* (1980). However, the introduction of methyl groups does not decrease the biodegradability of fatty amines with one long alkyl chain (Figure 6.2) which deviates from the generalizations described by Pitter and Chudoba (1990). Furthermore, it is clear that tertiary amines having two long alkyl chains are readily biodegradable (Figure 6.2).

Figure 6.3 shows that *N*-tallow-1,3-diaminopropane is readily biodegradable (Akzo result). This also applies to *N*-tallow-1,3-diaminopropane diacetic acid, thus showing that acids do not influence the biodegradation of amines. Figure 6.3 also illustrates the biodegradation curves of an amine oxide and an ethoxylated quaternary ammonium salt tested in closed bottle tests. The alkyldimethylamine oxide containing a $C_{12/14}$ hydrocarbon chain is readily biodegradable (Akzo result). Polyoxyethylene(15)tallowmethylammonium chloride is not readily biodegradable but when exposed to microorganisms in a closed bottle test for a longer period this compound is completely mineralized (Akzo result). Finally, Masuda *et al.* (1976) reported low biodegradation percentages of various pyridinium quaternary ammonium salts obtained in their MITI test. Biodegradability data on other cationic surfactants are scarce.

In conclusion, nearly all quaternary ammonium salts tested reach 60% biodegradation in screening tests which implies a total mineralization of these compounds. However, the test period necessary for total mineralization varies

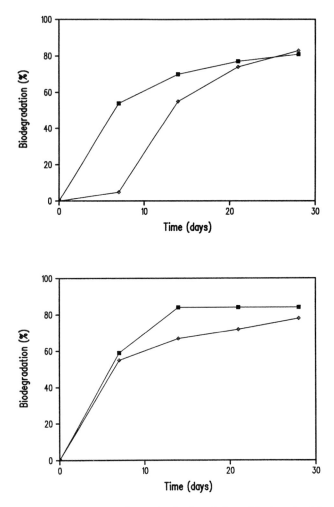

**Figure 6.2** Biodegradation curves of hexadecylamine (below, ■), hexadecyldimethylamine (below, ◊), cocodimethylamine (above, ■) and dicocomethylamine (above, ◊) obtained in closed bottle tests.

strongly. Larson (1983) demonstrated that results from the screening tests tend to be conservative and underestimate the biodegradation potential in the environment. Therefore, the readily biodegradable quaternary ammonium salts will not accumulate in most (eco)systems.

This statement has been confirmed by the half-life period of octadecyltrimethylammonium chloride which is 2.2 days in acclimated river water (Rapid Creek) (Larson, 1983). The half-life period of another alkyltrimethyl

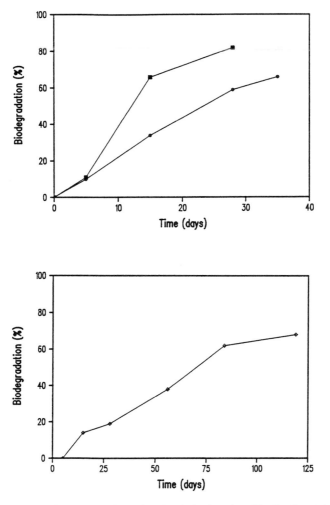

**Figure 6.3** Biodegradation curves of an alkyldimethylamineoxide (■), *N*-tallow-1,3-diamino-propane (□) and polyoxyethylene(15)-tallowmethylammonium chloride (◊) obtained in (pro-longed) closed bottle tests. *N*-Tallow-1,3-diaminopropane is tested in the presence of silica gel.

quaternary ammonium salt in sea water is 6 days (Vives Rego *et al.*, 1987). Dioctadecyldimethylammonium and ditallowdimethylammonium chloride are slowly biodegraded in the Sturm and the closed bottle test, respectively (Table 6.1). However, dioctyldimethylammonium chloride is significantly bio-degraded at a concentration of 0.5 mg/l in acclimated river water (Rapid Creek). The rate of biodegradation is enhanced by increasing the level of particulate matter. The estimated half-life periods of dioctadecyldimethylammonium chloride in river water with and without sediment are 4.9 and 13.8 days, respectively

(Larson, 1983; Larson and Vashon, 1983). These half-life periods are calculated from the carbon dioxide produced thus ensuring that true biodegradation has been determined. The biodegradation in biological wastewater treatment systems will be discussed in the next section.

### 6.2.2   Influence of toxicity on the biodegradation

Some quaternary ammonium salts are applied as biocides and could inhibit growth of microorganisms capable of degrading quaternary ammonium salts. Therefore, the inevitable presence of 'high' concentrations of quaternary ammonium salts in OECD screening tests might be the start of a fatal setup of a biodegradation experiment. Indeed, the non-biodegradability of quaternary ammonium salts with long alkyl chains in some MITI and Sturm tests is caused by inhibitory effects of the test compounds at 'high' initial concentrations (Dean Raymond and Alexander, 1977; Larson, 1983). Therefore, the results of Masuda *et al.* (1976) also reflect the different biocidal properties of the quaternary ammonium salts to microorganisms capable of degrading these compounds (Table 6.1). It is generally accepted that the toxicity of quaternary ammonium salts increases with increasing chain length (Dean-Raymond and Alexander, 1977).

The inhibitory effect of quaternary ammonium salts with long alkyl chains may be reduced by the addition of equimolar amounts of anionic surfactants. Under this condition insoluble anionic/cationic complexes are formed. Hence, the concentration of the quaternary ammonium salts in the water phase is reduced thus enabling biodegradability testing in screening tests. Several authors (Itoh and Naito, 1982; Games *et al.*, 1982; Larson, 1983) reported high biodegradation percentages of long-chain alkyl quaternary ammonium salts in the presence of these anionics (Figure 6.4, Table 6.1). 'High' initial concentrations of quaternary ammonium salts necessary in screening tests may be replaced by low concentrations of radiolabelled organic compounds. Using radiolabelled octadecyltrimethylammonium chloride, Larson (1983) demonstrated the ready biodegradation of this quaternary ammonium salt which is recalcitrant in standard screening tests (Figure 6.4).

N-Tallow-1,3-diaminopropane is also recalcitrant in a closed bottle test due to inhibitory effects of the test compound. Hence, this compound has been tested in the presence of silica gel to reduce the concentration in the water phase, thus preventing inhibitory effects. During the test period the test compound adsorbed releases slowly from the silica gel (Akzo result).

These results clearly illustrate that the non-biodegradability and the discrepancies of results found in standard screening tests (Table 6.1) are related to the antimicrobial action of cationic surfactants. The concentration of organic compounds in the screening tests is 2–20 mg/l, whereas the concentrations in nature are of the order of a few $\mu$g/l. Therefore, the use of radiolabelled material, anionic surfactants or silica gel allows a more realistic assessment of the biodegradation of these organic compounds in nature.

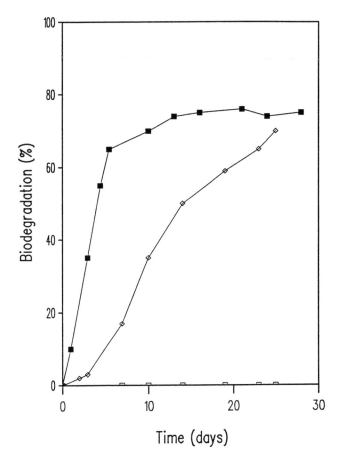

**Figure 6.4** Biodegradation of octadecyltrimethylammonium chloride in a Sturm test in the presence ($\diamond$) and absence ($\square$) of equal concentrations of linear alkylbenzenesulphonate and in Rapid Creek river water using radiolabelled octadecyltrimethylammonium chloride ($\blacksquare$) (redrawn from Larson, 1983).

### 6.2.3 Generalizations on biodegradability of quaternary ammonium salts

To minimize the influence of inhibitory effects of quaternary ammonium salts on biodegradability testing, the oxidation rates of various quaternary ammonium salts by a culture grown on hexadecyltrimethylammonium bromide may be used (van Ginkel and Kolvenbach, 1991). This culture has been adapted to hexadecyltrimethylammonium bromide, which results in high oxidation rates of various quaternary ammonium salts. The optimum concentration for the oxidation of the quaternary ammonium salts by this culture has been determined to assess the biodegradability without interference of inhibitory effects. Results obtained at this optimum concentration are in agreement with closed bottle test results, with the exception of alkyltrimethylammonium salts with long alkyl chains. The alkyl chain length of these compounds affects the oxidation rate only slightly. From these oxidation rates and the test results listed in the Table 6.1 the following generalizations on the aerobic biodegradability of quaternary ammonium salts emerge:

- an increase of the alkyl chain length to a limited extent reduces the biodegradability of the quaternary ammonium salts;
- the resistance to biological degradation is largely caused by increasing numbers of long alkyl chain;
- a benzyl group instead of a methyl group does not affect the biodegradability.

The 'recalcitrance' of dialkyldimethylammonium chloride presumably results from steric hindrance at the nitrogen atom. Yoshimura *et al.* (1980) concluded the same when comparing the biodegradability of various primary, secondary and tertiary long-chain amines. However, none of the cationic surfactants discussed should be regarded as persistent in nature (Section 6.2.1).

## 6.3   Behaviour of quaternary alkyl ammonium salts in waste water treatment plants

### 6.3.1   *Activated sludge plants*

Domestic waste water is predominantly treated in activated sludge systems. These activated sludge systems consist of a primary settling tank, a bioreactor and a clarifier. The clarifier, a solids separator, is used to remove sludge from the effluent stream. Most of the sludge removed is returned to the bioreactor where a high concentration of biomass is achieved primarily through controlled sludge wasting (Figure 6.5). The wasted activated sludge and settled primary sludge are often treated in an anaerobic digester and subsequently often used for agricultural purposes.

The use of quaternary ammonium salts as, e.g. fabric softeners, is expected to lead to the release of these compounds in activated sludge plants. The concentration of quaternary ammonium salts entering activated sludge plants has been calculated by Holman (1981) and Gerike *et al.* (1978). The levels calculated for Germany and the USA are 1.4 and 1.0 mg/l, respectively. These calculated concentrations agree with concentrations measured in raw sewage (Kupfer, 1982; Topping and Waters, 1982).

The presence of cationic surfactants in raw sewage has resulted in extensive studies of the behaviour of quaternary ammonium salts in activated sludge plants by various authors (Pitter and Svitalkova, 1961; Brown, 1976; May and Neufahrt, 1976; Gerike *et al.*, 1978, 1984). Cationic surfactants present in domestic waste water can leave the activated sludge plants together with excess sludge due to adsorption and/or with clarified water, and/or biodegrade in the reactor. Losses are not expected to occur by direct volatilization to the air. In the

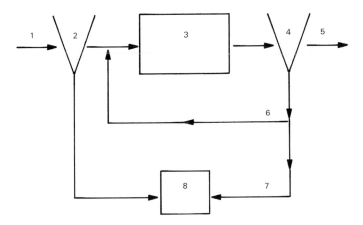

**Figure 6.5** Schematic representation of an activated sludge plant. The activated sludge plant consists of an influent supply (1), primary settling tank (2), bioreactor (3), clarifier (4), effluent outlet (5), sludge return (6), activated sludge waste (7) and digester (8).

case of adsorption the cationic surfactants may be degraded anaerobically in the digester or end up on agricultural soil. For a detailed account of the behaviour of cationic surfactants in wastewater purification plants, the reader is referred to the review of Boethling (1984).

### 6.3.2    Adsorption of cationic surfactants onto particles

Cationic surfactants strongly adsorb onto surfaces of solids which are mostly negatively charged. Solids present in domestic waste water are removed from this waste water by sedimentation in the primary settling tank. Therefore, only part of the cationic surfactants present in raw sewage is passed onto the aerobic bioreactor. The actual concentration of cationic surfactants reaching the bioreactor is dependent on its efficiency of removal during the primary sedimentation. According to a monitory study in full-scale activated sludge plants approx. 10% is removed in the primary settling tank (Topping and Waters, 1982). Huber (1987) also reported on the removal of cationic surfactants in primary settling tanks. In this case 20–40% of the cationic surfactants is removed.

In the bioreactor a decrease of >95% is reached within 2 h as a consequence of adsorption onto the activated sludge particles. This removal percentage has been measured by several authors (Games et al., 1982; Sullivan, 1983).

### 6.3.3    Removal of cationic surfactants in activated sludge reactors
(bioreactors)

The behaviour of organic compounds in activated sludge plants (bioreactors) has primarily been investigated in the OECD Confirmatory test (CAS) or in the

OECD semi-continuous activated sludge test (SCAS). The influent concentrations of cationic surfactants used in the CAS and SCAS tests vary from 10 to 20 mg/l, which is much higher than the actual concentrations in raw sewage. Nevertheless, the removal of cationic surfactants in these tests has on numerous occasions turned out to be >95% (Table 6.2). In addition, the removal of dialkylimidazolium salts has been determined in the Zahn–Wellens test (OECD) and turned out to be >90% (Domsch, pers. comm.). Because many removal percentages are achieved within 3 to 24 h, adsorption onto sludge particles seems to be the responsible mechanism. Subsequently, the cationic surfactants adsorbed may be mineralized by the activated sludge. Overall removal percentages of 87.5–95% determined in monitoring studies (Kupfer, 1982; Topping and Waters, 1982) of full-scale and pilot-scale activated sludge plants confirm the results of the CAS and SCAS tests.

The overlap of biodegradation and adsorption in activated sludge systems has been studied in some additional experiments in order to estimate the extent of biodegradation in activated sludge plants. Since only low levels of alkyltrimethylammonium chloride have been found in the activated sludge, these results suggest that this compound has been removed primarily by biodegradation (Games et al., 1982). To show unequivocally whether or not biodegradation of alkyltrimethylammonium salts is also responsible for the observed removal, an SCAS test with $^{14}C$ radiolabelled octadecyltrimethylammonium chloride has been performed. In this test $^{14}CO_2$ was produced within a few hours. From these results Games et al. (1982) have calculated a half-life

**Table 6.2** Removal/biodegradation of cationic surfactants in simulation tests of activated sludge plants selected from Swisher (1987)

| Test compound | Test | Removal (%) | References |
|---|---|---|---|
| Hexadecyltrimethylammonium bromide | CAS | 91–98 | Gerike, 1982 |
| Hexadecyltrimethylammonium bromide | CAS | 98–99 | Brown, 1976 |
| Hexadecyltrimethylammonium bromide | CAS | 100 | Pitter and Svitalkova, 1961 |
| Octadecyltrimethylammonium chloride | SCAS | 98 | Games et al., 1982 |
| Didecyldimethylammonium chloride | CAS | 95 | Gerike, 1982 |
| Dioctadecyldimethylammonium chloride | CAS | 95 | Gerike, 1982 |
| Dioctadecyldimethylammonium chloride | CAS | 91–93 | May and Neufahrt, 1976 |
| Dodecylbenzyldimethylammonium chloride | CAS | 96 | Gerike, 1982 |
| Tetradecylbenzyldimethylammonium chloride | CAS | >70 | Fenger et al., 1973 |
| Cocobenzyldimethylammonium chloride | CAS | 94 | Janicke and Hilge, 1979 |
| Hexadecylpyridinium bromide | CAS | 100 | Pitter and Svitalkova, 1961 |
| Dodecyldimethylamine oxide | CAS | 94 | Ruiz Cruz and Dobarganes, 1978 |
| Tetradecyldimethylamine oxide | CAS | >99 | Brown, 1976 |
| 1-Methyl-1-$C_{18}$-alkylamidoethyl-2-alkylimidalinium methosulphate | CAS | 111 | Schöberl et al., 1988 |

CAS, continuous flow activated sludge systems, also known as the OECD Confirmatory test.
SCAS test, semi-continuous activated sludge test.

period of about 2.5 h for octadecyltrimethylammonium chloride. This rate of biodegradation is fast enough to support biodegradation in activated sludge plants. This finding is in agreement with the ready biodegradability of alkyltrimethylammonium salts.

Alkylbenzyldimethylammonium salts are also readily biodegradable (Table 6.1). The ready biodegradability of these quaternary ammonium salts is also evident in an SCAS test performed with a $^{14}$C radiolabelled compound. After an adaptation period of a few days 80% of the $^{14}$C added to the SCAS unit has been converted to $^{14}CO_2$ (Krzeminski et al., 1973). Therefore, under normal operation conditions, the major proportion of alkytrimethylammonium salts and benzylalkyldimethylammonium salts adsorbed on activated sludge particles reaching the bioreactor is likely to be removed by biodegradation.

Sullivan (1983) studied the biodegradation of $^{14}$C radiolabelled ditallowdimethylammonium chloride in a semi-batch activated sludge reactor. The semi-batch reactor is fed daily with a synthetic sewage without removing effluent from the reactor. The sharp decrease of ditallowdimethylammonium chloride in the water phase is the result of adsorption. The biodegradation of this dialkyl quaternary ammonium salt is shown by a $^{14}CO_2$ recovery of approx. 40%.

In summary due to a rapid adsorption of cationic surfactants onto sludge particles and a subsequent complete biodegradation of both alkylbenzyldimethylammonium and alkyltrimethylammonium salts in the bioreactor, the overall removal of these compounds in activated sludge plants is higher than 95%. The almost complete removal of dialkyldimethylammonium salts in activated sludge plants is expected to be due to an initial rapid adsorption onto waste activated sludge solids. The adsorbed dialkyldimethylammonium salts, in turn, are at least partly biodegraded in sewage treatment plants.

### 6.3.4  Anaerobic biodegradation

In view of the adsorption of cationic surfactants onto the sludge particles and the subsequent treatment of primary sludge and activated sludge in a digester, the assessment of the anaerobic biodegradation is important. Nevertheless, information on the anaerobic biodegradation is scarce. Janicke and Hilge (1979) demonstrated that the concentration of quaternary ammonium salts does not or only slightly decrease in an anaerobic digester. Quaternary ammonium salts adsorbed on sludge particles are therefore not biodegraded under the anaerobic conditions in a digester. The possible biodegradation under anaerobic conditions has been tested in an anaerobic screening test (Battersby and Wilson, 1989). In this test, hexadecyltrimethylammonium bromide present at a concentration of 200 mg/l inhibited the methane production of the digester sludge used as inoculum. The inhibitory effects may therefore be responsible for the lack of biodegradation of hexadecyltrimethylammonium chloride in this anaerobic screening test.

## 6.3.5   Influence of cationic surfactants on biological processes

There is much evidence to show that quaternary ammonium salts have little or no effect on wastewater treatment systems at the concentrations found in sewage water. In an SCAS test using synthetic sewage the addition of 20 mg/l octade-cyltrimethylammonium chloride has no effect on the removal of organic compounds in the synthetic sewage (Games *et al.*, 1982). Furthermore, Sullivan (1983) reported no harmful effects on the performance of a semi-batch culture at 2.7 mg/l dialkyldimethylammonium salts. These results have been confirmed by the determination of the effect of a quaternary ammonium salt on the respiration rate of activated sludge. $EC_{50}$ values of tetradecyldimethylbenzylammonium chloride of either acclimated and unacclimated sludge ranged from 10 to 37 mg/l, respectively (Fenger *et al.*, 1973). $EC_{50}$ values are the concentrations of test substances which reduce the respiration rate by 50%. Comparing the actual concentration in the effluent of activated sludge plants and $EC_{50}$ values no adverse effects are expected. Moreover, the $EC_{50}$ values and some SCAS test results are obtained in the absence of anionic surfactants which are known to reduce the inhibitory effects of quaternary ammonium salts (Itoh and Naito, 1982; Larson, 1983).

Nitrification, i.e. conversion of ammonia to nitrate by nitrifying bacteria, is an important sewage treatment plant function. Nitrifiers are generally more susceptible than bacteria degrading organic compounds to inhibitors. The nitrification in a CAS test has been slightly influenced by 8 mg/l distearyldimethylammonium chloride in the influent (Gerike *et al.*, 1978). Therefore quaternary ammonium salts present in raw sewage will not inhibit nitrification in an activated sludge plant.

Also anaerobic formation of biogas in a digester is not inhibited by the quaternary ammonium salts at concentrations occurring in the activated sludge (Janicke and Hilge, 1979). Moreover, the toxicity to digester sludge has been determined according to a method described by Owen *et al.* (1979). The biogas production of digester sludge is not inhibited by ditallowdimethylammonium chloride at a concentration of 200 mg/l (Akzo result). These results show that there are no adverse effects on the performance of the digester due to the presence of cationic surfactants.

## 6.4   Biodegradation routes of quaternary ammonium salts

### 6.4.1   Tetramethylammonium chloride

Although tetramethylammonium chloride is not a cationic surfactant the biodegradation route of this compound will be discussed because this route has been extensively studied and tetramethylammonium chloride is structurally related to alkyltrimethylammonium salts. Various microorganisms can grow aerobically on tetramethylammonium salts as sole source of carbon and energy. The isolated strains belong to different genera such as *Pseudomonas* sp. (Hampton and Zatman,

1973). The initial oxidation of tetramethylammonium chloride results in the formation of formaldehyde and trimethylamine. This first step catalysed by a mono-oxygenase effects the following reaction (Hampton and Zatman, 1973).

$$(CH_3)_4N^+ + O_2 + NADH + H^+ \rightarrow (CH_3)_3N + H_2CO + NAD^+ + H_3O^+$$

A mono-oxygenase is an enzyme that catalyses oxidation of a substrate by removal of hydrogen which combines with molecular oxygen and by inserting an oxygen atom into the substrate. Reduced nicotiamide adenine dinucleotide (NADH) is an essential oxidizing–reducing co-enzyme. The catabolism of trimethylamine may proceed by two different pathways. The first route involves the oxidation of trimethylamine to trimethylamine-$N$-oxide which is catalysed by an inducible NADH-dependent mono-oxygenase (Large *et al.*, 1972; Colby and Zatman, 1973). Subsequently, the $N$-oxide is cleaved to dimethylamine and formaldehyde (Large, 1971; Myers and Zatman, 1971; Colby and Zatman, 1973). This reaction is catalysed by a non-oxidative demethylase. The second pathway is employed by microorganisms synthesizing a trimethylamine dehydrogenase which converts the trimethylamine directly to formaldehyde and dimethylamine (Colby and Zatman, 1971; Meiberg and Harder, 1978). Dehydrogenases are the most common oxidizing enzymes which catalyze the oxidation of a substrate by removing hydrogen from it.

$$(CH_3)_3N + NAD^+ + H_2O \rightarrow (CH_3)_2NH + CH_2O + NADH + H^+$$

Both pathways are shown in Figure 6.6. The following intermediate, dimethylamine, in aerobic methylotrophic microorganisms, in turn is usually attacked by a secondary amine mono-oxygenase that cleaves dimethylamine to methylamine and formaldehyde (Jarman *et al.* 1970; Colby and Zatman, 1973). More recently a dimethylamine dehydrogenase was purified from a *Hyphomicrobium* X grown anaerobically on dimethylamine as sole carbon source (Meiberg and Harder, 1978). Finally, a primary amine dehydrogenase, which has been measured using PMS instead of NAD, degrades methylamine yielding ammonia and formaldehyde (Eady and Large, 1968, 1971; Hampton and Zatman, 1973). The formaldehyde split off by the sequential reactions is oxidized to formate by a formaldehyde dehydrogenase found in methylotrophic microorganisms. Subsequently, a formate dehydrogenase catalyses the oxidation of formate to carbon dioxide (Anthony, 1982).

### 6.4.2 Ethyltrimethylammonium chloride

Ghisalba and Kuenzi (1983) isolated various strains capable of utilizing trimethylethylammonium chloride as sole carbon and energy source. This study demonstrates the ubiquitous presence of quaternary ammonium salt utilizing bacteria in nature. Trimethylethylammonium chloride is initially attacked by a mono-oxygenase yielding either acetaldehyde or methanal and a tertiary amine. It is not clear whether the cleavage of the $C_2$-unit takes place in the first, the

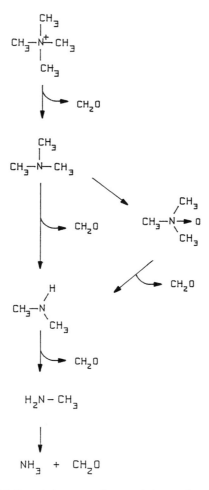

**Figure 6.6** Degradation routes of tetramethylammonium chloride.

second or third degradation step. However, it is evident that the last degradation step is not the cleavage of ethylamine because this compound does not serve as a growth substrate for the bacterium isolated.

### 6.4.3   Alkyltrimethylammonium salts

Only three articles deal with the metabolism of alkyltrimethyl ammonium salts. On the basis of paper chemistry, Macrell and Walker (1978) envisaged two potential points of attack. As depicted in Figure 6.7 these are:

–   a fission of the C–N bond in which the alkyl chain or a methyl group is cleaved from a tertiary amine;

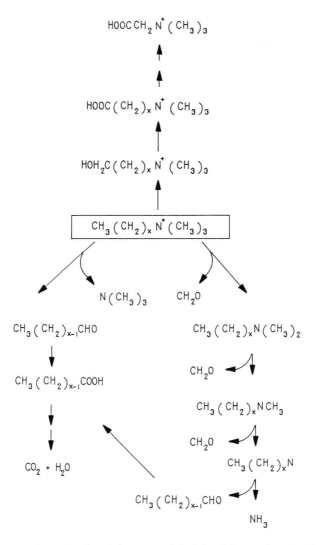

**Figure 6.7** Possible degradation routes of alkyltrimethylammonium chloride.

– an omega-oxidation in which the far end of the alkyl chain is first oxidized to a carboxylic acid. Biodegradation can then proceed via beta oxidation.

These authors, however, were not able to enrich quaternary ammonium salt degrading microorganisms because of the toxicity of these compounds to the microorganisms. Consequently, Macrell and Walker (1978) were not able to support the biodegradation routes with experimental data. Results obtained with a

Xanthomonad capable of biodegrading decyltrimethylammonium chloride support both degradation mechanisms (Dean-Raymond and Alexander, 1977). Using GC–MS 9-carboxynonyl- and 7-carboxyheptyltrimethyl-ammonium chloride have been detected in the supernatant liquor during growth of the organism on this quaternary alkyl ammonium salt. However, respiration experiments suggest a C–N cleavage yielding trimethylamine and decanal, which in turn is oxidized to the respective carboxylic acid. A monooxygenase catalysing this cleavage has been detected in a *Pseudomonas* sp. capable of utilizing hexadecyltrimethylammonium chloride as sole carbon and energy source (van Ginkel *et al.*, 1992). This bacterium oxidized the alkyl chain and as a consequence excreted trimethylamine.

In conclusion, biodegradation of quaternary alkyl ammonium salts probably proceeds via a fission of C–N bonds, yielding alkanals and amines. This initial oxidation is in accordance with the theory that closely packed nitrogen atoms resist biodegradation and known biodegradation routes of nitrogen derivatives.

### 6.4.4   *Possible formation of recalcitrant intermediates*

The degradation routes of quaternary ammonium salts outlined in Figure 6.7 do not result in the formation of recalcitrant intermediates. The aldehydes cleaved are oxidized to the respective carboxylic acids. Carboxylic acids are readily degraded via beta-oxidation to water and carbon dioxide. Formaldehyde resulting from the cleavage of a methyl group is readily biodegraded by many $C_1$-metabolizing microorganisms. The tertiary amines formed are also readily biodegradable as shown by Yoshimura *et al.* (1980). The oxidation of the far end of the alkyl chain of the alkyltrimethylammonium salt would lead to the formation of betaine. This compound is in turn biodegraded under both aerobic and anaerobic conditions (Kortstee, 1970; King, 1984; Heythuysen and Hansen, 1989; Fendrich *et al.*, 1990).

Games *et al.* (1982) stated that in their experiments using radiolabelled octadecyltrimethylammonium chloride no stable intermediates are formed. Quantitative recovery of the radiolabel has been obtained from the [14]C-methyl and [14]C-alkyl labelled octadecyltrimethylammonium chloride when added to activated sludge. [14]C-labelled carbon dioxide accounts for approximately 90% and 65%, respectively, whilst 10% and 35% respectively, are associated with activated sludge. The radiolabelled material associated with the activated sludge is assumed to represent material incorporated into microorganisms.

### References

Anthony, C. (1982) *The Biochemistry of Methylotrophs*. Academic Press, London.
Battersby, N.S. and Wilson, V. (1989) Survey of the anaerobic biodegradation potential of organic chemicals in digesting sludge. *Appl. Environ. Microbiol.* **55**, 433–439.
Boethling, R.S. (1984) Environmental fate and toxicity in wastewater treatment of quaternary ammonium surfactants. *Water Res.* **18**, 1061–1076.

Brown, D. (1976). The assessment of biodegradability. A consideration of possible criteria for surface active substances, in *Proceedings of VII International Congress on Surface-active Substances*, 4, USSR national committee on surface active substances, Moscow, pp. 44–57.

Colby, J. and Zatman, L.J. (1971) The purification and properties of a bacterial trimethylamine dehydrogenase. *Biochem. J.* **121**, 9–10.

Colby, J. and Zatman, L.J. (1973) Trimethylamine metabolism in obligate and facultative methylotrophs. *Biochem. J.* **132**, 101–112.

Dean-Raymond, D. and Alexander, M. (1977) Bacterial metabolism of quaternary ammonium compounds. *Appl. Environ. Microbiol.* **33**, 1037–1041.

Eady, R.R. and Large, P.J. (1968) Purification and properties of an amine dehydrogenase from *Pseudomonas* AM1 and its role in growth on methylamine. *Biochem. J.* **106**, 245–255.

Eady, R.R. and Large, P.J. (1971) Microbial oxidation of amines. Spectral and kinetic properties of primary amine dehydrogenase of *Pseudomonas* AM1. *Biochem. J.* **123**, 757–771.

Fendrich, C., Hippe, H. and Gottschalk, G. (1990) *Clostridium halophilium* sp. nov. and a marine species degrading betaine in the Stickland reaction. *Arch. Microbiol.* **154**, 127–132.

Fenger, B.H., Mandrup, M., Rohde, G. and Kjaer Sorensen, J.C. (1973) Degradation of cationic surfactants in activated sludge pilot plants. *Water Res.* **7**, 1195–1208.

Games, L.M., King, J.E. and Larson, R.J. (1982) Fate and distribution of a quaternary ammonium surfactant, octadecyltrimethylammonium chloride (OTAC) in wastewater treatment. *Environ. Sci. Technol.* **16**, 483–488.

Gerike, P. (1982) Uber den biologischen Abbau und die Bioelimination von kationischen Tensiden. *Tenside Surfact. Deterg.* **19**, 162–164

Gerike, P., Fischer, W.K. and Jasiak, W. (1978) Surfactant quaternary ammonium salts in aerobic sewage digestion. *Water Res.* **12**, 1117–1122.

Gerike, P., Holtman, W. and Jasiak, W. (1984) A test for detecting recalcitrant metabolites. *Chemosphere* **13**, 121–141.

Ghisalba, O. and Kuenzi, M. (1983) Biodegradation and utilization of quaternary alkylammonium compounds by specialized methylotrophs. *Experientia* **39**, 1264–1271.

Hampton, D. and Zatman, L.J. (1973) The metabolism of tetramethylammonium chloride by bacterium 5H2. *Biochem. Soc. Trans.* **1**, 667–668.

Heythuysen, J.H.F.G. and Hansen, T.A. (1989) Betaine fermentation and oxidation by marine *Desulfuromonas* strains. *Appl. Environ. Microbiol.* **55**, 965–969.

Holman, W.F. (1981) Estimating the environmental concentrations of consumer product components, in *Aquatic Toxicology and Hazard Assessment Fourth Conference ASTM STP 737*, ed. D.R. Branson and K.L. Dickson. American Society for Testing and Materials, Philadelphia, pp. 159–182.

Huber, L.H. (1987) Ecological behaviour of cationic surfactants from fabric softeners in aquatic environment. *J. Am. Oil Chem. Soc.* **61**, 377–382.

Itoh, S. and Naito, S. (1982) Studies on the biodegradation test method of chemical substances III. Ultimate biodegradabilities of cationic/anionic surfactant-complexes. *Yukagaku* **31**, 277–280.

Janicke, W. and Hilge, G. (1979) Biologisches Abbauverhalten von Anion/Kationtenside-Komplexen unter den aeroben und anaeroben Bedingungen der Abwasser-bzw Schlammbehandlung. *Tenside Surfact. Deterg.* **16**, 117–122.

Jarman, T.R., Eady, R.R. and Large, P.J. (1970) An enzymatically active P-420-type cytochrome involved in the mixed-function dimethyl oxidase system of *Pseudomonas aminovorans*. *Biochem. J.* **119**, 55–56.

Kappeler, T.U. (1982) The aquatic toxicity of distearyldimethylammonium chloride (DSDMAC) and its ecological significance. *Tenside Surfact. Deterg.* **19**, 157–161.

King, G.M. (1984) Metabolism of trimethylamine, choline, and glycine betaine by sulphate-reducing and methanogenic bacteria in marine sediments. *Appl. Environ. Microbiol.* **48**, 719–725.

Kortstee, G.J.J. (1970) The aerobic decomposition of choline by micro-organisms. I The ability of aerobic organisms, particularly coryneform bacteria, to utilize choline as sole carbon and nitrogen source. *Arch. Microbiol.* **71**, 235–244.

Krzeminski, S.F., Martin, J.J. and Brackett, C.K. (1973) The environmental impact of a quaternary ammonium bactericide. *Household Pers. Prod. Ind.* **10**, 22–24.

Kupfer, W. (1982) Spurenanalytik von kationischen Tensiden unter den speziellen Bedingungen im Wasser und Abwasser. *Tenside Surfact. Deterg.* **19**, 158–161.

Large, P.J. (1971) Non-oxidative demethylation of trimethylamine-N-oxide by *Pseudomonas aminovorans*. *Fed. Eur. Biochem. Soc. Lett.* **18**, 297–300.

Large, P.J., Boulton, C.A. and Crabbe, M.J.C. (1972) The reduced nicotinamide-adenine dinu-cleotide phosphate- and oxygen-dependent N-oxygenation of trimethylamine by *Pseudomonas aminovorans*. *Biochem. J. 128*, 137–138.

Larson, R.J. (1983) Comparison of biodegradation rates in laboratory screening studies with rates in natural water. *Residue Rev.* **85**, 159–171.

Larson, R.J. and Vashon, R.D. (1983) Adsorption and biodegradation of cationic surfactants in labo-ratory and environmental systems. *Dev. Ind. Microbiol.* **24**, 425–434.

Lewis, M.A. and Hamm, B.G. (1986) Environmental modification of the photosynthetic response of lake plankton to surfactants and significance to a laboratory-field comparison. *Water Res.* **20 (12)**, 1575–1582.

Lewis, M.A. and Suprenant, D. (1983) Comparative acute toxicities of surfactants to aquatic inverte-brates. *Ecotoxicol. Environ. Safety.* **7**, 312–322.

Lewis, M.A. and Wee, V.T. (1983) Aquatic safety assessment for cationic surfactants. *Environ. Toxicol. Chem.* **2**, 105–118.

Lewis, M.A., Taylor, M.J. and Larson, R.J. (1986) Structural and functional response of natural phy-toplankton and periphyton communities to a cationic surfactant with considerations on environ-mental fate, in *ASTM Spec. Techn. Publ., STP 920* pp. 241–268.

Mackrell, J.A. and Walker, J.R.L. (1978) The biodegradation of quaternary ammonium compounds. *Int. Biodeterior. Bull.* **14**, 77–83.

Masuda, F., Machida, S. and Kanno, M. (1976) Studies on the biodegradability of some cationic sur-factants, in *Proceedings of VII International Congress on Surface-active Substances*, 4, USSR national committee on surface active substances, Moscow, pp. 129–138.

May, A. and Neufahrt, A. (1976) Zum okologischen Verhalten von Kationtensiden, 3 Mitt.: Uber das Verhalten von Distearyldimethylammoniumclorid in Belebtschlamm-anlagen. *Tenside Surfact. Deterg.* **13**, 65–69.

Meiberg, J.B.M. and Harder, W. (1978) Aerobic and anaerobic metabolism of trimethylamine, dimethylamine and methylamine in *Hyphomicrobium* X. *J. Gen. Microbiol.* **106**, 265–276.

Miura, K., Yamanaka, K., Sangai, T., Yoshimura, K. and Hayashi, N. (1979) Application of the bio-logical oxygen consumption measurement technique to the biodegradation test of surfactants. *Yakagaku* **28**, 351–355.

Myers, P.A. and Zatman, L.J. (1971) The metabolism of trimethylamine-N-oxide by *Bacillus* PM6. *Biochem. J.* **121**, 10 pp.

Neufarth, A. and Pleschke, D. (1982) Studies of abiotic and biotic degradation processes on cationic surfactants XIII. *Jourdanos Com. Espanol Deterg.* **12**, 13–15

Owen, W.F., Stuckey, D.C., Healey, J.B., Young, L.Y. and McCarthy, P.L. (1979) Bioassay for monitoring biochemical methane potential and anaerobic toxicity. *Water Res.* **3**, 485–492.

Pitter, P. and Chudoba, J. (1990) *Biodegradability of Organic Substances in the Aquatic Environment. Section IV Aliphatic Amines and their Derivatives.* CRC Press, Boca Raton.

Pitter, P. and Svitalkova, J. (1961) Biodegradation of cationic agents in laboratory models of activat-ed sludge tanks. *Sb VSChT* **52**, 25–42.

Puchta, P., Krings, P. and Sandkühler, P. (1993) A new generation of softeners. *Tenside Surfact. Deterg.* **30(3)**, 186–192.

Ruiz Cruz, J. (1981) Pollution of natural water courses by synthetic detergents. XVII Influence of temperature and other variables on the biodegradation of cationic agents in river water. *Grasas Aceites* **32**, 147–153.

Ruiz Cruz, J. and Dobarganes, M.C. (1978) Pollution of natural waters by synthetic detergents XIII Biodegradation of nonionic surfactants in river water and determinations of their biodegradability by different test methods. *Grasas Aceites* **29**, 1–8.

Schöberl, P., Bock, K.J. and Huber L. (1988) Okologisch relevante Daten von Tensiden in Wasch-und Reinigungsmitteln. *Tenside Surfact. Deterg.* **25**, 86–98.

Simms, J.R., Woods, D.A., Walley, D.R., Keough, T., Schwab, B.S. and Larson, R.J. (1992) Integrated approach to surfactant environmental safety assessment: Fast bombardment mass spec-trometry and liquid scintillation counting to determine the mechanism and kinetics of surfactant biodegradation. *Anal. Chem.* **64**, 2951–2957.

Swisher, R.D. (1987) *Surfactant Biodegradation.* 2nd Edition, revised and expanded. Surfactant science series, Volume 18, Marcel Dekker, New York and Basel.

Sullivan, D.E. (1983) Biodegradation of cationic surfactants in activated sludge. *Water Res.* **17**, 1145–1151.

Topping, B.W. and Waters, J. (1982) Monitoring of cationic surfactants in sewage treatment plants. *Tenside Surfact. Deterg.* **19**, 164–169.

Valls, M., Bayona, J.M., Albaiges, J. and Mansour, M. (1990) Fate of cationic surfactants in the marine environment, II: photooxidation of long-chain alkylamines in aqueous media. *Chemosphere* **20**, 599–608.

van Ginkel, C.G. and Kolvenbach, M. (1991) Relations between structure of quaternary alkyl ammonium salts and their biodegradability. *Chemosphere* **23**, 281–289.

van Ginkel, C.G., van Dijk, J.B. and Kroon, A.G.M. (1992) Metabolism of hexadecyltrimethylammonium chloride in *Pseudomonas* strain B1. *Appl. Environ. Microbiol.* **58 (9)**, 3083–3087.

van Leeuwen, C.J., Roghair, C., de Greef, J. and de Nijs, T. (1990) Wasverzachter II resultaten van aanvullend onderzoek. $H_2O$, **23 (11)**, 294–299 (in Dutch).

Vives-Rego, J., Vaque, M.D., Sanchez Leal, J. and Parra, J. (1987) Surfactants biodegradation in sea water. *Tenside Surfact. Deterg.* **24**, 20–22.

Waters, J. and Kupfer, W. (1976) The determination of cationic surfactants in the presence of anionic surfactant in biodegradation test liquors. *Anal. Chim. Acta* **85**, 241–251.

Waters, J., Kleister, H.H., How, M.J., Barratt, M.D., Birch, R.R., Fletcher, R.J., Haigh, S.D., Hales, S.G., Marshall, S.J. and Pestell, T.C. (1991) A new rinse conditioner active with improved environmental properties. *Tenside Surfact. Deterg.* **28**, 460–468.

Wee, V.T. and Kennedy, J.M. (1982) Determination of trace levels of quaternary ammonium compounds in river water by liquid chromatography with conductometric detection. *Anal. Chem.* **54**, 1631–1633.

Yoshimura, K., Machida, S. and Masuda, F. (1980) Biodegradation of long chain alkylamines. *J. Am. Oil Chem. Soc.* **57**, 238–241.

# 7 Biodegradability of non-ionic surfactants

T. BALSON and M. S. B. FELIX

## 7.1 Introduction

Non-ionic surfactants can be described as being covalent compounds which have hydrophilic and hydrophobic moieties and which do not ionise when dissolved in water.

This general description of non-ionic surfactants applies to many product ranges and chemical types, such as alkanolamides, alkylpolyglucosides, sucroglycerides, esters, etc. This chapter will be concerned only with polyglycols, which are generally recognised to be synonymous with non-ionic surfactants. Polyglycols are based on alkylene oxides, mainly ethylene oxide (EO) but also propylene oxide (PO) and butylene oxide (BO), reacted onto an initiator containing one or more active protons.

Products based on an alkylene oxide, but which do not have both hydrophilic and hydrophobic moieties, such as the water-initiated polyethylene glycols or polypropylene glycols, are not regarded as surfactants.

In the literature the terminology employed to describe polyglycols is varied, with expressions such as ethoxylates, polyol, polyalkylene glycol, polyalkylene oxide, polyoxyalkylene, etc. being used to describe the whole or part (i.e. a block) of the surfactant. The IUPAC (International Union of Pure and Applied Chemistry) and CAS (Chemical Abstracts Services) use the terms oxirane and polyoxirane, respectively, to describe EO and its polymers, with methyloxirane and polymethyloxirane describing PO and its derivatives.

Since the first ethoxylates were prepared, in Germany in 1930 (Schoeller, 1930), there have been many different surfactants made with this chemistry based on the choice of initiator, choice of oxide(s) and feeding technique of the oxides. This will be covered in the ensuing sections.

The environmental fate of surfactants is an increasingly important facet of contemporary surfactant performance parameters. The post-war introduction of alkyl benzene sulphonates (ABS) in detergents, as a replacement for soap, led to unprecedented foam on rivers and in wastewater treatment plants. This was a direct consequence of the limited biodegradability of ABS as well as its excellent foaming properties even at very low concentrations. This led to the development of new products, such as linear alkyl benzene sulphonates, which were much more susceptible to microbial attack in a wastewater treatment plant.

This increased the regulatory awareness of the influence of surfactants on the aquatic environment and led directly to the legislation relating to the environmental compatibility of surfactants, especially those used in detergents (EEC Directive 73/404 with amendments 82/242 and 86/94).

Voluntary agreements not to use nonylphenol ethoxylates in some applications in certain European countries, and legislation related to the environmental classification requirements for products, dependent on the biodegradation, aquatic toxicity and bioaccumulation potential (EEC Directives 91/325 and 93/21), has led to a need to assess the environmental effects of these products and to the development of new chemical structures which can satisfy not only the existing performance criteria, but also the current and possible future regulatory requirements, such as risk assessment of existing and new chemicals.

## 7.2  Structure of polyglycol surfactants

All surfactants comprise hydrophobic and hydrophilic moieties, but polyglycol non-ionic surfactants are relatively unique in that it is easy to change, or fine tune, these units to give different properties to the molecule. This can be shown by looking at the different aspects of the chemistry of the polyglycol surfactant.

The hydrophobicity of a non-ionic polyglycol surfactant is normally derived from a carbon chain such as:

(1)  an alkyl group, e.g. a fatty alcohol;
(2)  a cyclic structure, e.g. phenol derivatives;
(3)  a polyether such as a PO or BO block;
(4)  a combination of the above such as nonylphenol.

The hydrophilicity of the surfactant comes mainly from:

(1)  an EO block;
(2)  the presence of hydroxyl groups;
(3)  from a combination of both, especially if the initiator has a high functionality.

In addition, the structure and performance of a polyglycol surfactant are also influenced by three other factors:

(1)  the functionality of the initiator which gives the surfactant its basic structure of being linear or multi-functional;
(2)  the choice of oxides used;
(3)  the feeding technique used for the oxides.

Possible oxide feeding techniques are as follows, but note that when PO is mentioned it can equally as well refer to BO:

(i)    regular block copolymers (PO block capped with EO block);
(ii)   reverse blocks (EO block capped with PO block);
(iii)  random blocks (mixed feed of EO and PO);

(iv)  any combination of the above, such as an initial block capped with random blocks, or the reverse.

The initiator and the alkylene oxides will be described in the following sections.

### 7.2.1   Preparation and properties of alkylene oxides and their polymers

This will be covered by briefly reviewing the synthesis of the alkylene oxides, the chemistry of alkoxylation reactions and then the properties of the alkylene oxides and how these influence the polyglycol surfactant.

#### 7.2.1.1   Synthesis of alkylene oxides.
Ethylene oxide is made by the direct oxidation of ethylene using a silver catalyst. The exothermic reaction is carried out under controlled conditions at high temperatures, about 250°C, and high pressure, 15 bar.

Propylene oxide is manufactured via two routes:

(1)  The chlorohydrin method, whereby propylene is reacted with hypochlorous acid to give propylene chlorohydrin which is then reacted with sodium hydroxide to yield PO with sodium chloride as a by-product.
(2)  The *tert*-butanol method, where iso-butane is oxidised to give a mixture of *tert*-butanol and peroxy-*tert*-butanol, the latter product is then reacted with propylene to give PO and *tert*-butanol.

Butylene oxide is made by the chlorohydrin method, in an analogous process to that of PO, derived from but-1-ene.

#### 7.2.1.2   Reactions of alkylene oxides.
Alkylene oxides react with compounds containing an active hydrogen in the presence of a catalyst at temperatures between 100 and 150°C dependent on the oxide and the final molecular weight of the product desired.

The alkoxylation reaction occurs in two stages. The initial reaction of an active proton with the alkylene oxide to yield a terminal hydroxyl function. This reaction can occur in the absence of a catalyst with highly active proton sources such as amines. Ring opening chain growth can then occur, in the presence of a catalyst, till the remaining alkylene oxides are reacted away. The normal catalyst used for an alkoxylation reaction is a base, typically potassium hydroxide, and under these basic conditions the anion, from the initiator or the newly formed alkoxylate, attacks preferentially, via second-order nucleophilic substitution reaction, SN2, the least hindered carbon atom of the alkylene oxide ring. This results in the formation of predominantly secondary terminal hydroxyl functions when PO or BO is used.

$$R{-}X{-} \quad + \quad \underset{\underset{O}{\diagdown\diagup}}{\overset{\overset{CH_3}{\diagup}}{CH{-}CH_2}} \quad \longrightarrow \quad R{-}X{-}CH_2{-}\underset{\underset{O{-}}{\diagdown}}{\overset{\overset{CH_3}{\diagup}}{CH}}$$

where X is O, S or N.

The relative reactivity of the alkylene oxides is such that EO is about five times faster than PO, which itself is about five times faster than BO and results in a tendency for the least reactive oxide to terminate the reaction when a mixed feed is used. This also implies that it is usually easier to completely cap an EO block with a higher alkylene oxide than *vice versa* because the higher reactivity of the primary hydroxyl anion will result in preferential polymerisation at the newly formed primary anion and thus incomplete capping of the secondary hydroxyl groups occurs. During the alkoxylation reaction any water present (from the base catalyst, the initiator and the reaction of the catalyst with the initiator) will act as an initiator for alkoxylation reactions forming a diol. If this is a problem the water can be removed, by distillation, prior to adding the oxide.

As with any polymerisation, the reaction of an alkylene oxide leads to a final polymer which has approximately a normal distribution of oligomers, as per Weibull and Nycander (1954). The distribution is relatively wide as shown by the typical analysis of a nonylphenol ethoxylate (with an average of 9 moles of EO) which can range from the oligomer containing 1 mole of EO to one with 20 moles of EO.

### 7.2.1.3 *Properties of alkylene oxides.*

The characters of the alkylene oxides change quite dramatically from EO, which is hydrophilic, through PO, partially hydrophilic, mainly hydrophobic but not lipophilic, to BO, lipophilic, and these give the main properties to the polyglycol as described below.

#### 7.2.1.3.1 *Characteristics of EO.*

The addition of EO to an initiator, or polyglycol, increases the hydrophilic character of the product as can be seen from observing the water solubility and aqueous cloud points of any ethoxylate series (*The Dow Chemical Company, 1988a*).

It will also raise the melting or freezing point of the product because the less sterically inhibited ethoxylate chains are capable of a more ordered structure in the lattice than other oxides. Thus the physical appearance of the ethoxylate moves from liquid through paste to solid.

This change in melting point by the addition of EO is clearly shown by comparing the melting points of polyethylene glycols as shown in Table 7.1 (*The Dow Chemical Company, 1988b*).

EO can be added to an initiator until very high molecular weights are reached. However, at increasingly high molecular weights the concentration of available hydroxyl functions is such that the reaction is very slow and can result in unacceptably long reaction times. This can be compensated for by using higher tem-

**Table 7.1**

| Product | Moles of EO | Melting point (°C) |
|---------|-------------|---------------------|
| E-200 | 4.1 | Supercools |
| E-300 | 6.4 | −10 |
| E-400 | 8.7 | +6 |
| E-600 | 13.2 | +22 |

peratures. Typically, the addition of EO to a hydrophobe creates a foaming surfactant.

*7.2.1.3.2 Characteristics of PO.* Adding PO to an initiator, or polyglycol, will increase its hydrophobicity, but not dramatically as observed from the aqueous cloud points of polypropylene glycols as shown below. This indicates the partial hydrophilicity of PO, due to the balance between the hydrophilic ether linkage and the hydrophobic propane unit.

Due to the methyl branch of PO, its derivatives cannot pack as easily into the lattice structure and so these products are invariably liquids and is illustrated by the pour points of the polypropylene glycol series as seen in Table 7.2 (*The Dow Chemical Company, 1988b*).

**Table 7.2**

| Product | Moles of PO | Pour point (°C) | 1% Aqueous cloud point (°C) |
|---------|-------------|-----------------|------------------------------|
| P-425 | 7 | −45 | >95 |
| P-1200 | 20 | −40 | 22 |
| P-2000 | 34 | −30 | 15 |
| P-4000 | 69 | −26 | 9 |

The formation of high molecular weight propoxylates is limited by the tendency of PO to isomerise to allyl alcohol in the presence of base and under the conditions of high temperature. In an analogy to the reaction of ethylene oxide, higher molecular weights imply longer reaction times and thus an increased probability of the isomerisation reaction. Allyl alcohol then becomes a new initiator for propoxylation and thus limits the molecular weights that can be made successfully. This is demonstrated by the polypropylene glycol series where the highest molecular weight sold commercially is 4000, whilst polyethylene glycols are sold commercially at 8000 and even much higher molecular weights (*The Dow Chemical Company, 1988b*). Propoxylates are hydrophobic but not lipophilic. However, if the amount of PO added is low, and the polyglycol initiator is lipophilic then the total molecule would probably also be lipophilic. The inclusion of PO in a non-ionic surfactant usually reduces its foaming tendency.

*7.2.1.3.3 Characteristics of BO.* Butylene oxide derivatives have some similarities to propoxylates and also yield liquid products, due to the pendant alkyl chain, when sufficient BO is incorporated into the molecule. However BO differs in that it is lipophilic because the character of the butane group has a much larger influence than that of the hydrophilic ether linkage. This is shown by the fact that medium molecular weight polybutylene glycols are not even dispersible in water. The inclusion of BO in a non-ionic surfactant also reduces the foaming tendencies but usually less BO than PO is required for a comparable reduction.

## 7.2.2   *Initiators used for polyglycol surfactants*

The initiator used to start a polyglycol must contain an active proton(s) such as alcohol, amine, thiol, etc. Although the initiator may only comprise a small percentage of the total molecule it can still have a large influence on the characteristics of the final product. The initiator can confer the following potential properties to a polyglycol:

(1)   it can provide hydrophobicity, and even lipophilicity, (such as a long-chain fatty alcohol) or lack of hydrophobicity (such as water, ammonia or methanol);
(2)   it can also provide hydrophilicity by having a high functionality, such as exists, for example, in sorbitol or higher ethylene amines;
(3)   the initiator can provide a reactive site, such as the unsaturation in allyl alcohol or other homologues;
(4)   it can bring hetero atoms to the polyglycol which give increased heat stability or corrosion inhibition, as well as the opportunity to detect the product easily in various matrices.

Any branching that exists in the initiator, besides affecting the biodegradability of the molecule, will also affect the pour point of the polyglycol particularly in the case of ethoxylates. Increased branching results in lower pour points.

## 7.2.3   *General properties of polyglycol surfactants*

The physical properties of polyglycols are dependent upon molecular weight, the initiator, the oxide(s) used and their feeding technique. Thus, variations in these parameters will affect properties such as physical appearance, viscosity, pour point, cloud point, etc. (Schick, 1967). There are however many physical and chemical properties of polyglycols, such as specific gravity, flash point and reactivity, which do not change, almost regardless of the initiator or oxide(s) used. The chemical similarities are all based on the fact that polyglycols contain only ether links and hydroxyl group(s) (unless the initiator contains a reactor group such as a double bond). The products are normally sold as 100% active but for convenience of handling, when higher levels of ethoxylation are used, it is possible to add water to convert a paste or solid to a solution.

In general the products are soluble in a wide range of solvents such as aromatics, chlorinated solvents, low molecular weight ketones and alcohols, glycol ethers, etc. In each case the solubility depends upon the level of ethoxylation and the hydrophobe used. However, the solubility of a polyglycol in a particular solvent can be improved by the addition of coupling agents such as glycol ethers. Non-ionic surfactants derived from alkylene oxides are relatively stable in aqueous solutions as indicated by their use in both acid and alkali-based cleaners. Polyglycols are also stable in the presence of metallic ions (Porter, 1991).

Polyglycols will slowly oxidise if in contact with oxygen or other oxidising agents. Oxidation of the hydroxyl group(s) leads directly to aldehydes and acids, but degradation will also begin at any C–H bond, especially where the carbon is alpha to an ether link, by initially removing a proton radical and generating a free radical, which results in further rearrangements, such as C–O bond breakage (Santacesaria *et al.*, 1991). This latter degradation, which only occurs noticeably at elevated temperatures such as during metal-to-metal lubrication, leads to the generation of smaller molecules which are soluble in the parent polyglycol and eventually evaporate. The extent of this degradation can be controlled by the addition of well-known radical inhibitors such as aromatic amines or sterically hindered substituted phenols.

The solubility of non-ionic surfactants depends primarily on the hydration of the ethoxylate chain and thus the solubility of a particular surfactant in water will increase with the amount of ethylene oxide. This is indicated by the empirical quantity used for all types of surfactants called the hydrophilic–lipophilic balance (HLB) of the molecule, where the HLB value, on a scale from 0 to 20, is defined for non-ionic surfactants as one-fifth of the weight percentage of the hydrophilic moieties (normally EO and hydroxyls) in the surfactant (Griffin, 1949). Knowledge of a surfactant's HLB value allows a simple comparison between different types of surfactants. Experience has demonstrated a correlation between the HLB value of a surfactant and its specific application such as a wetting agent or emulsifier. The use of the term HLB implies that the surfactant has both hydrophilic and lipophilic moieties. The former is always correct but the latter is not necessarily true. If the hydrophobe is a carbon chain then it is oil soluble, but if it is a PO block then it is not oil soluble, as shown by the fact that polypropylene glycols, even of high molecular weight, are not oil soluble and are more correctly described as being hydrophobic rather than lipophilic.

### 7.2.4 *Applications of polyglycol surfactants*

The whole range of commercial surfactants, including ionics as well as non-ionics, are so versatile that they are used as performance chemicals in almost all industrial applications, either as essential additives or processing aids (Karsa, 1987). Typical applications of the entire surfactant range include such activities as emulsification, demulsification, wetting, foaming, defoaming, dispersing, solubilising, etc. and may be found in such diverse operations as detergents, cosmetics, pharmaceuticals, textiles, mining, petroleum, paints, plastics, food, pulp and paper, agriculture and leather.

More particularly, non-ionic surfactants are used in a wide range of these applications, and industries, essentially because the flexibility inherent in the chemistry implies that the products may be tailor-made to suit the desired performance and application. The typical applications for non-ionic surfactants include some obvious examples such as detergents, pharmaceuticals, emulsifiers (for water-in-oil and reverse types), wetting agents (including low foaming types), foamers, dispersants, defoamers, etc. Some less obvious applications are as lubricants, deinking waste paper, drilling muds, emulsion stabilisers, solubilisers, etc. as well as being reactants for other surfactants such as alkyl ethoxy sulphates and silicon derivatives.

*7.2.4.1 Applications of nonylphenol ethoxylates.* The application have been reviewed many times, especially in company brochures (*The Dow Chemical Company, 1988a*), and they can be briefly described as covering the whole range of surfactant activities such as being emulsifiers (both water-in-oil and the reverse), wetting agents (especially for textiles and industrial detergents), foam control agents, stabilisers for latices, metal processing, agriculture and chemical intermediates.

*7.2.4.2 Applications of fatty alcohol ethoxylates.* These products are well established as having better environmental fitness than nonylphenol ethoxylates and have replaced them in many major applications, especially household detergents. One limiting factor for the fatty alcohol derivatives has been the relatively high melting point of these products which necessitates them being sold as either solids or aqueous solutions. The nonylphenol derivatives have less of a problem because the combination of a branched alkyl group and a benzene ring means that higher ethoxylation levels are obtainable before the product solidifies at ambient temperatures (*The Dow Chemical Company, 1988a*). The multiple applications which use these ethoxylates can be briefly described by industry sector such as detergents (mainly household but also industrial), textiles (scouring, dyeing, finishing), agriculture (emulsifiers, both water-in-oil and the reverse), water treatment (as emulsifiers), coatings (emulsifier and wetting agents for paints, stabiliser for latices) and paper (wetting agents).

*7.2.4.3 Applications of fatty alcohol alkoxylates.* These products can be regarded as being biodegradable alternatives to the block copolymers or low-foaming versions of fatty alcohol ethoxylates. They thus have similar characteristics such as being excellent wetting agents, emulsifiers and good foam control agents as well as having reasonable chemical stability such as in high or low pH environments.

For these reasons they have found application in such diverse areas as detergents (especially hard surface cleaners), metal treatment, textile lubricants, agricultural chemicals, rinse aid formulations and in defoaming applications.

*7.2.4.4 Applications of block copolymers.* The application of these products is guided by the aqueous cloud point and this is very much dependent upon the molecular weight of the product, the functionality of the initiator and the ratio of the EO to PO or BO block.

The aqueous cloud point is increased by adding EO but decreased by the addition of PO or BO. A decrease in molecular weight (EO and PO/BO weight ratio being constant) will also increase the aqueous cloud point.

The use of a PO block as the hydrophobe reduces the foaming characteristics of these surfactants. This can be shown by comparing the typical foam heights, from the Ross–Miles test, of standard wetting agents such as a fatty alcohol ethoxylate (8EO) and nonylphenol ethoxylate (9EO) with a block copolymer (Table 7.3).

**Table 7.3**

| Sample | Foam height (mm) | |
|---|---|---|
| | Immediately | After 5 min |
| PO block 30% EO | 50 | 4 |
| $C_{13}$ alcohol 8EO | 125 | 50 |
| Nonyl phenol 9EO | 130 | 95 |

The applications of the linear block copolymers are related to the amount of EO in the final molecule (as indeed it applies to any polyglycol surfactant). A level of 10–15 weight percent EO will yield a defoamer whilst 25–30% of EO gives a wetting agent, based on a PO block being the hydrophobe. The block copolymers have many favourable characteristics such as being good low-foam wetting agents, good foam control agents as well as having chemical stability and a low order of toxicity (human and aquatic). For these reasons they have found application (sometimes after esterification) in such diverse areas as detergents (especially alkaline-based cleaners), the production of mineral slurries, water treatment, latices stabilisers, emulsifiers, metal treatment, demulsification, textile lubricants, agricultural chemicals, rinse aid formulations and in many defoaming applications. When the wetting agent needs to have even lower foaming characteristics, this can be achieved by blending the defoamer in a 1:9 ratio with the wetting agent.

## 7.3  General biodegradability

Biodegradation is commonly defined as the destruction of chemicals by the metabolic activity of living organisms. Of particular relevance to surfactant biodegradation is the metabolic activity of microorganisms, such as bacteria, which can consume organic chemicals like surfactants as the only carbon food source. The oft quoted principle of biological infallibility says that given enough time and sufficient microbial variation any chemical can be degraded. This is a consequence of the simple morphological and physiological properties of bacteria that render them extremely flexible in adapting their capabilities with respect to alternative food sources. Further to this general description of biodegradation, there are more specific terms related to biodegradation which are of significance for the interpretation of the legislative requirements and are explained below.

### 7.3.1 Primary biodegradation

Primary biodegradation is said to have occurred when the surfactant structure is sufficiently altered, by microbial activity, such that it has lost its surfactant properties or when an analytical procedure, specific for the parent compound, shows no response. For polyglycols the usual test procedure is the Bismuth iodide Active Substance (BiAS) test which is an analytical procedure that detects molecules containing five or more consecutive alkoxide units. The principle of the method consists of the formation of a pseudo-crown ether complex between the bismuth iodide and polyether which is most stable when the alkoxide chain contains five or more ether units.

### 7.3.2 Ultimate biodegradation

Ultimate biodegradation is defined as the complete degradation of an organic chemical to carbon dioxide, water, mineral salts and biomass. This complete conversion is also sometimes referred to as mineralisation.

### 7.3.3 Inherent biodegradation

Inherent biodegradability is an arbitrary definition as given in the Organisation for Economic Cooperation and Development (OECD) guidelines for Testing of Chemicals which is designed to encompass compounds which have the potential to degrade in the environment. The test protocols used to define this category of compounds allow the use of an acclimated or adapted microbial population and prolonged incubation times (more than 28 days). The types and modes of acclimation are discussed in a subsequent section. The test protocols are listed in Table 7.4.

### 7.3.4 Ready biodegradability

This classification is, perhaps, the most important since it would appear to be a major requirement for the current and impending legislation. Ready biodegrad-

**Table 7.4** Test procedures for inherent biodegradability testing, OECD 302 series

|            | Name                                       | Measured parameter                        | Pass level |
|------------|--------------------------------------------|-------------------------------------------|------------|
| OECD 302A  | Semi-continuous activated sludge (SCAS) test | Dissolved organic carbon (DOC)          | 20%        |
| OECD 302B  | Zahn–Wellens test                          | DOC or chemical oxygen demand (COD)       | 20%        |
| OECD 302C  | MITI test (Level 2)                        | Biological oxygen demand (BOD), DOC       | 20%        |

**Table 7.5** Test procedures for ready biodegradability testing, OECD 301 series

| OECD test | Name                           | Measured parameter | Pass level           |
|-----------|--------------------------------|--------------------|----------------------|
| 301A      | DOC Die-Away                   | DOC                | >70% loss of DOC     |
| 301B      | $CO_2$ evolution (Modified Sturm) | Carbon dioxide  | >60% yield of $CO_2$ |
| 301C      | Modified MITI                  | BOD and DOC        | >60% BOD             |
| 301D      | Closed Bottle                  | BOD                | >60% BOD             |
| 301E      | Modified OECD Screening        | DOC                | >70% DOC             |
| 301F      | Manometric Respirometry        | BOD                | >60% BOD             |

ability is also a definition provided by the OECD and is a property given to compounds which succeed in passing one of the stringent tests according to OECD guidelines 310A–F. In these tests the innoculum consists of a low amount of bacteria derived from defined areas, for example municipal wastewater treatment plants, surface waters or soil to ensure that the microbial population is not acclimated to the compound to be tested. The rationale for this testing procedure is that reaching a pass level in these laboratory tests will ensure that the tested compound will degrade rapidly in the natural environment and should, therefore, not result in either persistence or bioaccumulation. The test protocols for these ready biodegradation tests are described in a little more detail with the appropriate analytical procedures in Table 7.5.

The tests are conducted for a maximum period of 28 days. A further limitation imposed on the tests is that the pass level should be reached within a '10-day window'. This means that the pass level for biodegradation should be reached within 10 days from the onset of biodegradation, which is defined as the 10% level of biodegradation.

### 7.3.5  Analytical methods for ready and inherent biodegradation tests

The analytical procedures for the test protocols described above are in the main non-specific. Three procedures are described. Two of the tests depend on the fact that biodegradation is primarily an oxidative process and consequently the uptake of oxygen, the biological oxygen demand (BOD), can be measured in a closed system per unit mass of test compound and can be compared with the theoretical oxygen demand of the tested compound. This procedure is required for the OECD tests 301C, D and F. A complementary procedure involves the measurement of the evolution of carbon dioxide as in the Modified Sturm test, OECD 301C. A further analytical procedure involves the measurement of the dissolved organic carbon (DOC). This method is based on the fact that the concentration of dissolved organic carbon in the medium and thus the test compound will be reduced as a result of its biodegradation and thus be reflected in the DOC measurement. A major problem associated with these non-specific analytical test procedures is that not all the test compound is oxidised for energy, but, is to a certain extent, at the same time, also converted into biomass. This effect is reflected in the lower pass level of 60% ($CO_2$ evolution and BOD) and 70% (DOC decrease).

### 7.3.6 The microorganisms

The microorganisms of most relevance when discussing the biodegradation of polyglycols are bacteria. However, non-bacterial organisms such as fungi have been identified which are also capable of biodegrading polyglycols. Since the test protocols defined by the regulatory authorities recommend innocula, derived from wastewater treatment plants, soil, or surface waters, which contain bacteria as the predominant species, the discussion will be restricted to this group of microorganisms.

Bacteria are simple unicellular organisms and exist in a huge variety of forms. It is precisely this simplicity in organisation and structure which permits changes to their metabolic abilities and allows them to utilise a wide variety of food sources. A bacterial system in the absence of a food source exists in a resting phase where the bacterial cells adapt to their environment and consume only a low amount of oxygen which they need for their endogenous metabolism. On addition of a food source there then exists a lag phase if the food source is unfamiliar to the microbial population. During this lag phase the organisms may synthesise enzymes, either intra- or extra-cellularly, required for the metabolism of the food. The addition of a well-recognised food source such as glucose does not require a lag phase and is usually followed by a logarithmic expansion in the population of the bacteria. This is known as the growth phase or log phase. During this activity the food source is converted to either biomass or is oxidised to carbon dioxide, mineral salts and water which is accompanied by oxygen uptake. The oxygen uptake slows as the food source is fully consumed and the population then returns to the resting state which is characterised by the absence of visible bacterial growth but is actually a result of a balance between dying, viable and still-dividing cells.

## 7.4 Mechanisms of biodegradation

Mechanisms used by bacteria for the biodegradation of chemicals are normally based on the conversion of available substrates into intermediates for the normal metabolic pathways. These reactions are usually enzyme catalysed and are located either extra- or intra-cellularly. However, it is important to note that the net effect of these chemical transformations is oxidation and a brief survey of the recognised biochemical routes of oxidation will be of help in understanding the differences in biodegradability of test compounds.

### 7.4.1  ω-Oxidation

Degradation of the hydrophobic part of the surfactant molecule is believed to occur via mechanisms similar to those for aliphatic hydrocarbons. Most bacterial species have the ability to degrade linear hydrocarbons and thus the process is conceived as occurring by means of constitutive enzymes. This explains the ready degradation of these materials. Bacterial attack on linear hydrocarbons can occur via oxidation of the terminal carbon atoms. This is known as ω-oxidation. Several pathways can be envisaged including the intervention of an oxygenase enzyme and then subsequent oxidation to the aldehydes and carboxylic acids. Alternatively, initial dehydrogenation followed by hydrolysis should also yield the same aldehyde structure. Studies carried out with $^{14}$C-labelled stearyl alcohol ethoxylates suggest that this is the primary mechanism of degradation (Nooi *et al.*, 1970). However, the alternative mechanism is one of initial hydrophobe–hydrophile scission and is described in a subsequent section. Both mechanisms probably occur in nature though the mixed bacterial strains found in domestic waste appear to favour the hydrophobe–hydrophile scission route (Kravetz, 1990). The subsequent steps in the biodegradation pathway occur via the oxidation cycle known as β-oxidation.

### 7.4.2  β-Oxidation

This process is commonly held to occur in the oxidation of fatty acids by living cells. In this reaction, pairs of carbon atoms are separated from the hydrocarbon chain by the moderation of a co-enzyme, acetyl co-enzyme A, leaving the hydrocarbon chain with a terminal carboxyl function free to re-enter the cycle as illustrated in Figure 7.1.

Such synthetases have a low specificity and are able to react with fatty acids with chain lengths from $C_6$ to $C_{20}$. Even-numbered carbon chains can be completely oxidised by this procedure. However, odd-numbered carbon chains eventually reach the propionyl group which is then further metabolised via an alternative propionyl coenzyme A cycle. It is clear that oxidation of a hydrocarbon via this route is most favoured by a linear carbon chain containing primary carbon atoms. The presence of secondary carbon atoms will prevent the oxida-

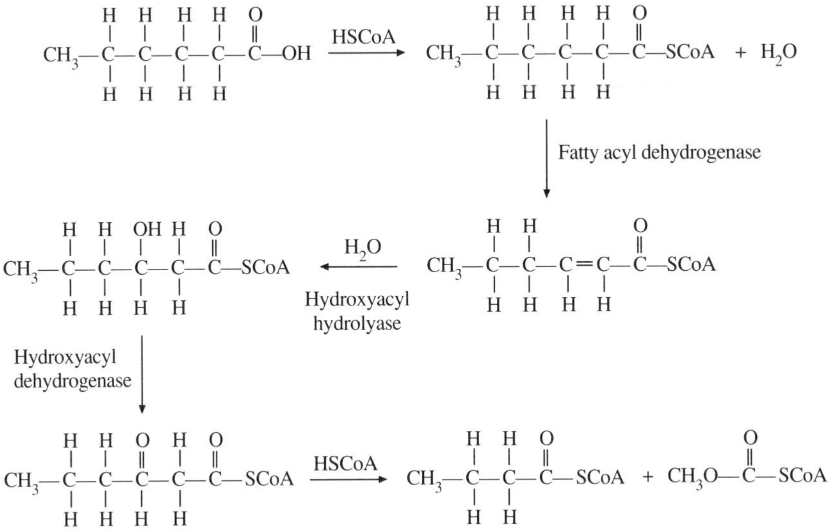

$$CH_3-C-C-C-C-C-OH \xrightarrow{\text{HSCoA}} CH_3-C-C-C-C-C-SCoA + H_2O$$

Fatty acyl dehydrogenase

$$CH_3-C-C-C-C-C-SCoA \xleftarrow[\text{Hydroxyacyl hydrolyase}]{H_2O} CH_3-C-C-C=C-C-SCoA$$

Hydroxyacyl dehydrogenase

$$CH_3-C-C-C-C-C-SCoA \xrightarrow{\text{HSCoA}} CH_3-C-C-C-SCoA + CH_3O-C-SCoA$$

**Figure 7.1** β-Oxidation cycle with acetyl coenzyme A (White *et al.*, 1978).

tion to a carboxylic acid function, stopping the cycle, and thus may require an alternative mechanism.

### 7.4.3 α-Oxidation

The presence of branching in a hydrocarbon chain inhibits the β-oxidation procedure if the pendant alkyl chain is on the β-carbon atom and thus requires that an alternative oxidation procedure referred to as α-oxidation be invoked. In this process the carbon atom is oxidised to a ketonic group and the carbon chain is shortened by oxidative decarbonylation (Emmanuel, 1978).

### 7.4.4 Alkoxylate chains

The preceding sections have provided mechanisms for the degradation of both branched and linear alkyl chains. However, an important structural characteristic of non-ionic polyglycols is the alkoxylate chains and any mechanistic discussion of degradation must of course cover these moieties. There are two ways in which a polyglycol chain based on ethylene oxide can be attacked. The chain can be degraded from the end, one glycol unit at a time, commonly referred to as the ω-hydrophile pathway. This process has been identified in species where the surfactant structure makes the hydrophobe resistant to attack, for example alkyl phenol ethoxylates or branched alcohol ethoxylates. Alternatively, the polygly-col chain can be broken at random into smaller glycol units (Haines and Alexander, 1975). However, most of the available evidence appears to point to the former process where the primary attack is at the terminus of the polymer (Pearce and Heydeman, 1980). The hydrolytic, and oxidative, mechanism via glycolic acid, and subsequent insertion into the oxidative dicarbonic acid cycle, via glyoxylic acid, oxalic acid and formic acid is illustrated in Figure 7.2.

However, some evidence exists that the mechanism of ethoxylate shortening may be more complex than a simple hydrolysis of ethylene glycol units (Watson and Jones, 1977). A complementary mechanism in which the terminal hydroxyl function is oxidised to a carboxylic acid and then the $C_2$ unit is lost as glyoxylic acid, without the intermediacy of glycolic acid, has also been implicated and may be the dominant mechanism (Kawai, 1987). The mechanistic details for propoxylates are less abundant and the ultimate degradation of these adducts is found to be poor. However, the mechanism is believed to be similar to that observed for ethoxylates in that a hydrolytic elimination of individual glycol units or an oxidation of the terminal hydroxyl group is operative. The presence of a secondary carbon atom in the PO backbone would inhibit the oxidative route and may explain the recalcitrance of these compounds. Specially isolated *Pseudomonas* and *Aerobacter* microbes have been used to develop a process to treat effluent from propylene oxide and polyol plants (Raja *et al.*, 1991). These species, when used in combination, showed more than 90% biodegradation (BOD) when exposed to PO- and glycol-containing waters. The mechanism pro-poses an oxidative sequence from propanediol to lactic acid, pyruvic acid and formic acid eventually resulting in complete mineralisation.

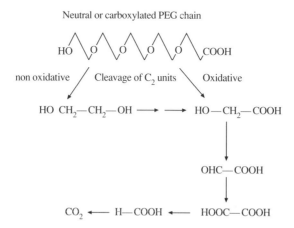

**Figure 7.2** Mechanism for the biodegradation of polyethylene glycols (Steber and Wierich, 1985).

## 7.5   Biodegradation of polyglycols

### 7.5.1   Nonylphenol ethoxylates (NPE)

The major product group within the generic alkylphenol ethoxylates is the nonylphenol derivative. The alternatives such as octylphenol, dinonylphenol and dodecylphenol are not frequently used because of economic and/or performance reasons.

For many years NPEs were the first choice for a non-ionic surfactant based on both performance and price. In their heyday they were well-liked and versatile surfactants, because they could be used in a wide variety of applications, covering the whole HLB range. Currently about 85 000 mtpa are sold in Western Europe. Their position of prominence has now been taken by the fatty alcohol ethoxylates due to biodegradability concerns as will be described later. NPEs are commercially available from a 2 mole ethoxylate up to a 50 mole ethoxylate, where the higher ethoxylates are frequently sold as 70% or 80% active solutions in water.

#### 7.5.1.1   Origin of the nonylphenol hydrophobe.   Nonylphenol is made by reacting nonene with phenol via the Friedel–Crafts reaction using aluminium chloride as catalyst. Nonene is either made by polymerising a smaller alkene, which leads to a branched alkene, for example propylene yields nonene or dodecene, isobutylene gives octene. Alternatively a larger alkene may be available from other reactions such as the Shell Higher Olefins Process (SHOP). The choice is normally trimerised propylene leading to a severely branched nonylphenol.

#### 7.5.1.2   Biodegradability of NPEs.   This subject will not be dealt with in great depth since the use of surfactants based on nonyl phenols is being phased out within the European Community and will continue to decrease. However, some general comments on the structural effects can be made. This class of compounds has been, and remains, a fruitful source of controversy and disagreement. It now appears that alkyl phenol ethoxylates undergo almost complete primary biodegradation given sufficient acclimation. Interestingly, some evidence exists to suggest that a global bacterial acclimation process with respect to alkyl phenol ethoxylates has occurred. Support for this hypothesis is provided by experiments which showed that biodegradation tests carried out in the 1970s showed better degradation than those carried out in the 1960s (Wickbold, 1974).

The presence of a highly branched alkyl chain derived from propylene impedes the $\beta$-oxidation of the hydrophobe. Furthermore, the steric hindrance imposed by the phenyl ring also reduces the participation of the hydrophobe–hydrophile scission mechanism and thus the overall ultimate biodegradability of these molecules is poor. The position of alkyl chain coordination to the benzene ring also appears to have an effect on the biodegradation rates, with the *ortho* substituted isomers being more resistant (Giger *et al.*, 1981). Extrapolating from the successful replacement of ABS by LAS, the biodegradation of linear alkyl phenol ethoxylates might be expected to be improved relative to the branched derivatives. The evidence for this supposition is not as conclusive as that for LAS. Moreover, the positional isomers with respect to the position of alkylation on the chain and also the location on the ring are important (Smithson, 1966; Marei *et al.*, 1976). What emerges is that the rate of primary biodegradation decreases on going from *para* to *meta* to *ortho* but is enhanced if the alkyl chain phenyl ring linkage is near the alkyl terminus.

### 7.5.2   Fatty alcohol ethoxylates (FAE)

This group of products is commercially the most important of the ethoxylates as more than 500 000 metric tonnes are sold in Europe each year, with the majority being used in detergents. Typically the alcohol used is in the range nonanol to octadecanol, or blends thereof, with an ethoxylation level of 2 to 30 commercially available.

#### 7.5.2.1   *Origin of the fatty alcohol hydrophobe.*   The alcohols used are in the range of 9 to 18 carbons, but most commonly are 12 to 15 or blends, distillation cuts, within that range, and are either natural (oleochemical) or synthetic (petrochemical), with the latter being either branched or linear or a combination of both. The fatty alcohol may be derived from several sources, either oleochemical or petrochemical. The type of fatty alcohol used, natural or synthetic, has led to much debate as to which is the most environmentally friendly hydrophobe (Schirber, 1991). The oleochemical derived fatty alcohols are made from either plant or animal fats and have acquired the mantle of being more environmentally friendly because they are based on renewable resources. They consist of linear carbon chains. This is in contrast to the rather negative perception of petrochemically derived fatty alcohols. It is debatable whether this negative perception is justified. A recent publication compared a 'natural' palm oil derived surfactant with a hydrocarbon derived material in terms of their environmental impact (Pittinger *et al.*, 1993). This revealed that although the extraction of the raw palm oil was cheaper in terms of energy, it generated significantly more solid waste. Petrochemical derived alcohols are manufactured by several methods resulting in different structural characteristics. They may be synthesised by the polymerisation of propylene resulting in a highly branched hydrophobe. Alternative methods are the Ziegler polymerisation of ethylene followed by hydrolysis. This yields an essentially linear alcohol, with an even number of carbon atoms. The other main method of synthesis is via the oxo-route in which linear $\alpha$-olefins are reacted with carbon monoxide and water to yield an alcohol. Since oxonation may occur at either carbon of the olefinic group a substantial amount of the mixture contains a methyl branch at the $\beta$ carbon atom. The

general consensus seems to accept that neither source can be truly regarded as environmentally superior to the other, but the debate still continues.

Natural alcohols used in surfactants, primarily in the 12–14 carbon atoms range, are normally derived from vegetable oils (such as coconut or palm kernel oils) where they are typically found as glyceride esters. These esters are hydrolysed to the acid and glycerol and the acid then reduced to the alcohol. The natural alcohols always have an even number of carbon atoms, but may have unsaturation.

### 7.5.2.2  Biodegradability of FAEs.

Alcohol ethoxylates have been developed as an alternative to the less eco-compatible nonyl phenol ethoxylates.

#### 7.5.2.2.1  Alkyl chain branching.

Considerable effort has been directed at biodegradation studies of alcohol ethoxylates and a few general rules have been formulated. The linear primary alcohol ethoxylates are characteristically readily biodegradable. These include those obtained from oleochemical derived initiators. The effect of branching on biodegradation is clearly demonstrated by the comparison of an essentially linear $C_{12}$–$C_{15}$ ethoxylate containing 9 EO with a highly branched $C_{13}$ alcohol containing 7 EO. The linear alcohol ethoxylate reached >80% biodegradation within 28 days. In contrast, the branched alcohol ethoxylate reached barely half that level (Kravetz et al., 1991). This is believed to be a consequence of the inhibition of the $\beta$-oxidation route described earlier. The effect of a small amount of methyl branching, when compared with a completely linear chain, is believed to be minimal. This was borne out by a recent study (Karsa et al., 1992) where fatty alcohols in the $C_{12}$–$C_{15}$ range with up to 100% 2-alkyl branching were studied. Results are summarised in Table 7.6. These results contrast with a similar study on fatty alcohol EO/PO copolymers (see Section 7.5.3). Very little information exists on the effect of alkyl chain length on biodegradation. However, some results appear to show that longer chain homologues exhibit faster primary biodegradation (Wickbold, 1974).

**Table 7.6** Influence of 2-alkyl branching on the primary biodegradation of fatty alcohol ethoxylates

| % 2-Alkyl branching in the hydrophobe | Moles EO in the ethoxylate (n) $R(OCH_2CH_2)_nOH$ | % Primary biodegradation |
|---|---|---|
| 0 | 3–20 | >95[a] |
| 20 | 5 | 99 |
|  | 9 | 98 |
| 35 | 7 | 95 |
| 60 | 7 | 93 |
| 100 | 4 | 97 |
|  | 10 | 85 |

[a] Literature values.
All measured values were the average of six samples.
Only the higher ethoxylate (10 EO) of the highly branched alcohol suggests an adverse influence of alkyl (mainly methyl) branching.

*7.5.2.2.2   Hydrophobe–hydrophile scission.*   Hydrophobe–hydrophile scission is the route which has been most widely identified as the means by which bacteria gain access to the hydrophobe portion of surfactants. In this case, scission of the internal ether link of the initiator to the alkoxylate chain yields a hydrophobe and a polyalkoxylate, thus achieving rapid primary biodegradation (Patterson *et al.*, 1970). The hydrophobe is then free to be oxidised by the routes described earlier. As mentioned before, evidence that $\omega$-oxidation is the biodegradation mechanism mode has also been presented. However, the scission mechanism is believed to predominate in strains of bacteria prevalent in domestic waste. The degree of branching present around the carbon forming the internal linkage can strongly influence this reaction. Initiators with a high degree of branching on the $\beta$-carbon atom can provide sufficient steric hindrance that will impede the biodegradation of these materials. Evidence for this hypothesis is provided in a publication which compared the primary biodegradability of a series of alcohol ethoxylates with hydrophobes derived from the oxo process containing differing amounts of $\beta$-carbon methyl substitution (Birch, 1984). In this case the primary biodegradability of an initiator containing 25% methyl branching was 100%. In contrast, a similar compound containing 50% methyl branching was degraded only 84%. Secondary linear alcohol ethoxylates revealed a greater resistance to primary degradation and a significant decrease in the amount of primary degradation was demonstrated on increasing the ethoxylate chain length (Birch, 1991). A similar trend is shown in the ultimate biodegradation. Further confirmation is also provided by studies on the biodegradation of 2-ethyl decyl ethoxylate. In this series no PEGS were found indicating the resistance to scission (Watson and Jones, 1979).

Both mechanisms of hydrophile–hydrophobe scission and $\omega$-oxidation probably occur in nature with the extent of each reaction being determined by the precise structure of the hydrophobe.

*7.5.2.2.3   Ethoxylates.*   It should be appreciated that although oxidative procedures for the biodegradation of ethylene glycols may be prevalent, the biodegradation of the alkyl hydrophobes, if linear, is considerably faster and occurs in preference to the glycol portion of the surfactants. The length of the ethoxylate chain has been shown to clearly influence the biodegradation of surfactants. Longer EO chains show increased bioresistance. This effect is rationalised in terms of the increased hydrophilicity and molecular dimensions limiting transport of the molecules through the cell walls. Extensive data illustrating this effect are collated in the excellent text by Swisher (1987) on surfactant biodegradation.

### 7.5.3   Fatty alcohol alkoxylates (FAA)

These alkoxylates are derived from the addition of alkylene oxides to the same fatty alcohols as described in the previous section 7.5.2.1.

The addition of the oxides can be the same four combinations as quoted in Section 7.2, but the main products encountered are either regular or reverse block copolymers. This group of products is similar to the FAEs, but the addition of the branched PO or BO molecule changes the potential biodegradability (as well as the physical properties) of the alkoxylate. The main characteristics of the FAAs are a combination of those of the linear block copolymers and the FAEs. Thus, the molecule does not have such a high molecular weight as the typical block copolymers, because the hydrophobic oxide is used to control the characteristics of the surfactant, such as low foaming, rather than provide the hydrophobicity.

*7.5.3.1  Biodegradability of FAAs.*    The inclusion of PO or BO in surfactants is now a common part of the surfactant manufacturer's palette. However, less biodegradability data are available than for the ethoxylates. Still less attention has been addressed to the subject of BO.

Both these molecules may be considered to have ambiguous characters with respect to their contributions to the HLB. While their structure includes the pendant methyl or ethyl groups which will contribute to the overall hydrophobicity of the molecule, the necessary inclusion of the ether linkage will also increase the hydrophilicity. This is less true for BO derivatives. The inclusion of PO or BO in a block copolymer can affect both the pour point and cloud points of the polymers.

However, information relating to the biodegradation of these polymers is sparse. A very recent study on alcohol initiated block copolymers of PO and EO has revealed the following effects. Branching in the alcohol hydrophobes diminishes the level of PO which can be tolerated before the primary biodegradation is affected. The most linear alcohol initiators can tolerate up to 3.5 moles PO (Taylor *et al.*, 1988). This study was carried out with the PO block included in between two EO blocks. No data are available for the location of the PO block with respect to the alcohol, although anecdotal evidence suggests that the attachment of EO to the alcohol, rather than PO, is preferable for the hydrophobe–hydrophile scission mechanism. The explanation for this may lie in the inefficiency of the hydrophobe–hydrophile scission mechanism when confronted by the pendant methyl group of PO.

A direct comparison of the influence of PO emerges on comparison of a $C_{12}/C_{18}$ alcohol reacted with 2.5 EO followed by 6 PO and the same alcohol reacted with 6 EO followed by 2 PO (Bock *et al.*, 1988). The former showed a degradation level of about 40% in two independent screening tests, thus falling short of the level required for a readily biodegradable description. In contrast the compound containing less PO revealed a biodegradation level of 70–80% in the same independent OECD screening procedures. This latter compound successfully reached the pass level for ready biodegradability. It is also educative to compare this data to a pure ethoxylate initiated from the same fatty alcohol. In this case the degree of biodegradation reached in similar OECD screening procedures was substantially higher than even the best EO/PO copolymer described above. This is further indication of the negative influence of PO on the biodegradability of fatty alcohol alkoxylates.

A recent study compared a series of fatty alcohol ethoxylates capped with varying amounts of PO and BO (Karsa *et al.*, 1992). This showed that the degree of 2-alkyl branching in the pure ethoxylate series had minimal effect on the primary biodegradability. In contrast, a series of fatty alcohol ethoxylates capped with an equivalent number of moles of PO was strongly affected by the degree of branching in the fatty alcohol. The products had exactly the same number of moles of both ethylene oxide and propylene oxide, i.e.

$$R—(OCH_2—CH_2)_n—(O—CH_2—CH)_m—OH$$
$$\underset{\displaystyle CH_3}{|}$$

where R is an alkyl group in the $C_{12}$–$C_{15}$ range, $n = 6$, and $m = 6.5$.

These products had different cloud points due to both variations in the average molecular weight of the hydrophobe and the fact that the more alkyl branching that occurs for a given molecular weight the more moles of ethylene oxide and propylene oxide are required to achieve both intermediate and final cloud points. That is to say that an alkyl branched alcohol is a more hydrophobic product than its linear analogue. Results are reported in Table 7.7.

**Table 7.7** Influence of 2-alkyl branching on the primary biodegradability of fatty alcohol EO/PO copolymers

| % 2-Alkyl branching in the alkyl hydrophobe | Moles EO/ moles PO | % Biodegradability |
|:---:|:---:|:---:|
| 20 | 6 EO + 6.5 PO | 97 |
| 35 | 6 EO + 6.5 PO | 93 |
| 60 | 6 EO + 6.5 PO | 80 |
| 100 | 6 EO + 6.5 PO | 10 |
| 0 | 8 EO + 5 PO | 95 |
| 60 | 8 EO + 5 PO | 73 |

A poly(oxypropyl) chain terminated alcohol ethoxylate effectively has a poly(oxyethylene) chain sandwiched between two hydrophobic groups and biodegradation is clearly retarded in the case of the derivatives of the more highly branched alcohol. This pattern is also observed when comparing a $C_{9,11}$-oxo-alcohol derivative with an iso-decanol analogue (Table 7.8). This implies that the primary degradation of fatty alcohol block copolymers occurs via a combination of two mechanisms, that of the $\omega$-hydrophile oxidation, and also the hydrophobe–hydrophile scission. If the $\omega$-hydrophile oxidation mechanism

**Table 7.8** Comparison of primary biodegradability of $C_{9,11}$-oxo alcohol and isodecanol based EO/PO copolymers

| Fatty alcohol | % 2-Alkyl branching | Moles EO/moles PO | % Biodegradability |
|:---|:---:|:---:|:---:|
| $C_{9,11}$-alkyl | 20 | 9.5 EO + 4.5 PO | 97 |
| $C_{10}$-alkyl | 100 | 10 EO + 5 PO | 70 |

is inhibited by the presence of PO then the level of 2-alkyl branching will dictate the extent of the hydrophobe–hydrophile scission. Moreover, if the degree of 2-alkyl branching is high then this mode of degradation is also inhibited resulting in poor primary biodegradability. A limit of 6–7 moles of PO has long been regarded as the limit placed on a linear fatty alcohol EO/PO block copolymer in order to still qualify for primary degradation. A recent study (Karsa *et al.*, 1992) of a $C_{12,14}$ linear fatty alcohol ethoxylate with 6 moles of ethylene oxide and propoxylated to various levels, i.e.

$$R(O-CH_2-CH_2)_6-(O-CH_2-\overset{\overset{\displaystyle CH_3}{|}}{CH})_x-OH$$

where R is a linear $C_{12,14}$-alkyl group, confirmed this general trend. Results are summarised in Table 7.9.

**Table 7.9** Influence of the moles of propylene oxide on the biodegradability of a fatty alcohol EO/PO copolymer

| Moles of propylene oxide (x) | % Biodegradability |
|---|---|
| 3 | 100 |
| 6 | 97 |
| 9.5 | 70 |

In contrast, a variation of the EO chain length between 3 and 9 had minimal effect on primary degradation. Again derivatives of a linear $C_{12,14}$ fatty alcohol were prepared with varying levels of poly(oxyethylene) and between 4 and 5 moles of propylene oxide, i.e.

$$R-(O-CH_2-CH_2)_y-(O-CH_2-\overset{\overset{\displaystyle CH_3}{|}}{CH})_{4-5}-OH$$

Results are summarised in Table 7.10.

**Table 7.10** Influence of the poly(oxyethylene) chain length on biodegradability of alcohol EO/PO copolymers

| Moles of ethylene oxide (y) | % Biodegradability |
|---|---|
| 3 | 100 |
| 6 | 100 |
| 7 | 95 |
| 9 | 97 |

This is also one of the few publications which has evaluated the biodegradability of butylene oxide tipped products. The data published are sparse but appear to indicate that the total moles of alkoxide in the molecule play a role in the primary biodegradability characteristics. In other words, the use of BO to obtain a particular HLB is preferable to PO, since less BO is required, and the use of less alkoxide favours the biodegradability of these molecules.

### 7.5.4   Alternative low-foam non-ionics to fatty alcohol EO/PO copolymers

The performance deficiencies perceived with the fatty alcohol EO/PO copolymers have resulted in both surfactant producers and formulators evaluating alternative species of low-foam non-ionics, examples of which include butylene oxide tipped and alkyl and aryl end-blocked fatty alcohol ethoxylates. These components exhibit improved performance characteristics, compared to the former.

Products were compared with two standards (Karsa et al., 1992) a biodegradable $C_{12,14}$-linear fatty alcohol EO/PO copolymer (cloud point 29°C) and a non-biodegradable EO/PO block copolymer (cloud point 34°C). These products were selected as they are widely used as components in rinse aid formulations, a major application area for such materials.

In the first study, butylene oxide tipped non-ionics were compared with these standards (Table 7.11). Butylene oxide is more hydrophobic than propylene oxide and hence fewer moles of butylene oxide are required to give equivalent final cloud points. Results in the B range suggest that total moles of alkylene oxide do begin to have an influence on biodegradability.

An alternative to alkylene oxide tipping of an ethoxylate is the reaction of the terminal hydroxyl group to give an alkyl or aryl 'end-blocked' non-ionic. The alkyl or aryl groups lower the cloud point of the parent ethoxylate and its

**Table 7.11** Biodegradability of fatty alcohol EO/BO copolymers

| Product | | Cloud point (1% aqueous solution) °C | % Biodegradability |
|---|---|---|---|
| $C_{12,14}$-alcohol EO/PO | | 29 | 94 |
| EO/PO block copolymer | | 32 | 20 |
| $C_{12,14}$-alcohol EO/BO | | | |
| copolymers | A1 | 67 | 99 |
| | A2 | 36 | 99 |
| | A3 | 22 | 99 |
| | B1 | 47 | 98 |
| | B2 | 32 | 87 |
| | B3 | 21 | 77 |

The number of moles of butylene oxide increases from A1 to A3 and from B1 to B3. A1–3 contain the same level of poly(ethylene oxide).
B1–3 also contain the same levels of poly(ethylene oxide) but, on average, 3 moles more than in the A1–3 series.

**Table 7.12** Biodegradability of aryl and alkyl 'end-blocked' alcohol ethoxylate

| Product | | Cloud point (1% aqueous solution) | % Biodegradability |
|---|---|---|---|
| $C_{12,14}$-alcohol EO/PO | | 26 | 94 |
| EO/PO block copolymer | | 34 | 20 |
| $C_{9,11}$-alcohol EO/aryl | C0 | 17 | 99 |
| end-blocked | C1 | 26 | – |
| | C2 | 37 | 89 |
| | C3 | 53 | 80 |
| $C_{9,11}$-alcohol EO/alkyl | D1 | 21 | 97 |
| end-blocked | D2 | 38 | 96 |
| | D3 | 43 | 88 |
| | D4 | 53 | 83 |

For a given alkyl or aryl blocking group, the number of moles of ethylene oxide increases from C0 to C4 and D1 to D4.

foaming capacity, provided that in the case of the alkyl group at least a $C_4$-chain length is used. It also depends on the 'degree of blocking' and in most cases at least 95% of the parent non-ionic should be reacted, particularly if the product is required to exhibit stability in caustic powder formulations or to chlorine release agents. However, it should be noted that in other applications, it can be advantageous to leave more of the base non-ionic unreacted, thus retaining some of the desired properties of the unblocked ethoxylate while achieving low-foam properties from the major component.

Results of the biodegradability testing of alkyl and aryl end-blocked non-ionics are given in Table 7.12. As expected, the alkyl end-blocked non-ionics exhibit similar primary biodegradation to the aryl analogues. In both series, relatively high levels of poly(ethylene oxide) can be incorporated into the non-ionic without loss of the required primary biodegradation. Hence, both reaction of fatty alcohol ethoxylates with butylene oxide or terminated with an alkyl or aryl group afford biodegradable intermediates in the correct cloud point range for use in rinse aids, machine dishwashing and other cleaning processes.

## 7.5.5  Linear block copolymers

In the previous sections the surfactant hydrophobe was a carbon chain, but in this section the hydrophobe is comprised of a branched carbon chain containing ether links, derived from a hydrophobic alkylene oxide, normally a PO block, which will contribute some hydrophilicity to the hydrophobe. An early reference is still useful for a basic understanding of these products (Schick, 1967).

### 7.5.5.1  Origin of the polyglycol hydrophobe.  Unlike the previous surfactants, which comprise a specific monool initiator reacted with just EO, i.e. a homopolymer, this group of products, made from the reaction of alkylene oxides onto an initiator, contains a wide range of possible chemistries based not only on

the functionality of the initiator but also the feeding technique of the oxides. The common factor that they all have is that the hydrophobe is relatively large, typically 90 carbon atoms and 30 oxygen atoms, and branched, from the pendant methyl and ethyl groups of PO and BO, respectively.

Typical initiators used are small molecules which are normally hydrophilic such as methanol (monool), propylene glycol (diol), glycerol (triol), ethylene diamine (tetraol) etc. The initiator (especially if it creates steric hindrance), the molecular weight of the surfactant and the ratio of EO to PO or BO will all play a role in biodegradation but for the purposes of this general discussion, all of these possible chemistries can be regarded as similar with a view to biodegradation. To this end the more common copolymer surfactant structure, regular block copolymers, will be discussed. These products were the ones used in various detergent formulations, such as caustic-based cleaners, until the recent derogation, from primary biodegradation regulations, was not renewed (Porter, 1991).

They are currently used in such non-detergent applications as wetting agents, de-inking waste paper, emulsifiers, demulsifiers and foam control agents.

Those block copolymers made from diol initiators, such as propylene glycol, will be reviewed. The first commercial products were made using PO blocks (based on propylene glycol) as the hydrophobe, but, due to the hydrophilicity associated with PO, these blocks had to have a molecular weight in excess of 1000 to achieve sufficient hydrophobicity. The most common molecular weight used for the PO block being 1750, with the EO block cap being from about 250 to 1200 depending on the application.

*7.5.5.2  Biodegradation of copolymers.*   The hydrophile element of polyglycol surfactants is usually provided by the ethylene oxide chains of the molecules. The propylene oxide, and to a lesser degree the butylene oxide chains can also be considered to partially contribute to the hydrophilicity of the polymer via the polyether linkage. Although recognised as being some of the least toxic type of non-ionic surfactants, block copolymers fail to meet the requirements for ready biodegradability (Shell Detergents Technical Bulletin, 1983). The data for these type of surfactants are not extensive and there has, to the knowledge of the authors, been no systematic study of the biodegradability of these materials with respect to the structure or PO vs. EO content. The mechanism of biodegradation of the block copolymers is probably similar to those described earlier in Section 7.4.4 whereby the alkoxylate chains are broken down from the terminus and lose individual glycol units. The high molecular weights of the block copolymers imply that the concentration of terminal hydroxyl functions is limited and thus the biodegradation of these materials is expected to be poor. Furthermore, the high molecular weights of the materials also make the transport through the cell walls of bacteria difficult and thus limit the extent of intracellular degradation. The inclusion of PO further reduces the biodegradability of these materials for the reasons elucidated earlier. A recent publication detailed the primary biodegradability of some block copolymers as varying between 5 and 58%, the higher values being obtained for the compounds with larger EO contents (Von Hettche and Klahr, 1982; Bock *et al.*, 1988). A similar trend was reported in the screening tests for ready biodegradability even though they all failed to reach the pass levels.

Primary biodegradation for block copolymers is based upon the BiAS method which detects an alkoxylate block of 5 or more alkylene oxide units. As a typical block copolymer contains a minimum of about 40 consecutive alkylene oxide units (30 PO, 10 EO) it is not surprising that block copolymers do not obtain primary biodegradation status. An indication of the lack of primary biodegradability associated with block copolymers is given by the fact that they were derogated from the EC directive relating to the biodegradability of surfactants used in detergents (EEC Directive 73/404 with amendments 82/242 and 86/94).

### 7.5.6   Fatty amine ethoxylates

Fatty amines and their ethoxylated adducts have been studied recently in closed bottle tests and a variant of the closed bottle test referred to as the 'prolonged closed bottle test' (van Ginkel *et al.*, 1993). This publication proposed a similar mechanism of biodegradation to that of the linear alcohol ethoxylates. The biodegradation curves are interpreted as indicating a rapid oxidation of the alkyl chain followed by a slower breakdown of the ethoxylated amines. Moreover, evidence for faster degradation of the alkyl chain of ethoxylated amides versus ethoxylated amines is also presented. This latter result is rationalised in terms of a rapid enzymatic hydrolysis of the amidic linkage in contrast to the oxidative scission of a C–N bond. Interestingly, this publication also states that the fatty tallow and oleyl bis(2-hydroxyethyl) amines are toxic to the inoculum used. This necessitated the use of a silica gel to reduce the concentration of the test compound in the aqueous phase. Under these conditions biodegradation levels of greater than 60% are reported. In contrast the ethoxylated amines did not reach the pass levels required for ready biodegradation.

### Acknowledgements

The authors wish to thank DOW EUROPE S.A. for permission to publish this chapter and their colleague Dr. Urs Friederich for his advice during its preparation.

### References

Birch, R.R. (1984). *J. Am. Oil Chem. Soc.* **61**, 340.
Birch, R.R. (1991). *Riv. Ital. Sostanze Grasse*, **LXVIII**, 433.
Bock, K.J., Huber, L. and Schöberl, P. (1988). *Tenside* **25**, 86.

*EEC Directive* 73/404 with amendments 82/242 and 86/94.

*EEC Directive* 91/325 and 93/21.

Emmanuel, B. (1978). *Biochim. Biophys. Acta.*, **528**, 239.

Giger, W., Stephanou, E. and Schaffner, C. (1981). *Chemosphere* **10**, 1253.

Griffin, W.C. (1949). *J. Soc. Cosmetic Chem.*, **1**, 311.

Haines, J.R. and Alexander, M. (1975). *Appl. Microbiol.*, **29**, 621.

Karsa, D.R. (1987). *Industrial Applications of Surfactants*, Royal Society of Chemistry, London.

Karsa, D.R., Adamson, J. and Hadfield, R.P. (1992). *Chimicaoggi* 39.

Kawai, F. (1987). *CRC Crit. Rev. Biotech.*, **6**, 273.

Kravetz, L. (1990). *Agricultural and Synthetic Polymers,* American Chemical Society, Washington, DC, p. 96.

Kravetz, L., Salanitro, J.P., Dorn, P.B. and Guin, K.F. (1991) *J. Am. Oil Chem. Soc.*, **68**, 610.

Marei, A., Kassem, T.M. and Gebril, B.A. (1976). *Indian J. Technol.* **14**, 447.

Nooi, J.R., Testa, M.C., and Wilemse, S. (1970). *Tenside*, **7**, 61.

Patterson, S.J., Scott, C.C. and Tucker, K.B.E. (1970). *J. Am. Oil Chem. Soc.* **47**, 37.

Pearce, B.A. and Heydeman, M.T. (1980). *J. Gen. Microbiol.*, 673.

Pittinger, C.A., Sellers, J.S., Janzen, D.C., Koch, D.G., Rothgeb, T.M. and Hunnicutt, M.L. (1993) *J. Am. Oil Chem. Soc.*, **70**, 1.

Porter, M.R. (1991). *Handbook of Surfactants*, Blackie, Glasgow.

Raja, L.M.V., Elvamuthy, G., Palaniappan, P. and Krishnan, R.M. (1991). *Appl. Biochem. Biotechnol.*, **28**, 827.

Santacesaria, E., Gelosa, D., Di Serio, M. and Tesser, R. (1991). *J. Appl. Polym. Sci.*, **42**, 2053.

Schick, M.J. (1967). *Nonionic Surfactants*, Marcel Dekker, New York.

Schirber, C.A. (1991). INFORM **2**, 1063.

Schoeller (1930). German Patent 548 201.

Shell Detergents Technical Bulletin DI 3.2.8, (1983).

Smithson, L.H. (1966). *J. Am. Oil Chem. Soc.* **43**, 568.

Steber, J. and Wierich, P. (1983). *Tenside* **20**, 183.

Steber, J. and Wierich, P. (1985). *Appl. Environ. Microbiol.*, **49**, 530.

Sturm, R.N. (1973). *J. Am. Oil Chem. Soc.* **50**, 159.

Swisher, R.D. (1987). *Surfactant Biodegradation*, Marcel Dekker, New York.

Taylor, C.G., Castaldi, F.J. and Hayes, B.J. (1988). *J. Am. Oil Chem. Soc.*, **65**, 1669.

The Dow Chemical Company. (1988a) DOWFAX 9N.

The Dow Chemical Company. (1988b) *The Polyglycol Handbook*.

van Ginkel, C.G. Stroo, C.A. and Kroon, A.G.M. (1993). *Tenside* **30**, 213.

Von Hettche, A. and Klahr, E. (1982). *Tenside* **19**, 127.

Watson, G.K. and Jones, N. (1977). *Water Res.*, **11**, 95.

Watson, G.K. and Jones, N. (1979). *Soc. Gen. Microbiol. Q.*, **6**, 78.

Weibull, B. and Nycander, B. (1954). *Acta. Chem. Scand.*, **8**, 847

White, A., Handler, P., Smith, E.L., Hill, R.L. and Lehman, L.R. (1978). *Principles of Biochemistry*, McGraw-Hill, New York.

Wickbold, R. (1974). *Tenside* **11**, 137.

# 8 Biodegradability of amphoteric surfactants

## A. DOMSCH

### 8.1 Introduction

Compounds with both acidic and alkaline properties, i.e. the ability to release or bind protons, are described as amphoteric electrolytes or ampholytes. In acidic solutions they form cations, in alkaline solutions anions and in the mid-pH range 'zwitterions', i.e. molecules with two ionic groups with equivalent charges. Molecules to which this principle is applied on the hydrophilic group and which, at the same time, contain a hydrophobic fatty chain are known as amphoteric surfactants.

In this chapter amphoteric surfactants, or in short amphoterics, are described. There are two main groups manufactured and used commercially: real amphoterics and betaines. The key functional groups are the more or less quaternized nitrogen, derived from an amine, and the carboxylic group. The carboxylic group can be replaced by the sulphonate or the phosphate group resulting in sulphobetaines and phosphobetaines, respectively.

Amphoterics are surfactants with an ionic charge. Depending on the pH value they can change between anionic character, the isoelectric neutral stage or the cationic character. But, in the isoelectric neutral stage, they are ionic substances and not non-ionics.

The first main group of amphoterics is the group of real amphoterics, i.e. they have one ionic group which determines the character of the structure. Depending on the pH value either anionics or cationics are formed. At a high pH value the carboxyl group is ionized and a more anionic surfactant results; at a low pH value the amino group is ionized and a cationic surfactant resembles a quaternary ammonium compound. Between these extremes there is a certain pH range in which the molecule has a neutral charge. This pH is the isoelectric range depending on the alkalinity of the nitrogen atom and the acidity of the carboxylic function in the given structure.

The betaines are the second group among the amphoterics. Because of the fully quaternized nitrogen they are present in the form of 'zwitterions', only. They are inner salts, because they consist of two functional ionic groups with opposite electric charge in one molecule. The difference to the real amphoterics is that an increase in pH does not give anionic properties and the quaternization of the nitrogen is independent of the pH value. The structures of real amphoterics and betaines as well as the influence of the pH value on it are shown in Figure 8.1.

All betaines can be regarded as derivatives of 'betaine' — *N,N,N*-trimethyl-glycine — a natural substance occurring in the sugar beet *Beta vulgaris*. In the

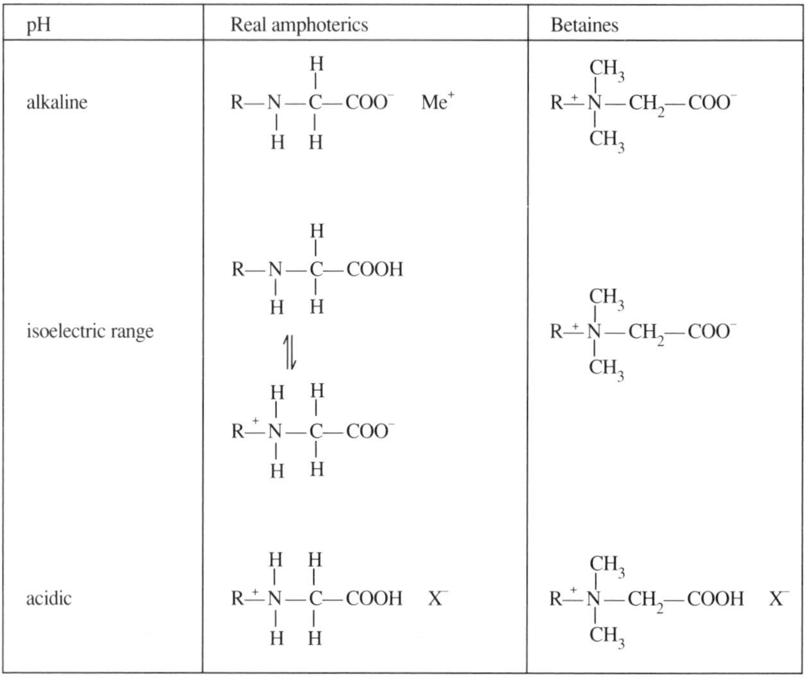

| pH | Real amphoterics | Betaines |
|---|---|---|
| alkaline | $\underset{\underset{H}{\mid}}{\overset{\overset{H}{\mid}}{R-N-C-COO^-}}$  Me$^+$ | $\underset{\underset{CH_3}{\mid}}{\overset{\overset{CH_3}{\mid}}{R-^{\pm}N-CH_2-COO^-}}$ |
| isoelectric range | $\underset{\underset{H}{\mid}}{\overset{\overset{H}{\mid}}{R-N-C-COOH}}$  $\Updownarrow$  $\underset{\underset{H}{\mid}}{\overset{\overset{H}{\mid}}{R-^{+}N-C-COO^-}}$ | $\underset{\underset{CH_3}{\mid}}{\overset{\overset{CH_3}{\mid}}{R-^{\pm}N-CH_2-COO^-}}$ |
| acidic | $\underset{\underset{H}{\mid}}{\overset{\overset{H}{\mid}}{R-^{\pm}N-C-COOH}}$  X$^-$ | $\underset{\underset{CH_3}{\mid}}{\overset{\overset{CH_3}{\mid}}{R-^{+}N-CH_2-COOH}}$  X$^-$ |

**Figure 8.1** Influence of pH value on structure of amphoterics.

simplest case one of the methyl groups is replaced by a long alkyl chain, forming the alkyl betaines.

Because of behaviour and a protein-like structure the amphoterics are dermatologically mild surfactants. They can form complexes with anionic surfactants and are able to reduce the irritative properties of these surfactants. Therefore the main use is in cosmetics and toiletries or hand dishwashing liquids as mild surfactants.

But beside mildness amphoterics are important surfactants in some detergents, especially in light-duty detergents and special wool care products. Special surface-active properties are the reason for these applications: amphoterics foam strongly and have an excellent capacity to disperse or emulsify oils and fats. They are very effective cleaning agents even in extreme pH ranges.

## 8.2  Structural elements and biodegradation in general

Amphoterics are noted for their particular compatibility with biological structures, a consequence of their protein-analogous structure. For this reason they are readily biodegradable in a dilute aqueous solution, assuming the fatty chains are essentially of natural origin (Eldib, 1977).

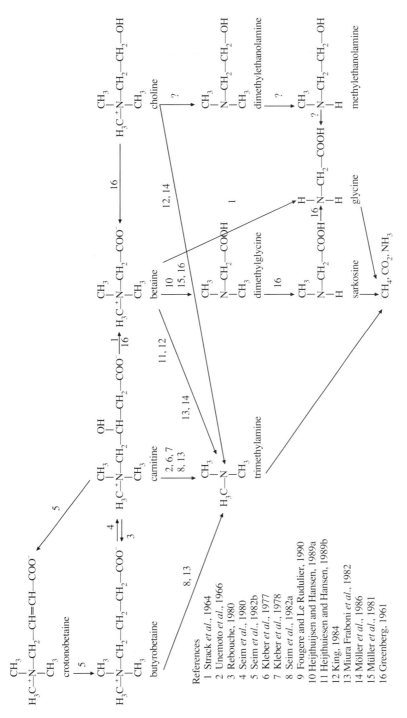

**Figure 8.2** Biological pathway of betaine and related compounds.

References
1 Strack et al., 1964
2 Unemoto et al., 1966
3 Rebouche, 1980
4 Seim et al., 1980
5 Seim et al., 1982b
6 Kleber et al., 1977
7 Kleber et al., 1978
8 Seim et al., 1982a
9 Fougere and Le Rudulier, 1990
10 Heijthuijsen and Hansen, 1989a
11 Heijthuiesen and Hansen, 1989b
12 King, 1984
13 Miura Fraboni et al., 1982
14 Möller et al., 1986
15 Müller et al., 1981
16 Greenberg, 1961

The fatty chain is degraded by the mechanism of $\omega$- and $\beta$-oxidation, as described, e.g. for linear alkylbenzene sulphonate (Steber, 1979; Berth *et al.*, 1984). This mechanism is generally valid for all surfactants based on natural oils and fats.

After $\omega$-oxidation a low molecular weight hydrophilic molecule remains. In the case of $\omega$-oxidation of alkyl betaine (see structure I in Figure 8.4) betaine (*N,N,N*-trimethylglycine) results. For this naturally occurring compound the degradation to methane, carbon dioxide and ammonia is well known.

For betaine and related compounds some biological pathways are described in the literature (see Figure 8.2). Different authors examined the ability of micro-organisms to decompose biological compounds containing quaternary nitrogen, like betaine, choline, carnitine, and butyrobetaine. With this the decomposition of alkyl betaines and related compounds can be understood, assuming that metabolites are a part of the pathway.

In contrast to anionic and non-ionic surfactants there is no very simple analytical method available to determine the surface-active properties, like methylene blue reaction or bismuth reaction for anionics and non-ionics, respectively. This could be the reason why very limited information about the primary degradation is published. There is a polarographic method, only, with which one can analyse the active content of amphoterics after a biodegradation process (Linhardt, 1972). In the case of polycarboxyglycinates HPLC has been used to measure the loss of active substance during degradation, which is now regarded as primary degradation.

Amphoterics in general have a certain antimicrobiological activity because of the quaternized nitrogen. This has to be taken into consideration for all biodegradation tests. Acclimatization to the bacteria is therefore necessary in order to get correct figures for the ultimate degradability.

Comparing the different types of amphoterics described below it can be assumed that every type is degraded more or less according to the same mechanism of decomposition, if the conditions of the test are constant as well as the lipophilic part of the molecule. Results on the basis of the OECD test 302B (Zahn–Wellens test) are shown in Figure 8.3 for alkylamidobetaine, alkylamphoacetate, alkylamphopropionate, and dimethylacetamido betaine, all with the same cocoalkyl chain as the lipophilic part of the molecule, illustrating nearly the same degradation kinetics.

## 8.3   Alkyl betaines

### 8.3.1   Chemical structure

Alkyl betaines can be regarded as derivatives of betaine (*N,N,N*-trimethylglycine) in which one methyl group is replaced by a long alkyl chain. In the preparation process equivalent quantities of long-chain tertiary amines and sodium chloroacetate are heated in aqueous solution at 70–80°C for several

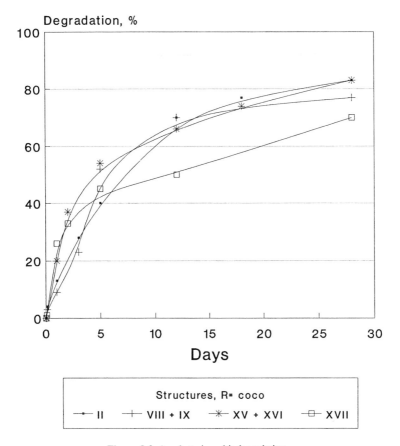

**Figure 8.3** Amphoterics – biodegradation.

hours. The general formula of the reaction is given in Figure 8.4. The final product is an aqueous solution of the alkyl betaine and contains an equimolar amount of sodium chloride. For most applications this is accepted. It is also possible to carry out the reaction in an alcoholic solution. In this case sodium chloride is precipitated and can be removed by filtration. Another possibility is the removal by electrodialysis. However, both processes are avoided because they

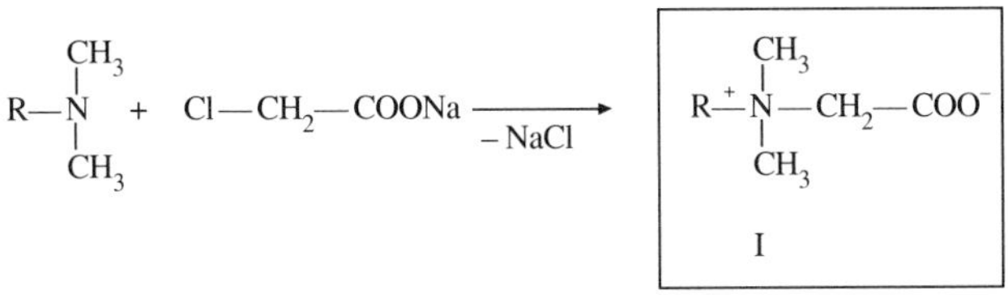

$$
R-\underset{\underset{CH_3}{|}}{\overset{\overset{CH_3}{|}}{N}} \;+\; Cl-CH_2-COONa \;\xrightarrow[-\,NaCl]{}\; R\overset{+}{-}\underset{\underset{CH_3}{|}}{\overset{\overset{CH_3}{|}}{N}}-CH_2-COO^-
$$

I

**Figure 8.4** Reaction scheme of alkyl betaines.

are more complicated than the alternative route for obtaining salt-free amphoterics. This is described in Section 8.8.

The group R in the molecular structure can be the alkyl chain derived from coconut oil or palm kernel oil ($C_8$–$C_{18}$), tallow ($C_{16}$–$C_{18}$ and $C_{18'}$), palm oil ($C_{16}$–$C_{18}$ and $C_{18'}$), or pure $C_{12}$-chain. This gives the different types of commercial products.

### 8.3.2  Properties, application

Alkyl betaines have very good stability against water hardness and are excellent lime soap dispersing agents. They are also stable at very low and high pH values and compatible with anionic, cationic, and non-ionic compounds, especially surfactants. There is complex formation between alkyl betaines and anionic surfactants, which depends upon the isoelectric point. The amphoterics are also stable in systems containing high amounts of electrolytes. With these properties and good detergency the alkyl betaines are widely used in hard surface cleaners and special textile detergents.

### 8.3.3  Primary degradation

The primary degradation of alkyl betaines (structure I in Figure 8.4) can be observed by measuring the loss of the surface-active substance during biodegradation using a polarographic method (Linhardt, 1972). The result is given in Table 8.1. In general, alkylbetaines show good primary degradability.

**Table 8.1**  Alkyl betaines — biodegradation

| Substance | Method of degradation | Analysis | Degradation in % | Evaluation of degradation | Reference |
|---|---|---|---|---|---|
| I, R = $C_{12}$ | CBT | Polarography | 100 | primary | Fernlay, 1978 |
| I, R = $C_{12}$ | CBT | $O_2$ | 55 | non readily | Fernlay, 1978 |
| | STURM | $CO_2$ | 91 | readily | Fernlay, 1978 |
| I, R = $C_{14}$ | CBT | $O_2$ | 58 | non-readily | Fernlay, 1978 |
| | STURM | $CO_2$ | 84 | readily | Fernlay, 1978 |
| I, R = $C_{16}$ | CBT | $O_2$ | 45 | non-readily | Fernlay, 1978 |
| | STURM | $CO_2$ | 84 | readily | Fernlay, 1978 |
| I, R = $C_{14-15}$ | CBT | $O_2$ | 52 | non-readily | Fernlay, 1978 |
| | STURM | $CO_2$ | 81 | readily | Fernlay, 1978 |
| I, R = cocoalkyl | CBT | $BOD_{30}$/COD | 57 | non-readily | Gerike, 1988 |
| | MOST | | >70 | readily | Henkel |
| | CBT | $BOD_{30}$/COD | >60 | readily | Henkel |
| I,R = cocoalkyl | OECD 303A | | 95 | ultimately | Hoechst |

I, structure I in Figure 8.4; CBT, Closed Bottle test; MOST, Modified OECD Screening test 301E; STURM, Sturm test; OECD 303A, OECD Method 303A = Coupled Units test; $O_2$, oxygen uptake; $CO_2$, $CO_2$ evolution; $BOD_{30}$, biological oxygen demand in 30 days; COD, chemical oxygen demand; Henkel, Safety Data Sheet of Dehyton AB 30, Henkel KGaA, Düsseldorf; Hoechst, Hoechst AG, Frankfurt, priv. comm.

### 8.3.4  Ultimate degradation

The degradation of alkyl betaines (structure I) is described with the Closed Bottle test as well as with the H. Sturm test (Fernlay, 1978). Both tests are described in more detail in Chapter 3. The data are shown in Table 8.1. A small difference, but probably not a significant one, is the difference between the naturally derived C-chain ($C_{14}$) and the synthetic one ($C_{14-15}$): the natural chain degrades in a given time to a higher extent. As a conclusion, alkyl betaines can be regarded as non-readily biodegradable, but after a certain time of adaption they are readily biodegradable, too.

## 8.4  Alkylamido betaines

### 8.4.1  Chemical structure

The synthesis of alkylamido betaines is carried out in two steps. The first step is the condensation of a fatty acid or their esters (especially the methyl ester or the corresponding triglyceride) with dimethylaminopropyl amine. The second step is the reaction of this intermediate with sodium chloroacetate. The reaction product is an aqueous solution of alkylamidopropyl betaine containing an equimolar amount of sodium chloride. The sodium chloride can be removed, but very seldom for commercial products. The reaction scheme is given in Figure 8.5. The link between the lipophilic and the hydrophilic part is an amide

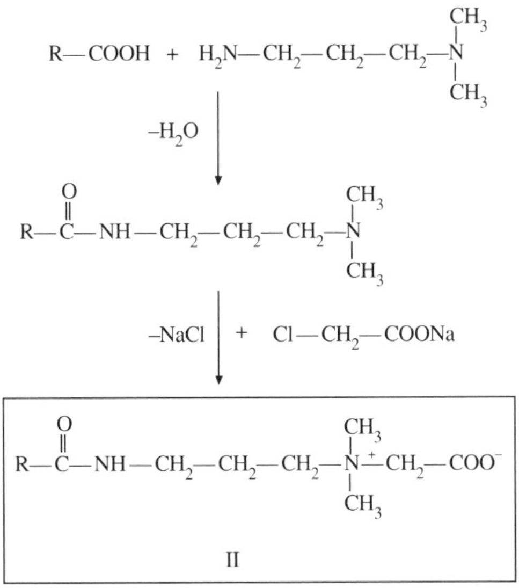

$$R-COOH \ + \ H_2N-CH_2-CH_2-CH_2-N\begin{smallmatrix}CH_3\\|\\|\\CH_3\end{smallmatrix}$$

$$\downarrow \ -H_2O$$

$$R-\overset{\overset{O}{\parallel}}{C}-NH-CH_2-CH_2-CH_2-N\begin{smallmatrix}CH_3\\|\\|\\CH_3\end{smallmatrix}$$

$$\downarrow \ -NaCl \quad + \quad Cl-CH_2-COONa$$

$$R-\overset{\overset{O}{\parallel}}{C}-NH-CH_2-CH_2-CH_2-\overset{\overset{CH_3}{|}}{\underset{\underset{CH_3}{|}}{N^+}}-CH_2-COO^-$$

II

Figure 8.5  Reaction scheme of alkylamido betaines.

functional group, corresponding to the amide link in proteins. This results in different properties compared with alkyl betaines.

The standard products in the range of alkylamido betaines are produced on the basis of R = cocoalkyl (see structural formula II in Figure 8.5) and mixed coco/oleoalkyl derivative or pure $C_{12}$-derivative.

### 8.4.2   Properties, application

The basic properties are similar to those of the alkyl betaines, i.e. the alkylamido betaines show stability against electrolytes, acids, alkali and water hardness. But the dermatological behaviour is much better. Therefore, the main use is in cosmetics and toiletries, but also in hand dishwashing liquids and special textile detergents.

The addition of alkylamido betaines to anionic surfactants, mainly sodium lauryl ether sulphates, increases the viscosity of the blend. With all of these properties the alkylamido betaines are now the most important second surfactant in shampoos, shower gels and foam baths or liquid soaps of higher quality.

### 8.4.3   Primary degradation

The degradation of alkylamido betaines (structure II in Figure 8.5) can be observed by measuring the loss of surface-active substance during biodegradation using the Orange II Method (Boiteux, 1984). The result is given in Table 8.2. In general, alkylamido betaines show good primary degradability.

### 8.4.4   Ultimate degradation

The inherent biodegradation of cocoamidopropyl betaine (structure II) was determined with the Zahn–Wellens test (OECD 302B, according to the German Standard DIN 38 412 Part 25–Static test). The general test procedure is described in Chapter 3. The result for the cocoalkyl derivative is given in Table 8.2. In conclusion the alkylamido betaines can be regarded as inherently biodegradable.

## 8.5   Sulphobetaines and hydroxysulphobetaines

### 8.5.1   Chemical structure

In the past the true sulphobetaines have been obtained by the reaction of tertiary amines with propane sultone (structure III in Figure 8.6). However, the sultones are regarded as carcinogenic, therefore this way is no longer used.

Today the usual procedure for preparation of sulphobetaines is the reaction of tertiary amines with chlorohydroxypropane sulphonic acid. The latter is obtained

**Table 8.2** Alkylamidopropyl betaine — biodegradation

| Substance | Method of degradation | Analysis | Degradation in % | Evaluation of degradation | Reference |
|---|---|---|---|---|---|
| II, R = Cocoalkyl | OECD Conf. | Orange II | 10 days: 98 | primary | Goldschmidt |
| | | Orange II | 7 h: 80–100 | primary | Boiteux, 1984 |
| II, R = cocoalkyl | CBT | $BOD_{30}$/COD | 84 | readily | Gerike, 1988 |
| | CBT | BOD/COD | 7 days: 30 | readily | De Waart and van der Most, 1986 |
| | | | 14 days : 51 | | |
| | | | 21 days : 80 | | |
| | | | 28 days : 82 | | |
| | MOST | | > 70 | readily | Henkel |
| | MOST | DOC | 90–94 | readily | Hoechst |
| | STURM | | | readily | |
| II, R = soy alkyl | MOST | DOC | 71 | readily | Hoechst |
| II, R = cocoalkyl | OECD 302B | COD | 3 h : 4 | inherently | |
| | | | 24 h : 13 | | |
| | | | 3 days : 28 | | |
| | | | 5 days : 40 | | |
| | | | 12 days : 70 | | |
| | | | 18 days : 77 | | |
| | | | 28 days : 83 | | |
| | OECD 302B | COD | 1 day : 24–27 | inherently | Z and S |
| | | | 2 days : 67–75 | | |
| | | | 3 days : 78–85 | | |
| | | | 6 days : 95 | | |
| | | | 7 days : 91–100 | | |
| II, R = cocoalkyl | OECD 303A | | 30 days : 71 | ultimately | Goldschmidt |

II, structure II in Figure 8.5; OECD Conf., OECD Confirmatory test; OECD 302B, OECD Method 302 B = Zahn–Wellens test = Static Test DIN 38412, part 25; OECD 303A, OECD Method 303A = Coupled Units test; CBT, Closed Bottle test; MOST, Modified OECD Screening test 301E; STURM, Shake Flask $CO_2$ Evolution System (Sturm test); COD, chemical oxygen demand; $BOD_{30}$, biological oxygen demand in 30 days; DOC, dissolved organic carbon; Orange II, Orange II Method acc. to Renault and Girard, 1963; Goldschmidt, Th. Goldschmidt AG, Essen, priv. comm.; Henkel, Safety Data Sheet of Dehyton K, Henkel KGaA, Düsseldorf; Hoechst, Hoechst AG, Frankfurt, priv. comm.; Z and S, Zschimer & Schwarz GmbH, Lahnstein, priv. comm.; no ref., REWO Chemische Werke GmbH, Steinau a. d. Straße.

by the reaction of epichlorohydrin with sodium hydrogensulphite. The reaction scheme is shown in Figure 8.6. This product group (structures IV and V in Figure 8.6) should be described correctly as hydroxysulphobetaines to separate them very clearly from the former version described in the first paragraph.

Commercial products are available as hydroxysulphobetaines with R = cocoalkyl and synthetic $C_{14-15}$-alkyl (structures IV and V in Figure 8.6) as an aqueous solution with molar equivalents of sodium chloride. For two special sulphobetaines the biodegradation is described in the literature (Larson, 1979), but no commercial products of this type are available. For the structures see

240

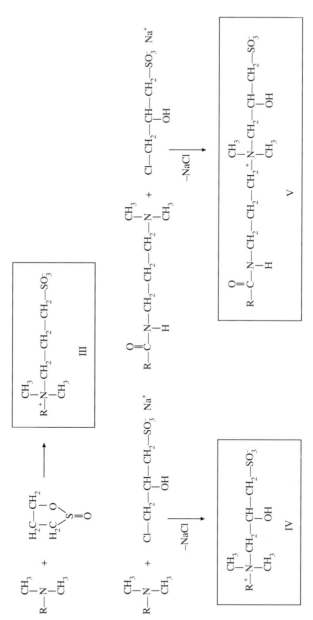

**Figure 8.6** Reaction scheme of sulphobetaines and hydroxysulphobetaines.

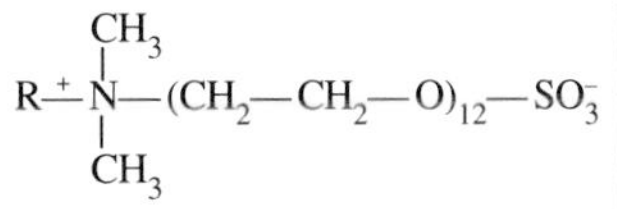

$$R \overset{+}{\underset{\underset{CH_3}{|}}{\overset{\overset{CH_3}{|}}{N}}} - (CH_2 - CH_2 - O)_{12} - SO_3^-$$

VI

$$R_1 \overset{+}{\underset{\underset{CH_3}{|}}{\overset{\overset{CH_3}{|}}{N}}} - R_2 - \underset{\underset{OH}{|}}{CH} - R_2 - SO_3^-$$

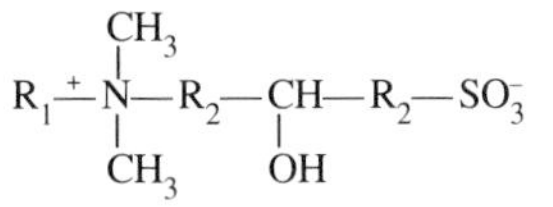

$R_1 = C_{10-18}$–alkyl

$R_2 = C_{1-4}$–alkyl

VII

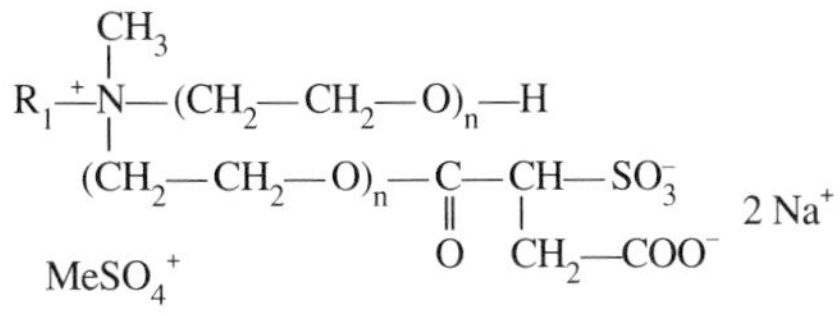

$$R_1 \overset{+}{\underset{\underset{(CH_2-CH_2-O)_n}{|}}{\overset{\overset{CH_3}{|}}{N}}} - (CH_2 - CH_2 - O)_n - H$$

$(CH_2 - CH_2 - O)_n - \underset{\underset{O}{\|}}{C} - \underset{\underset{CH_2-COO^-}{|}}{CH} - SO_3^-$  2 Na$^+$

$MeSO_4^+$

VIII

**Figure 8.7** Structure of special sulphobetaines.

Figure 8.7. Another sulphobetaine is based on ethoxylated primary amine, ethoxylated, quaternized and sulphosuccinated (see structure VIII in Figure 8.7).

### 8.5.2 *Properties, application*

With good wetting properties and good stability against electrolytes the hydroxy-sulphobetaines are of interest especially for household products. In the past their use was limited, but now there is increased interest in the use of these products .

### 8.5.3 *Primary degradation*

The primary degradation of sulphobetaines (structure III) and hydroxysulphobe-taines (structure IV) was determined by the loss of polarographic activity (Linhardt, 1972) in the Closed Bottle test. Alkyl sulphobetaine as well as alkyl hydroxysul-phobetaine are described (Fernlay, 1978). The data are given in Table 8.3. The primary degradation appears to be complete, but the 90% value for laurylsulpho-betaine gives a hint that the ultimate biodegradation is expected to be limited.

**Table 8.3** Sulphobetaines and hydroxysulphobetaines — biodegradation

| Substance | Method of degradation | Analysis | Degradation in % | Evaluation of degradation | Reference |
|---|---|---|---|---|---|
| III, R = $C_{12}$ | CBT | Polarography | 90 | primary | Fernlay, 1978 |
| III, R = $C_{16}$ | CBT | Polarography | 97 | primary | Fernlay, 1978 |
| IV, R = $C_{14-15}$ | OECD 301E | Polarography | 96 | primary | Fernlay, 1978 |
|  | CBT | Polarography | 100 | primary | Fernlay, 1978 |
| III, R = $C_{12}$ | CBT | $O_2$ | 25 | non-readily | Fernlay, 1978 |
|  | STURM | $CO_2$ | 49 | non-readily | Fernlay, 1978 |
| III, R = $C_{16}$ | CBT | $O_2$ | 26 | non-readily | Fernlay, 1978 |
|  | STURM | $CO_2$ | 56 | non-readily | Fernlay, 1978 |
| IV, R = $C_{14-15}$ | CBT | $O_2$ | 40 | non-readily | Fernlay, 1978 |
|  | STURM | $CO_2$ | 40 | non-readily | Fernlay, 1978 |
| V, R = cocoalkyl | STURM | $CO_2$ | 33 | non-readily |  |
|  |  | DOC | 47 | non-readily |  |
| VI, R = $C_{16}$ |  | COD | 33 days : 20 | non-readily | Larson, 1979 |
|  |  | DOC | 33 days : 25 | non-readily |  |
|  | Batch | DOC | 1 day : 32 | non-readily |  |
| VII, R = cocoalkyl |  | $CO_2$ | 42 days : 41 | non-readily | Larson, 1979 |
|  |  | DOC | 42 days : 49 | non-readily |  |
|  | Batch | DOC | 1 day : 57 | non-readily |  |
| VIII, R = cocoalkyl | OECD 302B | DOC | 28 days : 41 | non-inherently |  |

III, sulphobetaine, structure III in Figure 8.6; IV, V, hydroxysulphobetaine, structures IV and V in Figure 8.6; VI, VII, VIII, structures VI, VII, VIII in Figure 8.7; CBT, Closed Bottle test; OECD 301E, OECD Screening test; STURM, H. Sturm test; OECD 302E, OECD Method 302B; Batch, batch activated sludge; $O_2$, oxygen uptake; $CO_2$, $CO_2$ evolution; COD, chemical oxygen demand; DOC, dissolved organic carbon; no ref., REWO Chemische Werke GmbH, Steinau a. d. Straße.

### 8.5.4    Ultimate degradation

The ultimate biodegradation for both groups (structures III, IV, and V) is measured with the Closed Bottle test and the Sturm test. Data obtained are given in Table 8.3. Obvious differences can be recognized between sulphobetaines and hydroxysulphobetaines. The former products have only a very limited degradation, but the hydroxysulphobetaines show a better degradation behaviour nearer to the standard betaines. In conclusion, it can be stated that sulphobetaines as well as hydroxysulphobetaines are not readily biodegradable. Therefore, additional research and development work is necessary to overcome this situation. Two additional substances (structures VI and VII) are described in the literature. Both are evaluated as non-readily biodegradable.

In a patent application (Wentler et al., 1981) 18 different non-commercial ethoxylated types of sulphobetaines are described. For all structures the degradation is measured by the Sturm test. Only if the ester function is part of the molecule could ready biodegradability be confirmed. Ether linkages and higher degrees of ethoxylation did not show a good degradability.

The sulphobetaine based on the quaternized ethoxylated amine (structure VIII) has a very limited biodegradation like the traditional sulphobetaine.

## 8.6  Alkylamphoacetates

### 8.6.1  Chemical structure

This group of surfactants is based on fatty alkyl imidazolines obtained by the condensation of fatty acids or their esters (methylesters or triglycerides of fatty acids) with aminoethylethanol amine. The reaction scheme is given in Figure 8.8.

As intermediate an imidazoline ring structure is formed which was formerly assumed to be present in the final product. Intensive analysis leads to the interpretation that the ring structure is opened by the influence of hydrolysing conditions (Hein et al., 1978; Takano and Tsuji, 1983b; Rieger, 1984).

With 1 mole sodium chloroacetate the monoacetate, and with an excess of sodium chloroacetate the diacetate, will be formed. Numerous possible reaction products are described as a result of the first and the second step of the reaction (Hein et al., 1978, 1980; Schwarz et al., 1979; Takano and Tsuji, 1983a,b; Rieger, 1984; Zongshi and Zhuangyu, 1993). Therefore the usual commercial products are complex mixtures.

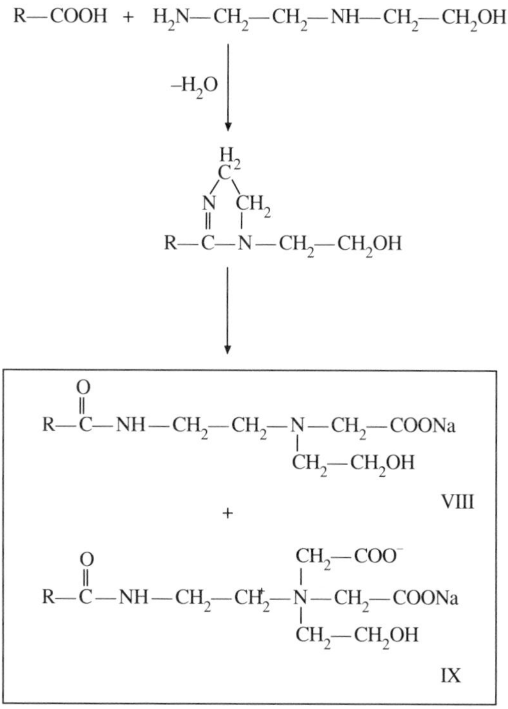

**Figure 8.8** Reaction scheme of alkylamphoacetates.

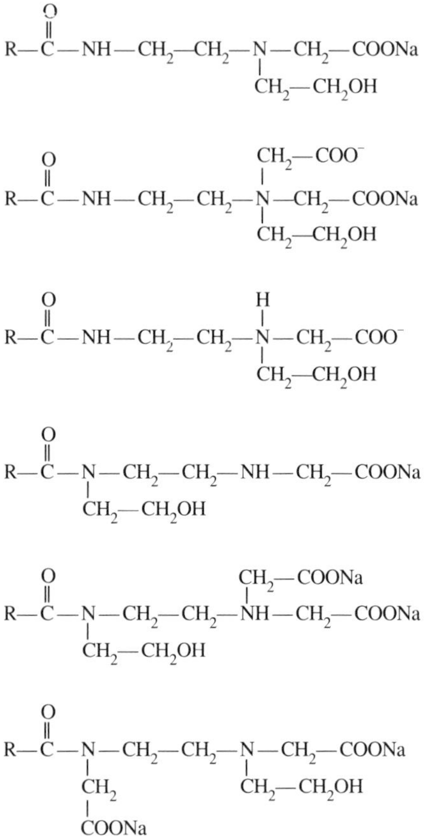

$$R—\overset{\overset{\displaystyle O}{\|}}{C}—NH—CH_2—CH_2—\underset{\underset{\displaystyle CH_2—CH_2OH}{|}}{N}—CH_2—COONa \quad \text{(VIII)}$$

$$R—\overset{\overset{\displaystyle O}{\|}}{C}—NH—CH_2—CH_2—\underset{\underset{\displaystyle CH_2—CH_2OH}{|}}{\overset{\overset{\displaystyle CH_2—COO^-}{|}}{N}}—CH_2—COONa \quad \text{(IX)}$$

$$R—\overset{\overset{\displaystyle O}{\|}}{C}—NH—CH_2—CH_2—\underset{\underset{\displaystyle CH_2—CH_2OH}{|}}{\overset{\overset{\displaystyle H}{|}}{N}}—CH_2—COO^-$$

$$R—\overset{\overset{\displaystyle O}{\|}}{C}—\underset{\underset{\displaystyle CH_2—CH_2OH}{|}}{N}—CH_2—CH_2—NH—CH_2—COONa$$

$$R—\overset{\overset{\displaystyle O}{\|}}{C}—\underset{\underset{\displaystyle CH_2—CH_2OH}{|}}{N}—CH_2—CH_2—\overset{\overset{\displaystyle CH_2—COONa}{|}}{NH}—CH_2—COONa$$

$$R—\overset{\overset{\displaystyle O}{\|}}{C}—\underset{\underset{\underset{\displaystyle COONa}{|}}{\underset{\displaystyle CH_2}{|}}}{N}—CH_2—CH_2—\underset{\underset{\displaystyle CH_2—CH_2OH}{|}}{N}—CH_2—COONa$$

**Figure 8.9** Structures of alkylamphoacetates.

Because of the reaction with sodium chloroacetate alkylamphoacetates contain the corresponding equimolar amount of sodium chloride. The commercial products are aqueous solutions.

Products are available with a broad range of lipophilic groups. In most of the products R in the formula corresponds to cocoalkyl, but also tallowalkyl or palm oil-alkyl are possible. Depending on the molar ratio of the sodium chloroacetate different degrees of 'quaternization' at the nitrogen can be observed. The chemical description of the products differs appreciably from author to author in the literature. Alkylamphoacetates have been described as alkylamino carboxylic acids, hydroxyalkyl alkylamidoethyl glycinates, carboxyglycinates, amphoglycinates, imidazoline derivatives (because of the intermediate), cocoamphodiacetate, or cocoampho(mono/di)acetate.

Some examples of the structures found in the complex mixture of reaction products are given in Figure 8.9 (Hein *et al.*, 1978; Takano and Tsuji, 1983b; Rieger, 1984).

## 8.6.2   Properties, application

Alkylamphoacetates are stable against acidic and alkaline pH values; also the foaming behaviour is not influenced by change of the pH value. There is good stability in systems with high content of electrolytes, with hard water and lime soaps. They are also compatible with anionic, cationic, and non-ionic surfactants. But the most important property is the extreme mildness on the skin and mucous membranes compared with other types of surfactants.

Alkylamphoacetates are used in formulating high quality toiletries, especially baby baths, shower gels and liquid soaps. In combination with anionic surfactants they improve the mildness of a given basic surfactant. Alkylamphoacetates are regarded as the coming generation of milder surfactants used in personal care products.

## 8.6.3   Ultimate degradation

The biodegradation of cocoamphodiacetate (mixture of structures VIII and IX) was first determined with the Closed Bottle test (OECD 301D). The general test procedure is described in Chapter 3. The results for the cocoalkyl derivative are given in Table 8.4.

These data show a degradation of 52% in only 28 days. The consequence is to evaluate these substances again with a higher advanced test of the test hierarchy. As a follow-up test, the determination of the inherent degradability or the simulation of a sewage plant is recommended.

The inherent biodegradation has been determined with the Zahn–Wellens test (OECD 302B, according to the German Standard DIN 38 412 Part 25 — Static test). The general test procedure is described in Chapter 3. The results for the cocoalkyl derivative are given in Table 8.4. These results show that the alkylamphoacetates can be regarded as inherently biodegradable.

The same result can be found for the 1:1 mixture of cocoamphodiacetate with sodium lauryl sulphate (see Table 8.4). The complex between the amphoteric and the anionic surfactant does not significantly influence the biodegradation.

## 8.7   Polycarboxyglycinates

### 8.7.1   Chemical structure

This special group of glycinates consists of the reaction products of a fatty polyamine with sodium chloroacetate (Palicka, 1990, 1991). The general structure is given in Figure 8.10. The polyamine used will have a distribution of amino groups with an average value of four amino groups which gives the structure shown in Figure 8.10. After reaction with sodium chloroacetate an equimolar quantity of sodium chloride will be produced and will be present in aqueous solutions of commercial polycarboxyglycinates.

**Table 8.4** Cocoamphodiacetate — biodegradation

| Substance | Method of degradation | Analysis | Degradation in % | Evaluation of degradation | Reference |
|---|---|---|---|---|---|
| VIII + IX, R = cocoalkyl | CBT | $BOD_{30}/COD$ | 66 | readily | Gerike, 1988 |
| | CBT | $BOD_{30}/COD$ | > 60 | readily | Henkel |
| | MOST | | > 70 | readily | Henkel |
| VIII + IX, R = cocoalkyl | OECD 302B | COD | 3 h: 3<br>24 h : 9<br>3 days : 23<br>5 days : 52<br>12 days : 70<br>18 days : 74<br>28 days : 77 | inherently | |
| | OECD 302B | DOC | 6 h : 17<br>1 day : 38<br>2 days : 69<br>3 days : 76<br>4 days : 80<br>7 days : 81<br>9 days : 79 | inherently | Z and S |
| VIII + IX, R = cocoalkyl, 1:1 blended with sodium laurylether sulphate | OECD 302B | COD | 3 h : 0<br>24 h : 3<br>2 days : 15<br>5 days : 22<br>10 days : 47<br>15 days : 59<br>20 days : 68<br>28 days : 80 | inherently | |

VIII, IX, structures VIII and IX in Figure 8.8; OECD 301D, Closed Bottle test; CBT, Closed Bottle test; MOST, Modified OECD Screening test 301E; OECD 302B, Zahn–Wellens test = Static test DIN 38412, Part 25; $BOD_{30}$, biological oxygen demand in 30 days; COD, chemical oxygen demand; DOC, dissolved organic carbon; Henkel, Safety Data Sheet of Dehyton G, Henkel KGaA, Düsseldorf; Z and S, Zschimmer und Schwarz, Lahnstein, priv. comm.; no ref., REWO Chemische Werke GmbH, Steinau a. d. Straße.

The group R in the structure can be the alkyl chain derived from coconut oil, oleic acid, or tallow. The main product in this range is produced on the basis of an alkyl chain derived from tallow. Data in Sections 8.7.3 and 8.7.4 are all valid for the tallow derivative.

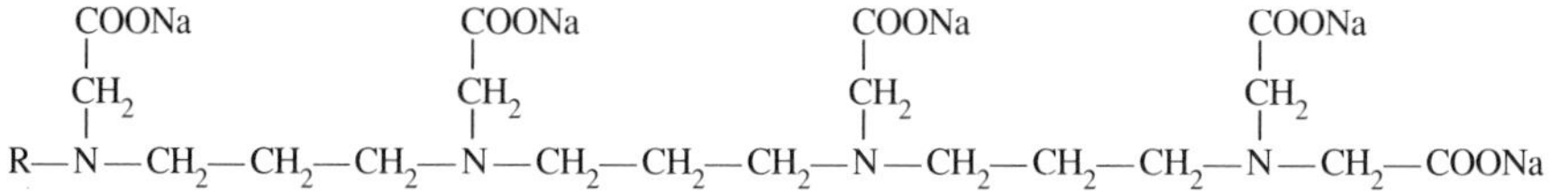

$$
\begin{array}{ccccccc}
COONa & & COONa & & COONa & & COONa \\
| & & | & & | & & | \\
CH_2 & & CH_2 & & CH_2 & & CH_2 \\
| & & | & & | & & | \\
R{-}N{-}CH_2{-}CH_2{-}CH_2{-}N{-}CH_2{-}CH_2{-}CH_2{-}N{-}CH_2{-}CH_2{-}CH_2{-}N{-}CH_2{-}COONa
\end{array}
$$

X

**Figure 8.10** Structure of polycarboxyglycinates.

## 8.7.2 Properties, application

From the application point of view this type of amphoteric is especially used in detergents because of high detergency, high sequestration ability, and dispersibility. This is combined with low irritation rates on skin and mucous membranes and the ability to stabilize enzymes in detergents. In personal care products they act as very mild surfactants.

## 8.7.3 Primary degradation

Results for the primary biodegradability of tallow polycarbocyglycinate (structure X) are given in Table 8.5. The method used was the Coupled Units test (according to OECD Method 303A, described in detail in Chapter 3) combined with HPLC to analyse the content of tallow polycarboxyglycinate. The primary degradability can be confirmed.

## 8.7.4 Ultimate degradation

The ultimate degradation of tallow polycarboxyglycinate (structure X) was determined with the Closed Bottle test (OECD Method 301D) and the Modified SCAS test (OECD Method 302A). Both tests are described in more detail in Chapter 3. The results of the measurements are given in Table 8.5. Both tests state that polycarboxyglycinates are inherently biodegradable.

## 8.8 Alkylamphopropionates

### 8.8.1 Chemical structure

Normally amphoterics contain equimolar quantities of sodium chloride because of the reaction with sodium chloroacetate. The removal of this is technologically complicated. For certain applications, where the presence of chlorides has to be avoided, an alternative product group of salt-free amphoterics is formed by the reaction of acrylic acid or its derivatives.

A salt-free carboxyethyl betaine is obtained by the electrophilic addition of acrylic acid, methyl acrylate, or ethyl acrylate on primary or secondary amines.

**Table 8.5** Polycarboxyglycinates — biodegradation

| Substance | Method of degradation | Analysis | Degradation in % | Evaluation of degradation | Reference |
|---|---|---|---|---|---|
| X, R = tallowalkyl | CUT | HPLC | > 90 | primary | Palicka, 1990, 1991 |
| X, R = tallowalkyl | CBT | BOD/COD | 5 days : 72.5 | readily | Palicka, 1990, 1991 |
| X, R = tallowalkyl | OECD 302A | DOC | 80 | inherently | Palicka, 1990, 1991 |

X, structure X in Figure 8.10; CUT, Coupled Units test; CBT, Closed Bottle test = OECD 301D, 5 day version; OECD 302A, Modified SCAS test; BOD, biological oxygen demand; COD, chemical oxygen demand; DOC, dissolved organic carbon.

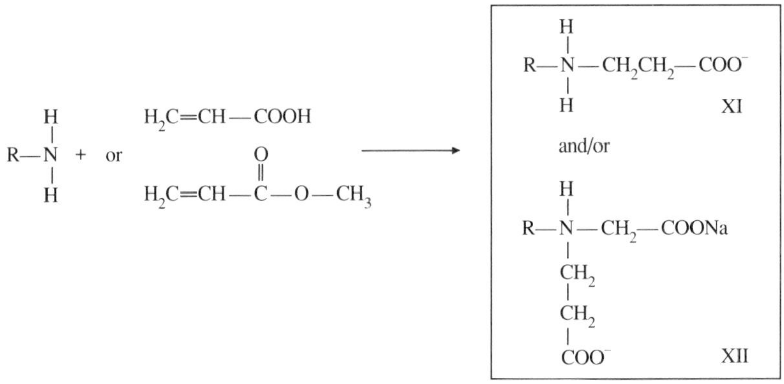

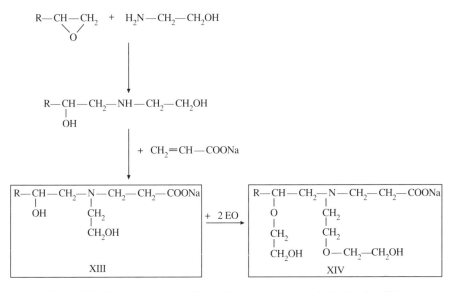

**Figure 8.11** Reaction scheme of alkylamphopropionates on the basis of amines (I).

This is shown in Figure 8.11. Depending on the amount of acrylic acid or its ester the mono- and diadducts are obtained. The first reaction products are the monoadducts, especially if a carbonate free amine is heated with an equimolar amount of methyl acrylate at 100°C, followed by vacuum distillation to remove the excess of acrylate. The ester adduct, a by-product, is hydrolysed with either alkali or acid. The substances formed can also be described as alanine derivatives (for monoadducts) or propionates (especially for diadducts).

Another possible preparation of alkylamphopropionates is the addition of methyl acrylate on $N$-2-hydroxyethyl-$N$-2-hydroxyalkyl-$\beta$-alanine. The reaction is described in Figure 8.12.

**Figure 8.12** Reaction scheme of alkylamphopropionates on the basis of amines (II).

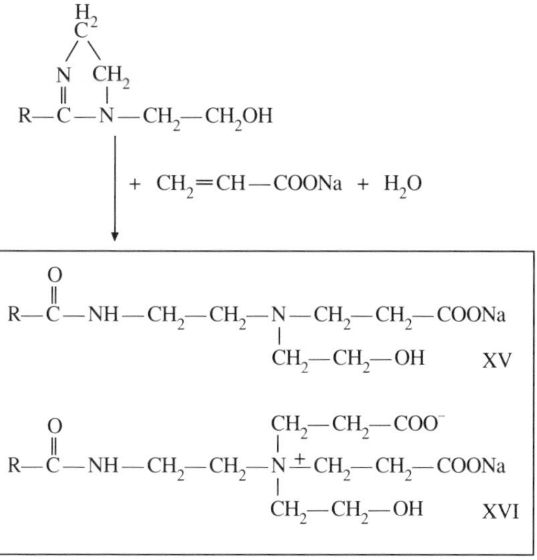

$$\begin{array}{c} \overset{H_2}{\underset{\diagup \diagdown}{C}} \\ N \quad CH_2 \\ \parallel \quad \mid \\ R-C-N-CH_2-CH_2OH \end{array}$$

$$\downarrow \quad + \; CH_2{=}CH-COONa \; + \; H_2O$$

$$\begin{array}{c}
\overset{O}{\overset{\parallel}{R-C}}-NH-CH_2-CH_2-\underset{\underset{CH_2-CH_2-OH}{\mid}}{N}-CH_2-CH_2-COONa \qquad XV \\[2em]
\overset{O}{\overset{\parallel}{R-C}}-NH-CH_2-CH_2-\overset{\overset{CH_2-CH_2-COO^-}{\mid}}{\underset{\underset{CH_2-CH_2-OH}{\mid}}{N^{\pm}}}CH_2-CH_2-COONa \qquad XVI
\end{array}$$

**Figure 8.13** Reaction scheme of alkylamphopropionates on the basis of amides.

A second group of salt-free betaines is produced by the addition of acrylic acid, methyl acrylate, or ethyl acrylate on the reaction product of aminoethylethanolamine and fatty acids (Hein *et al.*, 1980). The intermediate imidazoline ring is synthesized as described in Section 8.6. The synthesis in principle is shown in Figure 8.13.

The first group of alkylamphopropionates is available on the basis of R = cocoalkyl, lauryl, or stearyl. The second group (amide type) is represented by R = cocoalkyl.

A special substance in the alkylamphopropionates group is the reaction product of a secondary amine with sodium chloroacetate in the first step, later reacted with aminoethylethanolamine and finally with acrylic acid. The structure is shown in Figure 8.14.

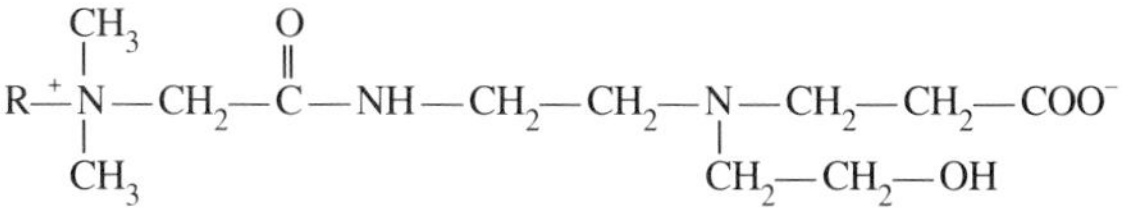

$$\underset{\underset{CH_3}{\mid}}{\overset{\overset{CH_3}{\mid}}{R-\overset{+}{N}}}-CH_2-\overset{O}{\overset{\parallel}{C}}-NH-CH_2-CH_2-\underset{\underset{CH_2-CH_2-OH}{\mid}}{N}-CH_2-CH_2-COO^-$$

XVII

**Figure 8.14** Structure of cocodimethylacetamido betaine.

### 8.8.2    Properties, application

All alkylamphopropionates are free of electrolytes, but all have a very good tolerance against high concentrations of salts or electrolytes. In some cases this could be higher than for other types of amphoterics, but most important is the high stability at very low and very high pH values. The dermatological properties of alkylamphopropionates are quite good.

The main use of alkylamphopropionates is in household products and other types of cleaners containing high amounts of acids or alkalis. They are also used in special cosmetic formulations where the presence of sodium chloride has to be avoided.

### 8.8.3    Ultimate degradation

Data are given for degradation of alkylamphopropionate on the basis of a primary amine (structures XI and XII) and $N$-alkyl-$\beta$-alanine derivatives (structures XIII and XIV). The methods used were an activated sludge test and a shake culture test. The results are presented in Table 8.6.

The degradation of $C_{12}$-$\beta$-alanine (structure XI) and $N$-stearyl-$\beta$-amino dipropionic acid (structure XII) was described on the basis of an activated sludge test (Eldib, 1977) and shows a good degradability. The degradation of $N$-(2-hydroxyethyl)-$N$-(2-hydroxyalkyl)-$\beta$-alanine (structure XIII) and its ethoxylated derivative (structure XIV) occurs to a high degree. The results are part of Table 8.6.

The inherent biodegradation has been determined for the alkylamphopropionate on the basis of an amidoamine (mixture of structures XV and XVI, and structure XVII) with the Zahn–Wellens test (OECD 302B, according to the German Standard DIN 38 412 Part 25 — Static test). The general test procedure is described in Chapter 3. The results for the imidazoline derived cocoalkyl derivatives (structures XV and XVI) are given in Table 8.6. With a degradability of 83% after 28 days this group of alkylamphopropionates can be classified as inherently biodegradable.

Cocodimethylacetamido betaine (structure XVII) was also tested for inherent biodegradability with the Zahn–Wellens test. Despite a certain antimicrobiological activity the degradation of this substance can be compared with standard betaines and can be stated as inherently biodegradable.

## 8.9    Imidazolinium betaines

### 8.9.1    Chemical structure

The addition of acrylic acid to an imidazoline should form an imidazolinium betaine. This was described by Schäfer et al. (cited in Hitz et al., 1983) for the model substance 1-hydroxyethyl-2-heptyl-imidazolinium-3-ethyl carboxylate. The synthesis is carried out by heating the imidazoline derivative for several

**Table 8.6** Alkylamphopropionates — biodegradation

| Substance | Method of degradation | Analysis | Degradation in % degradation | Evaluation of | Reference |
|---|---|---|---|---|---|
| XI, R=C12–18 | MOST | DOC | 79 readily | Hoechst | |
| XI, R=C12 | SCAS | Surface tension | 30 days: >95 | inherently | Eldib, 1977 |
| XII, R=C18 | SCAS | Surface tension | 30 days: >95 | inherently | Eldib, 1977 |
| XIII, R=C12–14 | SCT | | 8 days: 98 | | Takai et al., 1980 |
| XIV, R=C12–14 | SCT | | 2 days: 72,2 | | Takai et al., 1980 |
| | | | 8 days: 95,5 | | |
| XV + XVI, R=capryl | STURM | CO2 | | readily | |
| XV + XVI, R=cocoalkyl | OECD 302B | COD | 2 H: 0<br>1 day: 20<br>2 days: 37<br>5 days: 54<br>12 days: 66<br>18 days: 74<br>28 days: 83 | inherently | |
| XVII, R=cocoalkyl | OECD 302B | COD | 2h: 1<br>24 h: 26<br>2 days: 33<br>5 days: 45<br>12 days: 50<br>28 days: 70 | inherently | |

XI, C12-β-alanine, structure XI in Figure 8.11; XII, C18-β-amino dipropionic acid, structure XII in Figure 8.11; XIII, hydroxyalkyl-β-alanine, structure XIII in Figure 8.12; XIV, ethox. hydroxyalkyl-β-alanine, structure XIV in Figure 8.12; XV, XVI, amid based alkylamphopropionates, structures XV and XVI in Figure 8.13; XVII, cocodimethylacetamido betaine, structure XVII in Figure 8.14; MOST, Modified OECD Screening test 301E; SCAS, semi-continuous activated sludge test; SCT, shake culture test; STURM, Shake Flask $CO_2$ Evolution System (Sturm test); OECD 302B, Zahn–Wellens test; DOC, dissolved organic carbon; COD, chemical oxygen demand; $CO_2$, $CO_2$ evolution; Hoechst, Hoechst AG, Frankfurt, priv. comm.; no ref., REWO Chemische Werke GmbH, Steinau a. d. StraBe.

hours with a 5% excess of acrylic acid in the absence of water. More than 65% of the imidazolinium structure is retained, depending on the length of the fatty chain, but it is necessary that no water is present, otherwise the ring structure is opened. With excess of acrylic acid also the non-ring structure is obtained, depending on the temperature. The general reaction is described in Figure 8.15.

## 8.9.2 Degradation

If the imidazolinium betaines are used in water-free systems the ring structure will be retained (structure XVIII). No information concerning this group of amphoterics and the manner in which the imidazolinium ring degrades is available, but there are some data about the alkaline catalysed hydrolysis,

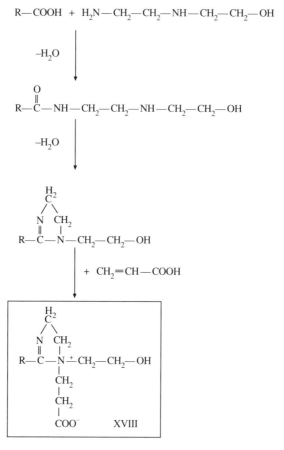

**Figure 8.15** Reaction scheme of imidazolinium betaines.

summarized by Watts (1990). Considering that the ring structure is transformed very fast to the linear structure, the mechanism according to which the imidazolinium ring degrades is not of importance.

In presence of water the ring will be opened and the alkylamphopropionate structure will result as described in Section 8.8.1. Therefore the results which are given for alkylamphopropionates would be expected to apply to imidazolinium betaines.

## Acknowledgements

The use of unpublished data kindly given by Dr. Volker Martin of Zschimmer and Schwarz GmbH, Lahnstein, Germany, Alwin K. Reng of Hoechst AG,

Frankfurt, Germany, and Dr. Hüttinger of Th. Goldschmidt AG, Essen, Germany, is greatly acknowledged.

## References

Berth, P., Gerike, P., Gode, P. and Steber, J. (1984) *Proceedings of the World Surfactants Congress* München. Kürle, Gelnhausen, pp. 227.

Boiteux, J.P. (1984) Dosage colorimétrique d'agents de surface amphotères et étude du comportement d'une alkyl amido bétaine en milieu naturel. *Riv. Ital. Sostanze Grasse* **61**, 491–495

De Waart, J. and Van der Most, M.M. (1986) Biodegradation test for microbicides. **22**, 113–120.

Eldib, L.A. (1977) Biodegradability of amphoteric detergents. *Soap Chem. Spec.* **41**, 77–80, 161, 163–165.

Fernlay, G.W. (1978) Zwitterionic surfactants: structure and performance, *J. Am. Oil Chem. Soc.* **55**, 98–103.

Fougere, F. and le Rudulier, D. (1990) Glycine betaine biosynthesis and catabolism in bacteroids of *Rhizobium meliloti*: Effect of salt stress. *J. Gen. Microbiol.* **136**, 2503–2510.

Gerike, P. (1988) Kosmetik-Grundstoffe unter Umweltschutzaspekten. *Parfüm. Kosmet.* **69**, 130–132.

Heijthuijsen, J.H.F.G. and Hansen, T.A. (1989a) Anaerobic degradation of betaine by marine Desulfobacterium strains. *Arch. Microbiol.* **152**, 393–396.

Heijthuijsen, J.H.F.G. and Hansen, T.A. (1989b) Betaine fermentation and oxidation by marine Desulfuromonas strains. *Appl. Environ. Microbiol.* **55**, 965–969.

Hein, H., Jaroschek, H.J. and Melloh, W. (1978) Beitrag zur Struktur amphoterer Tenside. *Fette Seifen Anstrichm.* **80**, 448–453.

Hein, H., Jaroschek, H.J. and Melloh, W. (1980) The structure of salt-free amphoteric surfactants, *Cosmet. Toilet.* **95**, 37–42.

Hitz, H., Schäfer, D., Schäfer, R. and Schäfer, W. (1983) Amphonyle (Amphotere Imidazolinium-Tenside) mit hohem Gehalt an Imidazolin-Ring. *Seifen Oele Fette Wachse* **109**, 20–21.

King, G.M. (1984) Metabolism of trimethylamine, choline, and glycine betaine by sulfate-reducing and methanogenic bacteria in marine sediments, *Appl. Environ. Microbiol.* **48**, 719–725.

Kleber, H.-P., Seim, H., Aurich, H. and Strack, E. (1977). *Arch. Microbiol.* **112**, 201–206.

Kleber, H.-P., Seim, H., Aurich, H. and Strack, E. (1978) *Arch. Microbiol.* **116**, 213–220.

Larson, R.J. (1979) Evaluation of biodegradation potential of xenobiotic organic chemicals. *Appl. Environ. Microbiol.* **38**, 1153–1163.

Linhardt, K. (1972) *Tenside*, **9**, 241.

Miura Fraboni, J., Kleber, H.-P. and Englard, S. (1982) Assimilation of gamma-butyrobetaine, and D- and L-carnitine by resting cell suspensions of *Acinetobacter calcoaceticus* and *Pseudomonas putida*. *Arch. Microbiol.* **133**, 217–221.

Möller, B., Hippe, H. and Gottschalk, G. (1986) Degradation of various amine compounds by mesophilic clostridia. *Arch. Microbiol.* **145**, 85–90.

Müller, E., Fahlbusch, K., Walther, R. and Gottschalk, G. (1981) Formation of *N,N*-dimethylglycine, acetic acid, and butyric acid from betaine by *Eubacterium limosum*. *Appl. Environ. Microbiol.* **42**, 439–445.

Palicka, J. (1990) Amphoterics in household detergents, in *Communicationes XXI Jornadas del Comite Espaniol de la Detergencia*, A.I.D., Barcelona, pp. 61–77.

Palicka, J. (1991) Amphoterics in household detergents, *J. Chem. Tech. Biotechnol.* **50**, 331–349.

Rebouche, C.J. (1980) In: *Carnitine Biosynthesis, Metabolism, and Functions*, ed. Frenkel, R.A. and McGarry, J.D. Academic Press, New York, pp. 57–67.

Renault, J. and Giraud, J.G. (1963) *Ann. Falsif. Exp. Chim.* **56**, 105–108.

Rieger, M.M. (1984) The structure of amphoterics derived from imidazoline. *Cosmet. Toilet.* **99**, 61–67.

Schwarz, G., Leenders, P. and Ploog, U. (1979) Zur Analytik von Kondensationsprodukten aus Fettsäuren bzw. deren Methylestern mit Aminoethylethanolamin. *Fette Seifen Anstrichm.* **81**, 154–158.

Seim, H., Ezold, R., Kleber, H.-P. and Strack, E. (1980) *Z. Allg. Mikrobiol.* **20**, 591–594.

Seim, H., Löster, H., Claus, R., Kleber, H.-P. and Strack, E. (1982a) Splitting of the C–N bond in carnitine by an enzyme (trimethylamine forming) from membranes of *Acinetobacter calcoaceticus*. *Fed. Eur. Microbiol. Soc. Microbiol. Lett.* **15**, 165–167.

Seim, H., Löster, H., Claus, R., Kleber, H.-P. and Strack, E. (1982b) Formation of γ-butyrobetaine and trimethylamine from quaternary ammonium compounds structure-related to L-carnitine and choline by *Proteus vulgaris*. *Fed. Eur. Microbiol. Soc. Microbiol. Lett.* **13**, 201–205.

Steber, J. (1979) *Tenside Deterg.* **16**, 140.

Strack, E., Aurich, H. and Grüner, E. (1964) *Z. Allg. Microbiol.* **4**, 164.

Takai, M., Hidaka, H., Ishikawa, S., Takada, M. and Moriya, M. (1980) New amphoteric surfactants containing a 2-hydroxyalkyl group: IV. Performance of amphoteric surfactant/soap blends. *J. Am. Oil Chem. Soc.* **57**, 183–188.

Takano, S. and Tsuji, K. (1983a) Analysis of cationic and amphoteric surfactants. IV Structural analysis of the amphoteric surfactants obtained by the reaction of 1-(2-hydroxyethyl)-2-alkyl-2-imidazoline with ethyl acrylate, *J. Am. Oil Chem. Soc.* **60**, 1798–1806.

Takano, S. and Tsuji, K. (1983b) Analysis of cationic and amphoteric surfactants. V Structural analysis of the amphoteric surfactants obtained by the reaction of 1-(2-hydroxyethyl)-2-alkyl-2-imidazoline with sodium monochloroacetate. *J. Am. Oil Chem. Soc.* **60**, 1807–1815.

Unemoto, T., Hayashi, M., Miyaki, K. and Hayashi, M. (1966) Formation of trimethylamine from DL-carnitine by *Serratia marcescens*. *Biochim. Biophys. Acta*, **121**, 220–222.

van Greenberg, L.L.M. (1961) *Metabolic Pathways*, Academic Press, New York.

Watts, M.M. (1990) Imidazoline hydrolysis in alkaline and acidic media — a review. *J. Am. Oil Chem. Soc.* **67**, 993–995.

Wentler, G.E., McGrady, J., Gosselink, E.P. and Cilley, W.A. (1981) EP 32 837 (Proctor and Gamble Comp.).

Zongshi, L. and Zhuangyu, Z. (1993) A Study on the Confirmation of the Structures of the Imidazoline Amphoteric Surfactant. Lecture held at the 10th Symposium of the GDCh Section Detergents, Potsdam, 1993.

# Index

activated sludge 97, 98
aerobic metabolism 40–42
AES *see* alcohol ether sulphates
alcohol EO/PO co-polymers *see* fatty
  alcohol alkoxylates
alcohol ether sulphates 171–175
  anaerobic biodegradability 175
  biodegradability pathways 173–175
  primary biodegradation 171–173
  ultimate biodegradation 172, 173
alcohol ethoxylates *see* fatty alcohol
  ethoxylates
alcohol sulphates *see* fatty alcohol
  sulphates
alkane sulphonates 141, 142
  anaerobic degradation 161
  biodegradation pathways 161
  primary biodegradability 159, 160
  ultimate biodegradability 159–161
alkylamido betaines 237, 238
  primary biodegradability 238
  properties and applications 238
  ultimate biodegradability 238
alkylamphoacetates 243–245
  chemical structure 243, 244
  properties and applications 245
  ultimate biodegradability 245
alkylamphopropionates 247–250
  biodegradability 247–250
  ultimate biodegradability 247–250
alkyklbenzene sulphonates 140, 141,
  149–158
  anaerobic biodegradability 157, 158
  biodegradation pathways 153, 157
  primary biodegradability 149–151
  ultimate biodegradability 151–153
alkylbetaines 234–237
  primary biodegradability 236
  properties and applications 236
  ultimate biodegradability 237
alkylene oxides and polymers, preparation
  of 206–211
alkyl ether phosphates 178, 179
  anaerobic biodegradability 179
  biodegradability data 178
  biodegradability pathways 178, 179
alkylimidazolines and derivatives *see*
  alkylyamphoacetates

alkyl phosphates
  anaerobic biodegradability 179
  biodegradability data 178
  biodegradability pathways 178, 179
alkyl trimethyl ammonium salts,
  biodegradation routes 198–200
amine ethoxylates 229
ampholytic surfactants *see* amphoteric
  surfactants
amphoteric surfactants 18, 23, 231–252
  biodegradability pathways 233
  influence of pH on structure 231, 232
anaerobic metabolism 40–42
anionic surfactants 134–179
  analysis 19, 102–106
  biodegradability legislation 16–18
  in surface waters 137, 138
  structure and applications 138, 139
AOS *see* α-olefin sulphonates
AS *see* fatty alcohol sulphates

bacteria
  adaptation of 53, 54
  attachment to solid surfaces/biofilms
    56–58
  co-metabolism by 54, 55
  genetic changes in 50–52
  influence of cationic surfactants 196
  population growth of 50–52
  use of mixed substrates 54, 55
BiAS method 21, 22
biodegradability
  anaerobic versus aerobic metabolis
    40–42, 74
  assessing the hydrophic chain 31–37
  assessment of 7, 8, 65–117
  definition of 5–7, 28–64, 66
  factors influencing 72
  influence of substrate concentration 52,
    53
  inherent 13, 14, 67, 89, 94, 95
  in the environment 42
  hydrophile degradation 37
  laboratory models 58, 59
  legislation 16–18, 127–132
  primary biodegradability 13, 65–117
  primary versus ultimate
    biodegradability 39, 40

biodegradability (*cont.*)
  ready biodegradability  13, 67, 88,
    89–91
  ultimate biodegradability  13, 65–86
biodegradability testing  10–15, 65–117
  analytical methods  100–114
  closed bottle test  94
  comparison of test method accuracy
    98–100
  control vessels  75, 76
  duration of test  76, 77
  early work  68–72
  Husmann unit  82–84
  inocula  73, 74, 83, 84
  manometric respirometry  92, 93
  MITI test  93, 94
  modified Sturm test ($CO_2$ evolution)
    91, 92
  OECD and EEC test for primary
    biodegradability  77–82
  porous pot equipment  83, 84
  reference compounds  77
  simulation methods  97, 98
  ultimate biodegradability methods
    87–89
bioelimination  66
BOD (definition)  67
butylene oxide and reactions  208, 209
  in fatty alcohol EO/BO co-polymers
    226

cationic surfactants  18, 183–200
  absorption onto particles  193
  anaerobic biodegradation  195
  biodegradability and analysis  22, 23,
    82, 86, 111–114, 183–200
  influence of biological process  196
  influence of toxicity on biodegradation
    190, 191
  in OECD screening tests  185
  removal in activated sludge reactors
    193–195
chromatographic methods of analysis  105,
  106, 109–111
  anionic surfactants  105, 106
  nonionic surfactants  109–111
closed bottle test  94
cobaltothiocyanate method for nonionic
  surfactants  109
COD (definition)  67
consortia  44, 45
cultures, pure versus mixed  42–44

10-day window (definition)  68
degradation phase (definition)  68
DOC (definition)  67
Dragendorf reagent  21

EEC Dangerous Chemicals Directive  130
EMPA method  *see* Zahn–Wellens
  method
end-blocked nonionics  227
enzyme biosynthesis  46–50
ethylene oxide and reactions  207, 208
ethylene oxide–propylene oxide block
  co-polymers  222–229
ethyl trimethyl ammonium chloride,
  biodegradation route  197, 198

FAA  *see* fatty alcohol alkoxylates
FAE  *see* fatty alcohol ethoxylates
fatty alcohol alkoxylates  222–226
fatty alcohol ethoxylates  220–222
  biodegradability  221, 222
  end-blocked  227
fatty alcohol sulphates  143, 144, 167–171
  anaerobic biodegradation  171
  biodegradability pathways  169–171
  primary biodegradation  167, 168
  ultimate biodegradation  167–169
fatty amine ethoxylates, biodegradability
  229
FES  *see* α-sulpho fatty acid esters

Husmann unit  82–84
hydroxysulphobetaines  *see*
  sulphobetaines

imidazoline derivatives *see*
  alkylamphoacetates
imidazolinium betaines  250–252
  biodegradability  251, 252

lag phase (definition)  68
legal requirements  127–132
  anionic surfactants  135–137
  EEC (current)  127–129
  future EEC requirements  130–132
  USA  129, 130

manomeric respirometry  92, 93
methyl ester sulphonates  *see* α-sulpho
  fatty acid esters
microbial nutrition and the carbon cycle
  28–30
MITI test  93, 94, 185–190
  applied to cationic surfactants  185–190

nonionic surfactants  204–229
  alkyl and aryl end-blocked  227
  analysis  20, 106–111
  biodegradability legislation  18,
    127–129
  biodegradation mechanisms  215–218
    α-oxidation  217

$\beta$-oxidation  216, 217
$\omega$-oxidation  216
    hydrophile degradation of  37
    inherent biodegradability  213–215
        analytical methods  214, 215
        primary biodegradability  213
        ready biodegradability  213–215
            analytical methods  214, 215
        ultimate biodegradability  213
nonyl phenyl ethoxylates  219, 220
    biodegradability  219, 220
NPE  *see* nonyl phenol ethoxylates

OECD and EEC tests for primary
    biodegradability  77–82
OECD confirmatory test  82
OECD screening test results for cationics
    185
$\alpha$-olefin sulphonates  142, 143, 162, 163
    pathways for biodegradation  162, 163
    primary biodegradation  162
    ultimate biodegradation  162

phosphate esters  *see* alkyl phosphates/
    alkyl ether phosphates
polycarboxyglycinates  245–247
    biodegradability  246
    primary biodegradation  247
    ultimate biodegradation  247
polyglycol surfactants
    applications of  210–212
    initiators for  209
    properties of  209, 210
    structure of  205, 206
porous pot equipment  83, 84
primary biodegradation  13, 65–117
    interpretation of results  126
    testing strategy  118, 119
propylene oxide and reactions  208
    in fatty alcohol alkoxylates  223–226

quaternary ammonium salts  191–200
    behaviour in waste water treatment
        plants  192, 193
    biodegradability  191, 192
    biodegradation pathways  196–200
    formation of recalcitrant intermediates
        200

risk assessment  24

SAS  *see* alkane sulphonates
screening tests (definition)  67
secondary alkane sulphonates  *see* alkane
    sulphonates
sewage treatment processes  8–10
    simulation test  12, 68

simulation tests (definition)  68
soap  1, 2, 139, 140
    aerobic degradation  149
    biodegradation pathway  148, 149
    primary and ultimate biodegradability
        146–149
    replacement in detergent products  2, 3
Sturm test, modified  91, 92
    applied to cationic surfactants  185–190
sulphobetaines and hydroxysulpho-
    betaines  238–242
    biodegradability  238–242
    chemical structure  238
    primary biodegradation  241, 242
    ultimate biodegradation  242
$\alpha$-sulpho fatty acid esters  143, 163–167
    anaerobic biodegradation  165–167
    biodegradation pathways  165
    primary biodegradation  163, 164
    ultimate biodegradation  163–165
sulphosuccinates  145, 175–178
    anaerobic biodegradation  178
    biodegradation pathways  176
    primary biodegradation  175–177
    ultimate biodegradation  176, 177
surfactants (definition)  1
    adaptation  45
    analysis  18
    as potential microbial nutrients  30, 31
    early development of  3–5
synthetic sewage  83

testing strategy and legal requirements
    118–132
    validation of results  124, 125
    pass levels  125, 126
tetramethyl ammonium chloride,
    biodegradation routes  196, 197
tetrapropylene benzene sulphonate
    replacement  15, 16
$ThCO_2$ (definition)  67
    $^{14}C$ labelled substances method  123,
        124
    headspace $CO_2$ method  124
ThOD (definition)  67
treatability (definition)  67

ultimate biodegradability  13, 65–68
    interpretation of results  126, 127
    methods of determination  88, 89
    testing strategy  119–123

Wickbold test/apparatus  20, 21, 107–109

Zahn–Wellens method (EMPA method)
    95–97